AF388994

Advanced Sensing and Control for Connected and Automated Vehicles

Advanced Sensing and Control for Connected and Automated Vehicles

Editors

Chao Huang
Haiping Du
Wanzhong Zhao
Yifan Zhao
Fuwu Yan
Chen Lv

MDPI • Basel • Beijing • Wuhan • Barcelona • Belgrade • Manchester • Tokyo • Cluj • Tianjin

Editors

Chao Huang
The Hong Kong
Polytechnic University
China

Haiping Du
University of Wollongong
Australia

Wanzhong Zhao
Nanjing University of
Aeronautics & Astronautics
China

Yifan Zhao
Cranfield University
UK

Fuwu Yan
Wuhan University
of Technology
China

Chen Lv
Nanyang Technological
University
Singapore

Editorial Office
MDPI
St. Alban-Anlage 66
4052 Basel, Switzerland

This is a reprint of articles from the Special Issue published online in the open access journal *Sensors* (ISSN 1424-8220) (available at: https://www.mdpi.com/journal/sensors/special_issues/ Sensing_Automated_Vehicles).

For citation purposes, cite each article independently as indicated on the article page online and as indicated below:

LastName, A.A.; LastName, B.B.; LastName, C.C. Article Title. *Journal Name* **Year**, *Volume Number*, Page Range.

ISBN 978-3-0365-3487-9 (Hbk)
ISBN 978-3-0365-3488-6 (PDF)

Cover image courtesy of Chao Huang

© 2022 by the authors. Articles in this book are Open Access and distributed under the Creative Commons Attribution (CC BY) license, which allows users to download, copy and build upon published articles, as long as the author and publisher are properly credited, which ensures maximum dissemination and a wider impact of our publications.
The book as a whole is distributed by MDPI under the terms and conditions of the Creative Commons license CC BY-NC-ND.

Contents

About the Editors

Chao Huang is currently a Research Assistant Professor in the Department of Industrial and Systems Engineering at the Hong Kong Polytechnic University. She received her bachelor's degree in control engineering from China University of Petroleum, Beijing, and her PhD from the University of Wollongong, Australia. Dr. Huang's research interests include system optimization and control, human–machine collaboration, fault diagnosis and fault tolerance, and automotive control and applications. Dr. Huang has published 33 journal papers, 20 conference papers, and 2 books.

Haiping Du received his PhD in mechanical design and theory from Shanghai Jiao Tong University, Shanghai, China, in 2002. He has been a Professor at the School of Electrical, Computer and Telecommunications Engineering, University of Wollongong, Australia, since 2016. He is a Subject Editor of the *Journal of The Franklin Institute*, an Associate Editor of *IEEE Transactions on Industrial Electronics* and *IEEE Control Systems Society* conference, and an Editorial Board Member for some international journals, such as the *Journal of Sound and Vibration, IET Control Theory and Application, IET Intelligent Transportation Systems, Mechatronics, Advances in Mechanical Engineering*, etc. Dr. Haiping Du's research interests include vibration control, vehicle dynamics and control systems, robust control theory and engineering applications, electric vehicles, robotics and automation, and smart materials and structures.

Yifan Zhao is a Reader in data science in the Centre for Life-cycle Engineering and Management (CLEM) at Cranfield University and the Academic Lead of the Through-life Engineering Services lab. He has over 18 years of experience in image/signal processing, computer vision, and artificial intelligence (AI) for degradation assessment and anomaly detection for complex systems. The applications cover critical asset health monitoring for construction sites, non-destructive testing and evaluation of high-value engineering components, driver behaviour monitoring for human-driven and automated vehicles, and medical data and image processing. He received his PhD in automatic control and system engineering from the University of Sheffield, UK, in 2007. He received his BEng and MSc degrees in automatic control engineering from the Beijing Institute of Technology, China, in 2000 and 2003, respectively. He is currently the academic PI of Fuel Coach (GBP 650,000, EEF8037), funded by the Department for Business, Energy & Industrial Strategy (BEIS) to investigate how AI can help reduce fuel consumption and greenhouse gas emissions in construction sites. He was PI on two Innovate-UK-funded projects ("The Learning Camera"—104794 (GBP 275,000) and "One Source of Truth"—105881 (GBP 675,000)) to develop innovations in critical asset digitalization and monitoring for the construction sector using computer vision and AI. As PI, in 2019 he completed a GBP 250,000 Lloyd-Registration-Foundation-funded (GA/100113) project RECBIT on AI for the counterfeit detection of electronic components. He was Co-I on "Distributed Intelligent Ultrasound Imaging System for Secure in-community Diagnostics (SecureUltrasound)", funded by an EPSRC GCRF grant (EP/R013950/1, GBP 1,000,000), where he lead the work on image analysis for evaluating and improving ultrasound image quality. He was the Co-I on the GBP 1,600,000 EPSRC project CogShift (EP/N012089/1) and a followed EPSRC IAA (EP/R511511/1) project in collaboration with Jaguar Land Rover (JLR), leading the research on characterizing driver behaviour using computer vision and AI. He has produced more than 150 publications, including 110 peer-reviewed journal papers, more than 30 conference papers, 3 book chapters, and 3 patents.

He has successfully supervised eight PhDs to completion and 45 MSc students. He is a fellow of the Higher Education Academy and a Senior Member of IEEE.

Chen Lv is currently an Assistant Professor at Nanyang Technology University, Singapore. He received his PhD from the Department of Automotive Engineering, Tsinghua University, China, in 2016. His research focuses on advanced vehicle control and intelligence, a field to which he has contributed over 100 papers and obtained 12 patents.

Editorial

Advanced Sensing and Control for Connected and Automated Vehicles

Chao Huang [1],*, Haiping Du [2], Wanzhong Zhao [3], Yifan Zhao [4], Fuwu Yan [5] and Chen Lv [6]

[1] Department of Industrial and Systems Engineering, The Hong Kong Polytechnic University, Hong Kong 999077, China

[2] Faulty of Engineering and Information Science, University of Wollongong, Wollongong, NSW 2522, Australia; hdu@uow.edu.au

[3] Department of Vehicle Engineering, Nanjing University of Aeronautics & Astronautics, Nanjing 213300, China; zwz@nuaa.edu.cn

[4] School of Aerospace, Transport and Manufacturing, Cranfield University, Bedford MK43 0AL, UK; yifan.zhao@cranfield.ac.uk

[5] Hubei Key Laboratory of Advanced Technology for Automotive Components, Wuhan University of Technology, Wuhan 430070, China; yanfw@whut.edu.cn

[6] School of Mechanical and Aerospace Engineering, Nanyang Technological University, Singapore 639798, Singapore; lyuchen@ntu.edu.sg

* Correspondence: hchao.huang@polyu.edu.hk

About the Editors

In recent years, connected and automated vehicles (CAV) have been a transformative technology that is expected to reduce emissions and change and improve the safety and efficiency of the mobilities. As the main functional components of CAVs, advanced sensing technologies and control algorithms, which gather environmental information, process data, and control vehicle motion, are of great importance. The development of novel sensing technologies for CAVs has become a hot spot in recent years. Thanks to the improved sensing technologies, CAVs are able to interpret sensory information to further detect obstacles, localize their positions, and navigate themselves and interact with other surrounding vehicles in the dynamic environment. Furthermore, leveraging computer vision and other sensing methods, in-cabin human body activities, facial emotions, and even mental states can also be recognized.

This Special Issue of *Sensors* aims at reporting on some of the recent research efforts on this increasingly important topic. The 12 accepted papers in this Issue cover vehicle position estimation [1], vehicle dynamic parameters estimation [2], cooperative collision warning systems [3], small object detection [4], impact identification of the driver's driving performance on executive control function [5], hybrid path planning for autonomous driving [6], trajectory tracking for autonomous driving [7], vehicle stability control [8], vehicle stability and ride comfort control [9], urban platooning protocol design for platoon [10], path planning algorithm for platooning [11], and self-driving architecture design for CAV platoon [12].

In the next paragraphs, a brief description of the content of each contribution forming the Special Issue is provided.

In [1], a data-driven object vehicle estimation scheme to solve measurement uncertainty and latency problems in radar systems is proposed. An accuracy model considers the different error characteristics depending on the zone. The accuracy model was used to solve the measurement uncertainty of radar. The authors also develop latency coordination for the radar system by analyzing the position error depending on the relative velocity. The authors claimed that proposed estimation method produces improved performance over the conventional radar estimation and previous methods.

In [2], a two-stage estimation method, consisting of multiple-models and the Unscented Kalman Filter, is proposed to estimate vehicle dynamic parameters. During the

Citation: Huang, C.; Du, H.; Zhao, W.; Zhao, Y.; Yan, F.; Lv, C. Advanced Sensing and Control for Connected and Automated Vehicles. *Sensors* **2022**, *22*, 1538. https://doi.org/10.3390/s22041538

Received: 11 January 2022
Accepted: 22 January 2022
Published: 16 February 2022

Publisher's Note: MDPI stays neutral with regard to jurisdictional claims in published maps and institutional affiliations.

Copyright: © 2022 by the authors. Licensee MDPI, Basel, Switzerland. This article is an open access article distributed under the terms and conditions of the Creative Commons Attribution (CC BY) license (https://creativecommons.org/licenses/by/4.0/).

first stage, the longitudinal vehicle dynamics model is used. Through vehicle acceleration/deceleration, this model can be used to estimate the distance between the vehicle centroid and vehicle front, the height of vehicle centroid, and tire longitudinal stiffness. The estimated parameter can be used in the second stage. During the second stage, a single-track vehicle model with roll dynamics is adopted. By making the vehicle have continuous steering, this vehicle model can be used to estimate tire cornering stiffness, the vehicle moment of inertia around the yaw axis, and the moment of inertia around the longitudinal axis. The results in [2] show that the proposed method is effective and vehicle dynamic parameters can be well estimated.

A vehicle-to-vehicle (V2V) cooperative collision warning system (CCWS) consisting of an ultra-wideband (UWB) relative positioning/directing module and a dead reckoning (DR) module with wheel-speed sensors is proposed in [3]. An over-constrained localization method is proposed to calculate the relative position and orientation with the UWB data more accurately. Vehicle velocities and yaw rates are measured by wheel-speed sensors. An extended Kalman filter (EKF) is applied based on the relative kinematic model to combine the UWB and DR data. Finally, the time to collision (TTC) is estimated based on the predicted vehicle collision position. The authors of [3] concluded that the proposed method significantly improves the positioning and directing, and the proposed system can efficiently provide collision warning.

As small object detection is very important for the understanding of traffic scene environments, [4] proposes a small object detection method in traffic scenes based on attention feature fusion. First, a multi-scale channel attention block (MS-CAB) is designed, which uses local and global scales to aggregate the effective information of the feature maps. Based on this block, an attention feature fusion block (AFFB) is proposed, which can better integrate contextual information from different layers. Finally, the AFFB is used to replace the linear fusion module in the object detection network and obtain the final network structure. The authors in [4] conclude that the proposed approach increases the mAP of all objects by 0.9 percentage points on the validation set of the traffic scene dataset BDD100K, and at the same time, increases the mAP of small objects by 3.5%.

To explore the relationship between the driver's driving performance and executive control function, the authors of [5] invite a total of 35 healthy subjects to take part in a simulated driving experiment and a task-cuing experiment. The subjects were divided into three groups according to their driving performance (aberrant driving behaviors, including lapses and errors) by the clustering method. Then, the performance efficiency and electroencephalogram (EEG) data acquired in the task-cueing experiment were compared among the three groups. The authors concluded that this research presented evidence of the close relationship between executive control functions and driving performance.

In [6], a hybrid path planning is proposed to avoid unsatisfying path generation and to improve the performance of autonomous driving by combining the potential field with the sigmoid curve. The repulsive and attractive potential fields are redesigned by considering the safety and the feasibility. Based on the objective of the shortest path generation, the optimized trajectory is obtained to improve the vehicle stability and driving safety by considering the constraints of collision avoidance and vehicle dynamics. The effectiveness is examined by simulations in multiobstacle dynamic and static scenarios. The authors claimed that the proposed method shows better performance on vehicle stability and ride comfortability than that of the traditional potential field-based method in all the examined scenarios during autonomous driving.

Trajectory tracking is a key technology for precisely controlling autonomous vehicles. A trajectory-tracking method based on model predictive control is proposed in [7]. Instead of using the forward Euler integration method, the backward Euler integration method is used to establish the predictive model. To meet the real-time requirement, a constraint is imposed on the control law, and the warm-start technique is employed. The authors of [7] concluded that the proposed the tracking performance of the proposed controller is much better than that of controllers using the forward Euler method.

In [8], studies on comprehensive three-dimensional vehicle dynamics modelling and stability control strategies in the event of a sudden tire blow-out are conducted. An integrated control framework for a combined yaw plane and roll-plane stability control is presented. The authors of [8] concluded that the proposed lower-level MPC can successfully improve the roll stability in the challenging scenario of a tire blow-out during a fishhook maneuver when the vehicle has a big load transfer.

A Unified Chassis Control (UCC) strategy for enhancing vehicle stability and ride comfort by the coordination of four In-Wheel Drive (IWD), Four-Wheel Independent Steering (4WIS), and Active Suspension Systems (ASS) is designed in [9]. A hierarchical control structure was adopted to realize the UCC, including high-level sliding mode control, fixed point CA, and a normal tire force robust tracking controller. The authors claimed that the proposed method can effectively realize the tire force distribution to control the vehicle body attitude and driving stability even in high-demanding scenarios.

When an existing vehicle platoon is applied to urban roads, many challenges are more complicated to address than highways. They include complex topology, various routes, traffic signals, intersections, frequent lane changes, and communication interference depending on a higher vehicle density. To address these challenges, [10] propose a distributed urban platooning protocol (DUPP) that enables high mobility and maximizes flexibility for driving vehicles to conduct urban platooning in a decentralized manner. DUPP performs forwarder selection using an analytic hierarchy process. The performance of the proposed DUPP is compared with that of ENSEMBLE, which is the latest European platooning project. The authors of [10] concluded that the proposed DUPP is well suited to dynamic urban environments by maintaining a vehicle platoon as stable as possible.

In [11], a path planning algorithm for the platooning of articulated cargo trucks has been developed. Using the Kalman filter, V2V communication, and a novel update-and-conversion method, each following vehicle can accurately compute the trajectory of the leading vehicle's front part for using it as a target path. The authors claimed that on severe driving scenarios, the proposed algorithm could provide lateral string stability and robustness for truck platooning.

A self-driving architecture combining the sensing, planning, and control for CAV platoons in an end-to-end fashion is proposed in [12]. This multi-task model can switch between two tasks to drive either the leading or following vehicle in the platoon. The architecture is based on an end-to-end deep learning approach and predicts the control commands, i.e., steering and throttle/brake, with a single neural network. The inputs for this network are images from a front-facing camera, enhanced by information transmitted via V2V communication. The authors claimed that the approach eliminates casual confusion for the following vehicle, which is a known limitation of end-to-end self-driving.

In summary, there is a huge potential for CAV in collision avoidance, safety improvement and driving stability improvement. The papers gathered in this Special Issue contributed by proposing solutions to the general problem of state and/or parameter estimation [1–3], obstacle detection [4], driver behavior estimation [5], and/or classical academic problems such as path planning [6], path tracking [7], and stability control for autonomous driving [8,9] and/or by suggesting applications of the combined use of sensors and advanced algorithms to platooning [10–12], thus showing the theoretical challenges and practical interest of this research topic. Finally, we wish to thank the authors, reviewers, and journal staff for their commitment and effort, which made it possible to complete this Special Issue on time.

Funding: This work received funding from the PolyU (UGC), via grant A0040253 associated with grant A0039179.

Conflicts of Interest: The authors declare no conflict of interest.

References

1. Choi, W.Y.; Yang, J.H.; Chung, C.C. Data-Driven Object Vehicle Estimation by Radar Accuracy Modeling with Weighted Interpolation. *Sensors* **2021**, *21*, 2317. [CrossRef] [PubMed]
2. Li, W.; Li, H.; Xu, K.; Huang, Z.; Li, K.; Du, H. Estimation of Vehicle Dynamic Parameters Based on the Two-Stage Estimation Method. *Sensors* **2021**, *21*, 3711. [CrossRef] [PubMed]
3. Wang, M.; Chen, X.; Jin, B.; Lv, P.; Wang, W.; Shen, Y. A Novel V2V Cooperative Collision Warning System Using UWB/DR for Intelligent Vehicles. *Sensors* **2021**, *21*, 3485. [CrossRef] [PubMed]
4. Lian, J.; Yin, Y.; Li, L.; Wang, Z.; Zhou, Y. Small Object Detection in Traffic Scenes Based on Attention Feature Fusion. *Sensors* **2021**, *21*, 3031. [CrossRef] [PubMed]
5. Yan, L.; Wen, T.; Zhang, J.; Chang, L.; Wang, Y.; Liu, M.; Ding, C.; Yan, F. An Evaluation of Executive Control Function and Its Relationship with Driving Performance. *Sensors* **2021**, *21*, 1763. [CrossRef] [PubMed]
6. Lu, B.; He, H.; Yu, H.; Wang, H.; Li, G.; Shi, M.; Cao, D. Hybrid path planning combining potential field with sigmoid curve for autonomous driving. *Sensors* **2020**, *20*, 7197. [CrossRef] [PubMed]
7. Huang, Z.; Li, H.; Li, W.; Liu, J.; Huang, C.; Yang, Z.; Fang, W. A New Trajectory Tracking Algorithm for Autonomous Vehicles Based on Model Predictive Control. *Sensors* **2021**, *21*, 7165. [CrossRef]
8. Li, B.; Huang, C.; Wu, Y.; Zhang, B.; Du, H. A Three-Dimensional Integrated Non-Linear Coordinate Control Framework for Combined Yaw- and Roll-Stability Control during Tyre Blow-Out. *Sensors* **2021**, *21*, 8328. [CrossRef]
9. Chen, X.; Wang, M.; Wang, W. Unified Chassis Control of Electric Vehicles Considering Wheel Vertical Vibrations. *Sensors* **2021**, *21*, 3931. [CrossRef] [PubMed]
10. Jeong, S.; Baek, Y.; Son, S. Distributed Urban Platooning towards High Flexibility, Adaptability, and Stability. *Sensors* **2021**, *21*, 2684. [CrossRef] [PubMed]
11. Lee, Y.; Ahn, T.; Lee, C.; Kim, S.; Park, K. A Novel Path Planning Algorithm for Truck Platooning Using V2V Communication. *Sensors* **2020**, *20*, 7022. [CrossRef]
12. Huch, S.; Ongel, A.; Betz, J.; Lienkamp, M. Multi-task end-to-end self-driving architecture for CAV platoons. *Sensors* **2021**, *21*, 1039. [CrossRef] [PubMed]

Article

A Three-Dimensional Integrated Non-Linear Coordinate Control Framework for Combined Yaw- and Roll-Stability Control during Tyre Blow-Out

Boyuan Li [1], Chao Huang [2,*], Yang Wu [1], Bangji Zhang [1] and Haiping Du [3]

[1] State Key Laboratory of Advanced Design and Manufacturing for Vehicle Body, Hunan University, Changsha 410082, China; bl995@uowmail.edu.au (B.L.); kisera@126.com (Y.W.); bangjizhang@hnu.edu.cn (B.Z.)
[2] Department of Industrial and Systems Engineering, The Hong Kong Polytechnic University, Hong Kong
[3] School of Electrical, Computer and Telecommunications Engineering, University of Wollongong, Wollongong, NSW 2522, Australia; hdu@uow.edu.au
* Correspondence: ch449@uowmail.edu.au

Abstract: A tyre blow-out can greatly affect vehicle stability and cause serious accidents. In the literature, however, studies on comprehensive three-dimensional vehicle dynamics modelling and stability control strategies in the event of a sudden tyre blow-out are seriously lacking. In this study, a comprehensive 14 degrees-of-freedom (DOF) vehicle dynamics model is first proposed to describe the vehicle yaw-plane and roll-plane dynamics performance after a tyre blow-out. Then, based on the proposed 14 DOF dynamics model, an integrated control framework for a combined yaw plane and roll-plane stability control is presented. This integrated control framework consists of a vehicle state predictor, an upper-level control mode supervisor and a lower-level 14 DOF model predictive controller (MPC). The state predictor is designed to predict the vehicle's future states, and the upper-level control mode supervisor can use these future states to determine a suitable control mode. After that, based on the selected control mode, the lower-level MPC can control the individual driving actuator to achieve the combined yaw plane and roll plane control. Finally, a series of simulation tests are conducted to verify the effectiveness of the proposed control strategy.

Keywords: tyre blow-out; yaw stability; roll stability; vehicle dynamics model; model predictive control

Citation: Li, B.; Huang, C.; Wu, Y.; Zhang, B.; Du, H. A Three-Dimensional Integrated Non-Linear Coordinate Control Framework for Combined Yaw- and Roll-Stability Control during Tyre Blow-Out. *Sensors* **2021**, *21*, 8328. https://doi.org/10.3390/s21248328

Academic Editor: Felipe Jiménez

Received: 12 October 2021
Accepted: 8 December 2021
Published: 13 December 2021

Publisher's Note: MDPI stays neutral with regard to jurisdictional claims in published maps and institutional affiliations.

Copyright: © 2021 by the authors. Licensee MDPI, Basel, Switzerland. This article is an open access article distributed under the terms and conditions of the Creative Commons Attribution (CC BY) license (https://creativecommons.org/licenses/by/4.0/).

1. Introduction

A sudden vehicle tyre blow-out may cause significant problems to vehicle stability and road safety. In the United States (US), the published statistical data shows 'tyre blow-out' caused more than 300,000 road accidents in the years 1992 to 1996 [1]. Based on the data from the report by the National Highway Traffic Safety Administration (NHTSA) in the US, tyre blow-outs caused 414 fatalities, 10,275 nonfatal injuries, and 78,392 crashes in 2003 [2]. In addition, tyre blow-outs also cause serious stability issues in electric industrial vehicles, such as forklift trucks [3,4].

The blow-out of one specific tyre makes the tyre pressure significantly decrease and causes a significant change to the vehicle's dynamic response. Various studies have proved that a tyre blow-out can be completed within 0.1 s and the tyre parameter change can be considered as a step change [5,6]. It is argued that the tyre deflation greatly affects cornering stiffness, radial tyre stiffness and rolling resistance [5,7]. In [8], actual experiments on 26 vehicles were carried out to study the vehicles' dynamic response to tyre blow-outs. The experiment results suggested that the increased rolling resistance of deflated tyres could generate longitudinal drag force and cause additional yaw moment to pull the vehicle away from the original path. The studies [6,9] also pointed out that the tyre cornering stiffness and radial stiffness decreased significantly after tyre blow-out. The assumption of the tenfold

5

drop of radial stiffness after tyre blow-out was verified by the tests on a 165SR13 D90 tyre, and the decreased tyre radial stiffness caused the tyre's instantaneous radius reduction and significantly increased the load transfer effect. It is suggested in [5] that tyre cornering stiffness and radial stiffness reduces by 25–40% after tyre deflation. Similarly, Wang et al. proposed a non-linear coordinate motion controller for the vehicle after tyre deflation by assuming the rolling resistance increased 30 times and the cornering stiffness reduced to 28% of the original value [9]. In addition, when a tyre blow-out happens, at the steering wheel more steering input is required to compensate for the increased total alignment moment caused by the deflated tyre [8], and the steering controller needs to be redesigned, for instance, with the human-machine adaptive shared control [10]. However, the steering control system design is not focused on in this study. Based on the review of the above studies, it can be summarised that the tyre blow-out mainly affects vehicle dynamics performance in three aspects: (1) the additional yaw moment is induced by the increased rolling resistance of the deflated tyre; (2) the changed tyre lateral force is caused by the decreased tyre cornering stiffness; (3) the decreased radial stiffness will cause a significant decrease of the wheel's instantaneous radius and induce a big load transfer effect.

In current literature, a number of studies have proposed different kinds of vehicle dynamics models to present the dynamics performance after tyre blow-out. In [6,7], the three-dimensional Engineering Dynamics Vehicle Simulation Model (EDVSM) is used to describe tyre blow-out behaviour. This comprehensive vehicle model has 15 degrees of freedom (DOF): 6 DOF for the vehicle body, 4 DOF for the suspension system, 4 DOF for the wheel rotation and one DOF for the steering wheel. Similarly, the high-order comprehensive commercial vehicle dynamics model veDYNA is applied in [5,11] to present the vehicle dynamics performance after tyre blow-out. However, the EDVSM and veDYNA vehicle models are all commercial products and the detailed mathematical equations of these models are not presented, so it is hard to carry out the theoretical study on the tyre blow-out modelling. The stability controller design after a tyre blow-out in studies [5,9,11] is only based on the yaw plane dynamics equation (only considering the changed rolling resistance and cornering stiffness after tyre blow-out), and the suspension motion and vertical dynamics have been neglected. When one specific tyre blows out, the suddenly decreased tyre radial stiffness will cause the reduction of the instantaneous tyre radius. This reduction will transfer to the suspension system and cause a big suspension deflection, load transfer and increase of the roll angle. This will cause a strong coupling effect on the yaw plane dynamics and should be considered in the controller design. Therefore, a three-dimensional full-vehicle dynamics model, which considers all six degrees of freedom of the vehicle body (longitudinal motion, lateral motion, vertical motion, yaw motion, roll motion and pitch motion) and integrating the suspension system and vehicle body dynamics system, is required to comprehensively present the dynamics response of a vehicle after one specific tyre blow-out for the stability controller design.

In the current vehicle industry, the tyre pressure monitoring system (TPMS) based on new in-tyre sensors and electronics is widely used to monitor the tyre pressure in real-time and detect tyre blow-out early [2,12]. Although some studies have proposed fault diagnosis and estimation approaches in the literature [13], we can simply assume the location of tyre blow-out is already known. After the blow-out of a specific tyre has been detected, various vehicle stability control systems are designed to improve the vehicle handling and stability. The control algorithms in the literature can be classified into three types: the steering-only control, the braking-only control and the integrated control. In [14], a steering-only control approach is presented, and the control system is triggered by an alarm generated by the TPMS. Chen et al. proposes the control strategy for an emergency automatic braking system when a tyre blows out [15]. Wang et al. developed a control optimization strategy for the yaw-plane motion by coordinating both the steering and braking based on a triple-step control method: steady-state controller, feedforward controller and feedback controller [9,11]. In [9,11], the longitudinal vehicle dynamics are neglected and the longitudinal velocity is assumed to be available from the estimation algorithm. However, the time-varying longitudinal velocity will greatly affect the vehicle handling and stability after a tyre blow-out and the effect of the changing longitudinal velocity on the controller design should not be neglected. In [5], the gain scheduling robust controller with respect to time-varying longitudinal velocity after tyre blow-out is proposed. The feedback control gain of a high-level controller can be real-time adjusted by the changing scaling factors determined by different values of current longitudinal velocity value and maximum and minimum velocity values.

The above studies [5,9–11] focus on the yaw-plane stability control during tyre blow-out and the main control targets are the side-slip angle and yaw rate. However, a tyre blow-out strongly affects the vehicle roll dynamics and the roll plane control targets should be also included in the controller design. Currently, a rollover can be mitigated by using the brakes [16–18], steering [19,20], antiroll bars [21] or a combination of different actuators [22,23]. Some of the current studies have discussed the combined control of yaw stability and roll stability. For example, Rajamani et al. carried on a study to explore the vehicle yaw and roll dynamics response in the steady-state turning manoeuvre [24]. It is concluded that in steady-state cornering, the roll angle and rollover index remain unchanged unless the longitudinal velocity or the cornering trajectory is changed. Alberding et al. propose a non-linear hierarchical control allocation algorithm for vehicle yaw stabilisation and rollover prevention by using differential braking, and this controller eliminates the roll controller by introducing the rollover prevention as a constraint in the control algorithm [25].

Model predictive control (MPC) can predict the vehicle's future state and is greatly advantageous in rollover prevention. In addition, MPC is suitable for dealing with multiple control targets within defined constraints. Yin et al. propose a non-linear MPC to achieve the path-tracking control by utilising the prediction horizon of MPC [26]. Similarly, Chen et al. also design an LQR lateral control method based on the optimal front tyre lateral force [27].

A recent study proposes a combined yaw and roll-stability control framework based on the MPC method [28]. In [28], however, only the control actuator of differential braking is utilised to achieve various control targets, which limits the control performance. In [29], an MPC control system is proposed by integrating lateral stability control, rollover prevention and longitudinal slip control. Furthermore, an integrated control system based on fuzzy differential braking is developed to improve the yaw and rollover stability of off-road vehicles [30]. The new emerging technology of electric vehicles with in-wheel motors can achieve four-wheel-independent-driving (4WID) and the driving or braking torque can be optimally controlled and allocated to the individual wheel and the control envelope is substantially enlarged. A number of studies have proposed utilising the 4WID function to achieve better dynamic stability control performance [31,32].

In this study, first a 14 DOF vehicle dynamics model including the yaw-plane motion, roll-plane motion, pitch-plane motion and suspension dynamics is proposed, which is utilised to present the impact of abruptly changed tyre rolling resistance, cornering stiffness and vertical stiffness after a tyre blow-out on vehicle dynamics performance. Then, based on the comprehensive dynamics model, a three-dimensional MPC control allocation framework for integrated yaw-plane stability and roll-stability control after tyre blow-out is proposed. Based on a 4WID electric vehicle, this control framework can optimally distribute the driving and braking torque of individual wheels and achieve cruise control, yaw-plane stability control and roll-stability control simultaneously. The proposed control framework has a two-layer control structure and has three control modes: cruise control mode, yaw stability control mode and roll-stability control mode. In the upper-level control strategy, a model predictor is proposed to predict the vehicle's future states, and a control mode supervisor can determine the suitable control mode based on the predicted states. In the lower level, a MPC controller is applied to allocate the control actuators based on the selected control mode.

The major contribution of our study can be summarised as follows:

(1) A comprehensive 14 DOF dynamic model is applied to describe the vehicle dynamics performance during tyre blow-out, which is less focused on in the literature.

(2) A new integrated yaw- and roll-stability MPC controller based on the 14 DOF model is proposed specifically for the tyre blow-out scenario.

The rest of this paper is organised as follows. First, a 14 DOF vehicle dynamics model is presented in Section 2 to describe the dynamics performance of a tyre blow-out. Then, in Section 3, the simulation results of the dynamics performance of the 14 DOF model and the 8 DOF model after tyre blow-out are compared with the EDVSM model which has been validated by actual experimental results. Section 4 describes the proposed integrated yaw-stability and roll-stability control framework based on MPC. Finally, the simulation results of vehicle control performance during tyre blow-out are presented to validate the proposed control framework.

2. Vehicle Dynamics Model Considering the Tyre Blow-Out Effect

2.1. Vehicle Body Dynamics Model

In this section, the comprehensive 14 DOF vehicle dynamics model is proposed to present the actual vehicle dynamics performance after tyre blow-out [33] and the detailed diagram description is shown in Figure 1.

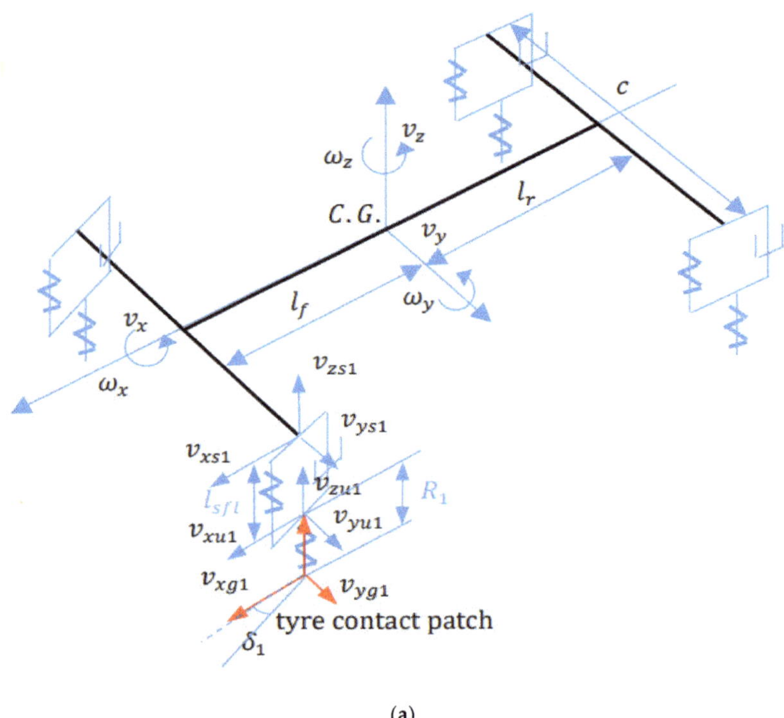

(a)

Figure 1. *Cont.*

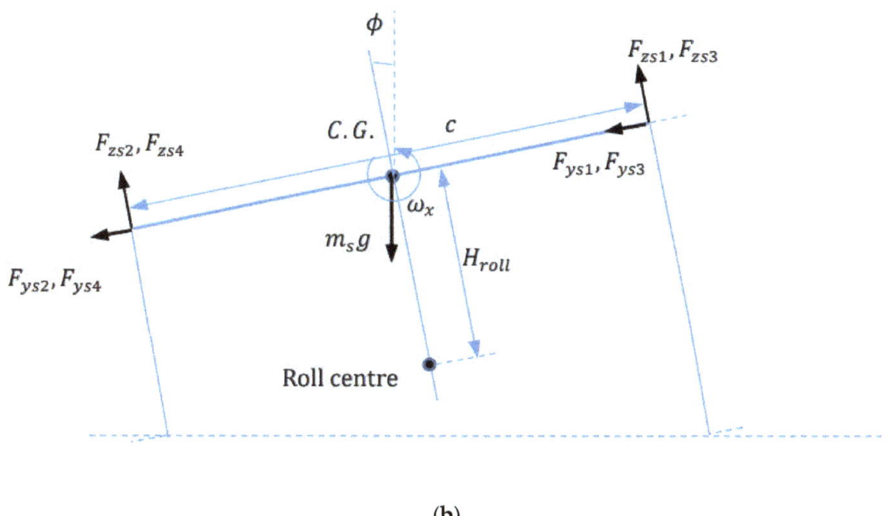

(b)

Figure 1. Schematic of 14 DOF vehicle dynamics model (**a**) yaw plane (**b**) roll plane.

The equations of motion of the vehicle sprung mass can be presented as following the six DOF model:

$$m_s\left(\dot{v}_x - \omega_y v_z - \omega_z v_y\right) = \sum(F_{xsi}) + m_s g \sin\theta \tag{1a}$$

$$m_s\left(\dot{v}_y + \omega_z v_x - \omega_x v_z\right) = \sum(F_{ysi}) - m_s g \sin\phi \cos\theta \tag{1b}$$

$$m_s\left(\dot{v}_z + \omega_x v_y - \omega_y v_x\right) = \sum F_{zsij} - m_s g \cos\phi \cos\theta \tag{1c}$$

$$J_x\dot{\omega}_x + (J_z - J_y)\omega_y\omega_z = \sum(M_{xi}) + m_s g H_{roll} \sin\phi + \frac{c(F_{zs1} - F_{zs2} + F_{zs3} - F_{zs4})}{2} \tag{1d}$$

$$J_y\dot{\omega}_y + (J_x - J_z)\omega_z\omega_x = \sum(M_{yi}) + l_r(F_{zs3} + F_{zs4}) - l_f(F_{zs1} + F_{zs2}) \tag{1e}$$

$$J_z\dot{\omega}_z + (J_y - J_x)\omega_x\omega_y = l_f(F_{ys1} + F_{ys2}) - l_r(F_{ys3} + F_{ys4}) + \frac{c(-F_{xs1} + F_{xs2} - F_{xs3} + F_{xs4})}{2} \tag{1f}$$

where m_s is the vehicle sprung mass and g is the acceleration gravity. J_x, J_y, J_z are inertial moments of pitch, roll and yaw, respectively. v_x, v_y, v_z are longitudinal velocity, lateral velocity and vertical velocity, respectively. $\omega_x, \omega_y, \omega_z$ are pitch rate, roll rate and yaw rate, respectively. $F_{xsi}, F_{ysi}, F_{zsi}$ represent the longitudinal force, lateral force and vertical force transferred to C.G. in the coordinate system attached to C.G. $i = 1, 2, 3, 4$, which presents the front left, front right, rear left and rear right wheel. F_{dzij} shows the load transfer force of each wheel. l_f is the front wheelbase and l_r is the rear wheelbase. c is the track width. h_f and h_r represent the distance between front and rear roll centres and C.G. M_{xi} and M_{yi} are roll moment and pitch moment transmitted to the sprung mass. H_{roll} is the distance between C.G. and vehicle roll centre of the sprung mass.

The roll angle ϕ, pitch angle θ and yaw angle ψ can be determined as the following equations:

$$\dot{\theta} = \omega_y \cos\phi - \omega_z \sin\phi \tag{2a}$$

$$\dot{\psi} = \frac{\omega_y \sin\phi}{\cos\theta} + \frac{\omega_z \cos\phi}{\cos\theta} \tag{2b}$$

$$\dot{\phi} = \omega_x + \omega_y \sin\phi \tan\theta + \omega_z \cos\phi \tan\theta \tag{2c}$$

The tyre force F_{xsi} and F_{ysi} can be determined by subtracting the unsprung mass weight and inertial force from the corresponding forces acting on the tyre contact patch:

$$F_{xsi} = F_{xgsi} + m_{ui}g\sin\theta - m_{ui}\dot{v}_{xui} + m_{ui}\omega_z v_{yui} - m_{ui}\omega_y v_{zui} \tag{3a}$$

$$F_{ysi} = F_{ygsi} - m_{ui}g\sin\phi\cos\theta - m_{ui}\dot{v}_{yui} + m_{ui}\omega_x v_{zui} - m_{ui}\omega_z v_{xui} \tag{3b}$$

where m_{ui} is the unsprung mass of an individual corner. v_{xui}, v_{yui}, v_{zui} are unsprung mass longitudinal velocity/lateral velocity/vertical velocity in a coordinate system attached to C.G. F_{xgsi}, F_{ygsi}, F_{zgsi} are tyre–road contact forces in the body-fixed coordinate system, which can be projected from the F_{xgi}, F_{ygi}, F_{zgi} (tyre force in the coordinate system fixed at the tyre contact patch) as:

$$\begin{bmatrix} F_{xgsi} \\ F_{ygsi} \\ F_{zgsi} \end{bmatrix} = \begin{bmatrix} 1 & 0 & 0 \\ 0 & \cos\phi & \sin\phi \\ 0 & -\sin\phi & \cos\phi \end{bmatrix} \begin{bmatrix} \cos\theta & 0 & -\sin\theta \\ 0 & 1 & 0 \\ \sin\theta & 0 & \cos\theta \end{bmatrix} \begin{bmatrix} F_{xgi} \\ F_{ygi} \\ F_{zgi} \end{bmatrix} \tag{4}$$

The roll moment M_{xi} and pitch moment M_{yi} can be determined by the following equations:

$$M_{x1} = F_{ys1}H_{roll} \tag{5a}$$

$$M_{x2} = F_{ys2}H_{roll} \tag{5b}$$

$$M_{x3} = F_{ys3}H_{roll} \tag{5c}$$

$$M_{x4} = F_{ys4}H_{roll} \tag{5d}$$

$$M_{yi} = -\left(F_{xsgi}R_i + F_{xsi}l_{si}\right) \tag{5e}$$

where R_i is the instantaneous length of tyre radius and l_{si} is the instantaneous length of strut. The vertical tyre force F_{zsi} can be determined according to the following equation:

$$F_{zs1} = k_{sf}x_{s1} + b_{sf}\dot{x}_{s1} - \frac{M_{ARB\_F}}{c} \tag{6a}$$

$$F_{zs2} = k_{sf}x_{s2} + b_{sf}\dot{x}_{s2} + \frac{M_{ARB\_F}}{c} \tag{6b}$$

$$F_{zs3} = k_{sr}x_{s3} + b_{sr}\dot{x}_{s3} - \frac{M_{ARB\_R}}{c} \tag{6c}$$

$$F_{zs4} = k_{sr}x_{s4} + b_{sr}\dot{x}_{s4} + \frac{M_{ARB\_R}}{c} \tag{6d}$$

where x_{si} is the suspension spring compression. k_{sf}, k_{sr} are suspension stiffness and b_{sf}, b_{sr} are suspension damping coefficient. The anti-roll moment from the anti-roll bar can be determined by:

$$M_{ARB\_F} = 0.5k_{ARB,f}(x_{s1} - x_{s2}) + 0.5b_{ARB,f}\left(\dot{x}_{s1} - \dot{x}_{s2}\right) \tag{7a}$$

$$M_{ARB\_R} = 0.5k_{ARB,r}(x_{s3} - x_{s4}) + 0.5b_{ARB,r}\left(\dot{x}_{s3} - \dot{x}_{s4}\right) \tag{7b}$$

where $k_{ARB,f}, k_{ARB,r}$ are the stiffness of anti-roll bar and $b_{ARB,f}, b_{ARB,r}$ are the damping coefficient of the anti-roll bar.

The jacking force F_{dzi} transmitted to the sprung mass through the struts can be calculated as:

$$F_{dz2} = -F_{dz1} = \frac{F_{ygs1}R_1 + F_{ygs2}R_2 + F_{ys1}l_{s1} + F_{ys2}l_{s2} - (F_{ys1} + F_{ys2})H_{roll}}{c} \tag{8a}$$

$$F_{dz4} = -F_{dz3} = \frac{F_{ygs3}R_3 + F_{ygs4}R_4 + F_{ys3}l_{s3} + F_{ys4}l_{s4} - (F_{ys3} + F_{ys4})H_{roll}}{c} \tag{8b}$$

2.2. Tyre Model

The non-linear Dugoff tyre model is used in this paper to present the tyre's non-linear characteristics and determine the tyre longitudinal force F_{xti} and lateral force F_{yti} [34,35], and is described by:

$$\lambda_i = \frac{\mu F_{zgi} \left[1 - \varepsilon_r v_{si} \sqrt{s_i^2 + \tan^2 \alpha_i} \right] (1 - s_i)}{2\sqrt{C_s^2 s_i^2 + C_{\alpha i}^2 \tan^2 \alpha_i}} \tag{9a}$$

$$f(\lambda_i) = \begin{cases} \lambda_i (2 - \lambda_i) \ (\lambda_i < 1) \\ 1 \ (\lambda_i > 1) \end{cases} \tag{9b}$$

$$F_{yti} = \frac{C_\alpha \tan \alpha_i}{1 - s_i} f(\lambda_i) \tag{9c}$$

$$F_{xti} = \frac{C_s s_i}{1 - s_i} f(\lambda_i) \tag{9d}$$

where μ is the tyre–road friction coefficient. C_s is the longitudinal cornering stiffness and $C_{\alpha i}$ is the lateral cornering stiffness of each wheel. ε_r is a constant value. The side-slip angle α_i and slip ratio s_i of the individual tyres can be calculated as the following:

$$\alpha_1 = \delta_1 - \tan^{-1}\left(\frac{v_{yg1}}{v_{xg1}}\right) \tag{10a}$$

$$\alpha_2 = \delta_2 - \tan^{-1}\left(\frac{v_{yg2}}{v_{xg2}}\right) \tag{10b}$$

$$\alpha_3 = \delta_3 - \tan^{-1}\left(\frac{v_{yg3}}{v_{xg3}}\right) \tag{10c}$$

$$\alpha_4 = \delta_4 - \tan^{-1}\left(\frac{v_{yg4}}{v_{xg4}}\right) \tag{10d}$$

$$v_{sfl} = \cos \delta_{fl} (v_{xg1}) + \sin \delta_{fl} (v_{yg1}) \tag{11a}$$

$$v_{sfr} = \cos \delta_{fr} (v_{xg2}) + \sin \delta_{fr} (v_{yg2}) \tag{11b}$$

$$v_{srl} = \cos \delta_{rl} (v_{xg3}) + \sin \delta_{rl} (v_{yg3}) \tag{11c}$$

$$v_{srr} = \cos \delta_{rr} (v_{xg4}) + \sin \delta_{rr} (v_{yg4}) \tag{11d}$$

$$s_i = \frac{\omega_i R_i - v_{si}}{max(\omega_i R_i, v_{si})} \tag{12}$$

Longitudinal velocity and lateral velocity at tyre contact patch v_{xgi} and v_{ygi} can be presented as the following equations:

$$v_{xgi} = \cos \theta (v_{xui} - \omega_y R_i) + \sin \theta (v_{zui} \cos \phi + \sin \phi (\omega_x R_i + v_{yui})) \tag{13a}$$

$$v_{ygi} = \cos \phi (v_{yui} + \omega_x R_i) - v_{zui} \sin \phi \tag{13b}$$

F_{zgi} is the vertical force acting on the tyre–ground contact patch, which can be calculated by the following equation:

$$F_{zgi} = k_{ti} x_{ti} \tag{14}$$

k_{ti} is the tyre vertical stiffness. x_{ti} is the tyre spring compression and the initial tyre compression:

$$x_{t0} = \frac{\dfrac{ml_r}{2(l_f + l_r)} + m_{ui}}{k_t} \tag{15}$$

The velocity of the tyre's instantaneous deflection can be calculated as the following:

$$\dot{x}_{ti} = v_{xui}\sin\theta - \cos\theta\left(v_{zui}\cos\phi + v_{yui}\sin\phi\right) \tag{16}$$

The instantaneous tyre radius can be calculated as:

$$R_i = \frac{R_0 - x_{ti}}{\cos\theta\cos\phi} \tag{17}$$

where R_0 is the nominal tyre radius.

The wheel dynamics equations can be presented as follows:

$$I_\omega\dot{\omega}_i = -R_i F_{xti} + T_i - M_{yi} \tag{18}$$

where I_ω is the wheel rotational inertia. ω_i is the wheel angular speed and T_i is the traction/brake torque of the individual wheel. M_{yi} is the rolling resistance moment, which can be presented by the following equation:

$$M_{yi} = F_{zgi}\left(K_{fi} + K_{fvi}v_{xgi}{}^2\right)R_i \tag{19}$$

where K_{fi}, K_{fvi} are the tyre rolling resistance coefficients of the individual tyre.

2.3. Suspension System

The instantaneous compression of the suspension spring x_{si} can be calculated by the following equation:

$$\dot{x}_{si} = -v_{zsi} + v_{zui} \tag{20}$$

where v_{zsi} is the vertical velocity of the strut mounting point of each wheel, which can be calculated as the following equation:

$$v_{zs1} = v_z + \frac{c\omega_x}{2} - l_f\omega_y \tag{21a}$$

$$v_{zs2} = v_z - \frac{c\omega_x}{2} - l_f\omega_y \tag{21b}$$

$$v_{zs3} = v_z - \frac{c\omega_x}{2} + l_r\omega_y \tag{21c}$$

$$v_{zs4} = v_z + \frac{c\omega_x}{2} + l_r\omega_y \tag{21d}$$

The unsprung mass vertical velocity v_{zui} can be calculated as:

$$m_u\dot{v}_{zui} = \cos\phi\left(\cos\theta\left(F_{zgi} - m_{ui}g\right) + \sin\theta F_{xgi}\right) - \sin\phi F_{ygi} - F_{dzi} - x_{si}k_{si} - \dot{x}_{si}b_{si} - m_{ui}\left(v_{zui}\omega_x - v_{xui}\omega_y\right) \tag{22}$$

Forces F_{xgi} and F_{ygi} can be obtained by the following equation:

$$F_{xgi} = F_{xti}\cos\delta_i - F_{yti}\sin\delta_i \tag{23a}$$

$$F_{ygi} = F_{yti}\cos\delta_i + F_{xti}\sin\delta_i \tag{23b}$$

The longitudinal and lateral velocities v_{xui}, v_{yui} of unsprung mass can be calculated as:

$$v_{xui} = v_{xsi} - l_{si}\omega_y \tag{24a}$$

$$v_{yui} = v_{ysi} + l_{si}\omega_x \tag{24b}$$

where longitudinal and lateral velocity of the strut mounting point of each wheel (v_{xsi} and v_{ysi}) can be calculated as the following equation:

$$v_{xs2} = v_{xs4} = v_x + \frac{c}{2}\omega_z \tag{25a}$$

$$v_{xs1} = v_{xs3} = v_x - \frac{c}{2}\omega_z \tag{25b}$$

$$v_{ys1} = v_{ys2} = v_y + l_f\omega_z \tag{25c}$$

$$v_{ys3} = v_{ys4} = v_y - l_r\omega_z \tag{25d}$$

The instantaneous length of the strut l_{si} can be calculated as:

$$l_{si} = l_{s0} - (x_{si} - x_{s0}) \tag{26}$$

where the initial strut length $l_{s0} = h - (R_0 - x_{t0})$ and the initial suspension deflection

$$x_{s0} = \frac{ml_r}{2\left(l_f + l_r\right)k_s} \tag{27}$$

2.4. The Effectiveness of Tyre Blow-Out

The tyre blow-out will cause the sudden increase of the rolling resistance of the deflated tyre and induce an additional yaw moment T_b. Based on [36], the yaw moment T_b caused by a tyre blow-out can be determined by the following equation:

$$T_b = 0.5c\left(F_{cfl} - F_{cfr}\right) \tag{28a}$$

where c is the tracking width of the vehicle, the location of the blow-out tyre is at the front axle or

$$T_b = 0.5c(F_{crl} - F_{crr}) \tag{28b}$$

where the location of the blow-out tyre is at the rear axle. F_{ci} presents the tyre rolling resistance force, which can be calculated as the following equation:

$$F_{ci} = K_{fi}F_{zgi} \tag{29}$$

It is suggested that the typical rolling resistance stiffness for the light vehicle is around 0.012 and 0.015 [36] and it is argued in [7] that this value increases thirty times after a tyre blow-out. Thus, in this study, the rolling resistance coefficient when the tyre is in a healthy condition is chosen as 0.014 and this value increases to 0.42 after tyre blow-out.

In addition, a tyre blow-out causes a sudden decrease of tyre vertical stiffness k_{ti} and a decrease of the tyre's instantaneous radius R_i. The following equation shows the effect of the changed tyre's vertical stiffness on the vehicle vertical tyre load and suspension system:

$$F_{zgi} = k_1 k_{ti} x_{ti} \tag{30}$$

where k_1 is the ratio of the changed vertical stiffness related to the tyre deflation. When the tyre is in a healthy condition, $k_1 = 1$; when the tyre has a blow-out, $k_1 = 0.28$ [5]. The changed vertical tyre force F_{zgi} caused by the tyre blow-out can induce a significant load transfer effect.

The tyre cornering stiffness also reduces to 28% of the original value [5], which will greatly affect the tyre cornering force:

$$B_y = k_2 B_y \tag{31}$$

where k_2 is the ratio of changed cornering stiffness related to the tyre deflation.

3. Simulation Performance of the Vehicle after Tyre Blow-Out

In this section, the simulation test is carried out to present the vehicle dynamics performance after tyre blow-out based on the suggested 14 DOF comprehensive vehicle dynamics model. For comparison purposes, the dynamics performance of the widely applied 8 DOF vehicle model which neglects the pitch dynamics motion, vertical dynamics motion and suspension dynamics motion of the four wheels is also presented [34]. Furthermore, the above simulation results are validated against the simulation results from the EDVSM tyre blow-out model proposed in a study [7], where the EDVSM model has been verified by the experimental results of tyre blow-out. The simulation test applies the same straight-line manoeuvre in [7], where the vehicle speed is 101 km/h at the time of a rear right tyre blow-out, and the driver steers and brakes to maintain vehicle control. The tyre–road friction coefficient is assumed as 1. The tyre blow-out happens at 3.8 s and the duration is 0.1 s. The vehicle model parameters applied in the simulation are the same as the values in [7], which is shown in Table 1.

Table 1. Vehicle parameters of 8 DOF model and 14 DOF model (same as [7]).

Vehicle Mass	m	1440 kg
Distance between front axle and C.G.	l_f	1.016 m
Distance between rear axle and C.G.	l_r	1.524 m
Track width	b	1.5 m
Pitch moment of inertia	J_x	900 kg.m^2
Roll moment of inertia	J_y	900 kg.m^2
Yaw moment of inertia	J_z	2000 kg.m^2
Height of C.G.	h	0.75 m
Front suspension stiffness	k_{sf}	35,000 N/m
Rear suspension stiffness	k_{sr}	35,000 N/m
Front suspension damping ratio	b_{sf}	2500 N.s/m
Rear suspension damping ratio	b_{sr}	2500 N.s/m
Vertical front tyre stiffness	k_{tf}	200,000 N/m
Vertical rear tyre stiffness	k_{tr}	200,000 N/m
Tyre cornering stiffness	C_α	30,000 N/m

In Figure 2a–d, the longitudinal velocity, yaw rate, longitudinal acceleration and lateral acceleration responses of the 14 DOF model and the 8 DOF model are compared with the results from EDVSM model. After the rear right tyre deflation at 3.8 s, there is a sudden change of the longitudinal velocity, yaw rate, longitudinal and lateral acceleration in both the 8 DOF model and 14 DOF model due to the generated additional yaw moment caused by the sudden increase of the wheel rolling coefficient. It can also be noted that the 14 DOF model shows very similar responses in longitudinal velocity and longitudinal acceleration as the EDVSM model response. The 8 DOF model has a smaller negative longitudinal acceleration response and consequently, the longitudinal velocity is much larger than EDVSM. There are some mismatches of vehicle dynamics responses, such as the yaw rate and lateral acceleration between the 14 DOF model EDVSM model, which required further investigation. The 8 DOF model shows a larger yaw rate and lateral acceleration responses compared with EDVSM model.

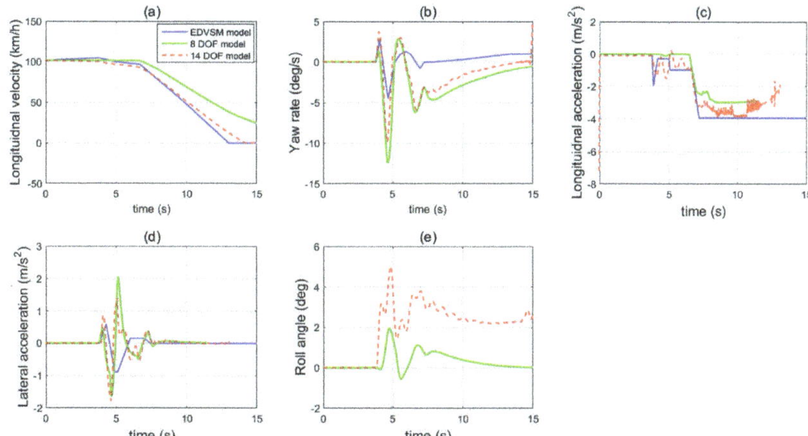

Figure 2. The vehicle state dynamics responses of 8 DOF model and 14 DOF model after tyre deflation (**a**) longitudinal velocity (**b**) yaw rate (**c**) longitudinal acceleration (**d**) lateral acceleration (**e**) roll angle.

Figure 3 also suggests that the instantaneous tyre radius of the deflated rear right tyre of the 14 DOF model at the beginning is smaller than the 8 DOF model due to the tyre compression. The 14 DOF model considers the sudden decrease of tyre vertical stiffness when the tyre blows out, which will induce the significant tyre instantaneous radius reduction. However, the 8 DOF model neglects the tyre vertical dynamics and suspension system and considers the tyre radius as a constant value, which cannot accurately present the tyre blow-out effect.

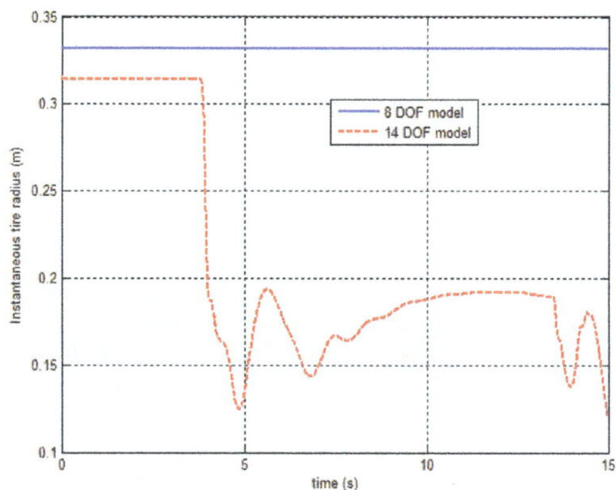

Figure 3. The tyre's instantaneous radius responses of 8 DOF model and 14 DOF model.

Figure 4 presents the load transfer effect of 8 DOF model and 14 DOF model after a rear right tyre blow-out. The initial vertical load of 8 DOF model is smaller than 14 DOF model since the 8 DOF model neglects the weight of unsprung mass including the wheel hub, wheel mass and suspension system. For the 8 DOF model, there is no obvious load transfer before and after tyre blow-out happens although the vertical load response of each wheel has a small oscillation during the tyre deflation. On the other hand, the 14 DOF model shows obvious load transfer effect after a front left tyre blow-out: at the beginning of the tyre blow-out, the tyre's vertical load of the rear right wheel decreases sharply and then a brief spike occurs, which is shown in Figure 4d and is very close to the simulation results from EDVSM model. After that, due to vehicle roll and pitch motion, the tyre's vertical loads of the front right and rear left wheel are increased and the tyre's vertical load of the front left is decreased as shown in Figure 4a–c. This vertical load transient response after a rear right tyre blow-out is very similar to the response described in [7].

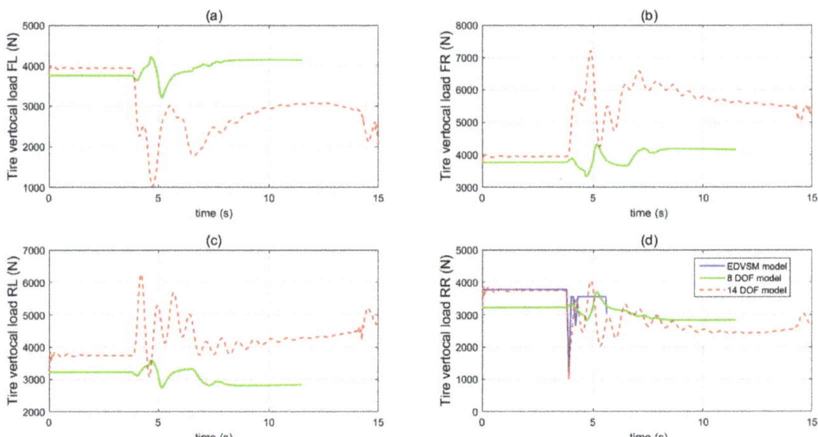

Figure 4. The vertical load responses of 8 DOF model and 14 DOF model (**a**) front left wheel (**b**) front right wheel (**c**) rear left wheel (**d**) rear right wheel.

Figures 5 and 6 compare the yaw-plane stability region of the14 DOF model and the 8 DOF model in different longitudinal velocity conditions. In this study, the yaw-plane stability is determined by the value λ_i in the Dugoff tyre model: if λ_i of an individual tyre is larger than 1, the vehicle is in a stable condition; if λ_i of the individual tyre is equal or smaller than 1, the vehicle is moving in an unstable condition. A group of simulation tests have been carried out to determine the stability transition point when $\lambda_i = 1$ and consequently the stability boundary can be determined. According to Figures 5 and 6, the stability region of 14 DOF model is generally smaller than 8 DOF model. This is mainly because the 14 DOF model considers the coupling effect of the vehicle roll and pitch motion on the yaw motion and the yaw-plane stability is compromised.

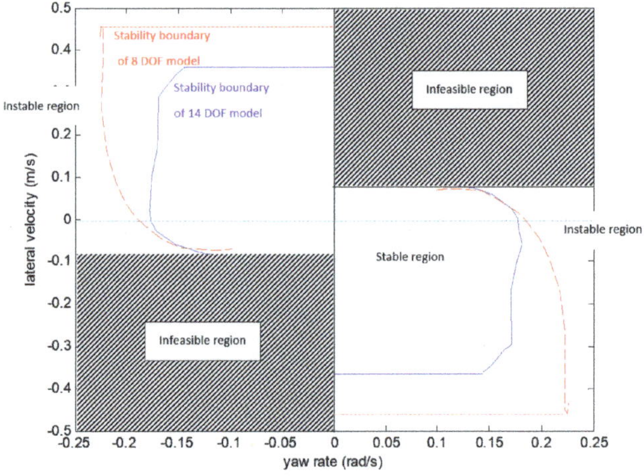

Figure 5. Compares the yaw-plane stability region of 14 DOF model and 8 DOF model ($v_x = 20$, $\mu = 1$).

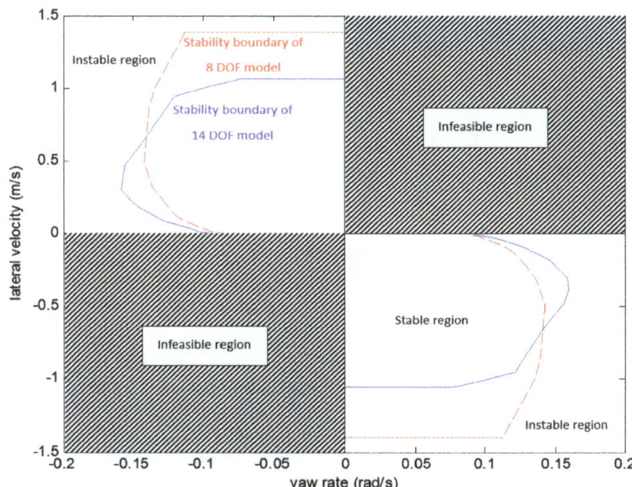

Figure 6. Compares the yaw-plane stability region of 14 DOF model and 8 DOF model ($v_x = 40$, $\mu = 1$).

In this section, the proposed 14 DOF model has been validated in the simulation to accurately present the dynamics performance of a tyre blow-out. The value λ_i in the Dugoff tyre model can be utilised to determine the stability region of the 14 DOF model in the yaw plane. The determined stability region is a very useful tool to select a suitable control mode for the integrated controller design in the following section.

4. Three-Dimensional Integrated Yaw-Plane Stability and Roll-Plane Stability 14 DOF MPC Control Framework

In this section, a three-dimensional non-linear coordinate control framework is designed to achieve the integrated control of yaw-plane stability and roll-plane stability when tyre blow-out. Based on 14 DOF model, The hierarchy of the whole 14 DOF MPC control framework consists of the vehicle states predictor, upper-level control supervisor and lower-level 4 DOF MPC controller. Based on predicted vehicle states from a model predictor, the upper-level control mode supervisor selects the most suitable control mode from the options of cruise control mode, yaw-plane stability control mode and roll-stability control mode. Then according to the selected control mode, the lower-level 4 DOF MPC algorithm is applied to allocate the desired control value to the individual actuator. The whole structure of the control framework is shown in Figure 7.

4.1. Vehicle States Predictor

A vehicle model predictor based on the model predictive algorithm is presented to determine the vehicle's future states. Based on some the vehicle's critical future states, such as the longitudinal velocity, lateral velocity and roll angle, the vehicle control mode can be selected in the upper-level control supervisor.

The major difficulty in implementing the 14 DOF model-based MPC control allocation is the complex model structure of 14 DOF model and the significant increase in computational time. In order to deal with this issue, the 14 DOF model in MPC can be simplified as 4 DOF by assuming some vehicle states are already known or can be directly measured. Therefore, in this section, the vehicle state predictor can be utilised to estimate the vehicle states which cannot be measured directly. Then, the estimated and measured vehicle states can be directly used as input information in the simplified 4 DOF MPC in the lower-level controller.

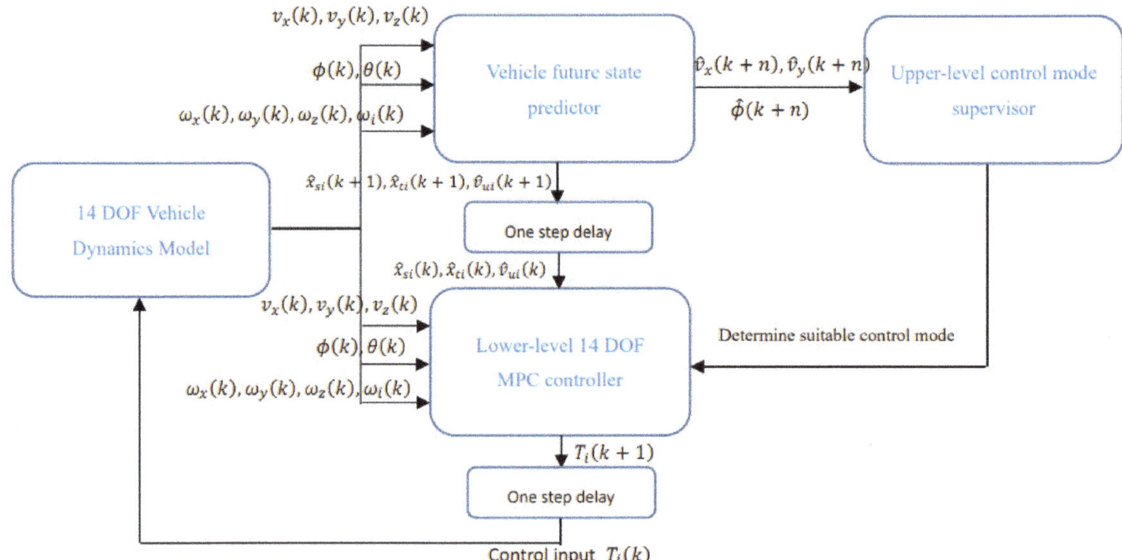

Figure 7. The structure diagram of proposed three-dimensional integrated stability control framework.

Assumption 1. *It is assumed that velocity longitudinal velocity v_x, lateral velocity v_y and vertical velocity v_z in C.G. can be easily estimated [37,38]. Vehicle roll angle ϕ, roll rate ω_x, pitch angle θ, pitch rate ω_y, yaw rate ω_z of C.G. and wheel angular velocity of each wheel ω_i are all easy to measure with various sensors. In addition, the tyre cornering stiffness change, tyre vertical stiffness change and rolling resistance change after tyre blow-out are all assumed to be known.*

The model predictive estimator algorithm can be presented in discrete time in this section. The vehicle states which are hard to measure and intended to be estimated are tyre compression x_{ti}, suspension spring compression x_{si} and vertical velocity of unsprung mass v_{zui}.

First, the velocity of suspension mounting points in the current time step can be calculated based on Equations (21) and (25) in discrete time:

$$v_{xs1}(k) = v_{xs3}(k) = v_x(k) - 0.5c\omega_z(k) \tag{32a}$$

$$v_{xs2}(k) = v_{xs4}(k) = v_x(k) + 0.5c\omega_z(k) \tag{32b}$$

$$v_{ys1}(k) = v_{ys2}(k) = v_y + l_f\omega_z(k) \tag{33a}$$

$$v_{ys3}(k) = v_{ys4}(k) = v_y - l_r\omega_z(k) \tag{33b}$$

$$v_{zs1}(k) = v_z(k) + 0.5c\omega_x(k) - l_f\omega_y(k) \tag{34a}$$

$$v_{zs2}(k) = v_z(k) - 0.5c\omega_x(k) - l_f\omega_y(k) \tag{34b}$$

$$v_{zs3}(k) = v_z(k) - 0.5c\omega_x(k) + l_r\omega_y(k) \tag{34c}$$

$$v_{zs4}(k) = v_z(k) + 0.5c\omega_x(k) + l_r\omega_y(k) \tag{34d}$$

The length of suspension strut in the current time step can be presented based on Equation (26):

$$l_{si}(k) = l_{s0} - (\hat{x}_{si}(k) - x_{s0}) \tag{35}$$

The instance tyre radius in the current time step can be calculated based on Equation (17):

$$R_i(k) = \frac{R_0 - \hat{x}_{ti}(k)}{\cos\theta(k)\cos\phi(k)} \tag{36}$$

It is noted in Equations (30)–(32), $v_x(k)$, $v_y(k)$, $v_z(k)$, $\omega_x(k)$, $\omega_y(k)$ and $\omega_z(k)$ are measured vehicle state values in current time step. $\hat{x}_{si}(k)$ and $\hat{x}_{ti}(k)$ are the estimated suspension spring compression and tyre compression in the current time step. The initial conditions x_{t0} and x_{s0} can be determined by Equations (14) and (26).

The longitudinal and lateral velocity of the unsprung mass in the current time step can be determined based on Equation (24):

$$v_{xui}(k) = v_{xsi}(k) - l_{si}\omega_y(k) \tag{37a}$$

$$v_{yui}(k) = v_{ysi}(k) + l_{si}\omega_x(k) \tag{37b}$$

The velocity on the tyre contact patch in the current time step can be calculated based on Equation (13):

$$\begin{aligned} v_{xgi}(k) = \cos\theta(k)\left(v_{xui}(k) - \omega_y(k)R_i(k)\right) + \\ \sin\theta(k)\left(\hat{v}_{zui}(k)\cos\phi(k) + \sin\phi(k)\left(\omega_x(k)R_i(k) + v_{yui}(k)\right)\right) \end{aligned} \tag{38a}$$

$$v_{ygi}(k) = \cos\phi(k)\left(v_{yui}(k) + \omega_x(k)R_i(k)\right) - \hat{v}_{zui}(k)\sin\phi(k) \tag{38b}$$

where $v_{zui}(k)$ is hard to measure and can be updated and estimated in every discrete-time iteration with the initial conditions of $v_{zui}(0) = 0$.

The lateral side-slip angle and longitudinal slip ratio of each wheel in discrete time can be determined based on Equations (10)–(12). The lateral tyre force of each wheel F_{yti} can be calculated based on Equation (9). The longitudinal tyre force F_{xti} at the current time step can be directly determined by:

$$F_{xti}(k) = \frac{T_i(k)}{R_i(k)} \tag{39}$$

The tyre forces applied on the wheel F_{xgi} and F_{ygi} can be determined by Equation (23). The tyre force transmitted to vehicle C.G. F_{xgsi} and F_{ygsi} can be determined based on Equation (4). The unsprung mass should be subtracted from forces F_{xgsi} and F_{ygsi}:

$$F_{xsi}(k) = F_{xgsi}(k) + m_u g \sin\theta(k) \tag{40a}$$

$$F_{ysi}(k) = F_{ygsi}(k) - m_u g \sin\phi(k)\cos\theta(k) \tag{40b}$$

Vehicle load transfer of each wheel $F_{dzi}(k)$ can be obtained from Equation (8). The vertical tyre force $F_{zgi}(k)$ can be determined based on Equation (14).

The estimated velocity of vehicle suspension in the current time step can be calculated as:

$$\hat{\dot{x}}_{si}(k) = -v_{zsi}(k) + \hat{v}_{zui}(k) \tag{41}$$

The estimated velocity of wheel radius change in the current time step can be determined according to Equation (15):

$$\hat{\dot{x}}_{ti}(k) = v_{xui}(k)\sin\theta(k) - \cos\theta(k)\left(\hat{v}_{zui}(k)\cos\phi(k) + v_{yui}(k)\sin\phi(k)\right) \tag{42}$$

The vertical acceleration of unsprung mass in the current time step can be determined according to Equation (22):

$$\begin{aligned} m_u\hat{\dot{v}}_{zui}(k) = \ & \cos\phi(k)\left(\cos\theta(k)\left(F_{zgi}(k) - m_u g\right) + \sin\theta(k)F_{xgi}(k)\right) - \sin\phi(k)F_{ygi}(k) \\ & -F_{dzi}(k) - x_{si}(k)k_s - \dot{x}_{si}(k)b_s - m_u\left(\hat{v}_{zui}(k)\omega_x(k) - v_{xui}(k)\omega_y(k)\right) \end{aligned} \tag{43}$$

Finally, the estimated values of $\hat{x}_{si}$, $\hat{x}_{ti}$ and $\hat{v}_{zui}$ in the next time step can be estimated by:

$$\hat{x}_{si}(k+1) = \hat{x}_{si}(k) + \dot{x}_{si}(k)(t(k+1) - t(k)) \tag{44a}$$

$$\hat{x}_{ti}(k+1) = \hat{x}_{ti}(k) + \dot{x}_{ti}(k)(t(k+1) - t(k)) \tag{44b}$$

$$\hat{v}_{zui}(k+1) = \hat{v}_{zui}(k) + \dot{v}_{zui}(k)(t(k+1) - t(k)) \tag{44c}$$

The predicted vehicle state values of v_x, v_y, ω_z, ϕ in the next n time steps can be determined by following equations:

$$\hat{v}_x(k+n) = v_x(k+n-1) + \dot{v}_x(k)(t(k+n) - t(k+n-1)) \tag{45a}$$

$$\hat{v}_y(k+n) = v_y(k+n-1) + \dot{v}_y(k)(t(k+n) - t(k+n-1)) \tag{45b}$$

$$\hat{\omega}_z(k+n) = \omega_z(k+n-1) + \dot{\omega}_z(k)(t(k+n) - t(k+n-1)) \tag{45c}$$

$$\begin{aligned}\hat{\phi}(k+n) = \ &\phi(k+n-1) \\ &+ (\omega_x(k) + \omega_y(k)\sin(\phi(k+n-1))\tan(\theta(k)) \\ &+ \omega_z(k)\cos(\phi(k+n-1))\tan(\theta(k)))\end{aligned} \tag{45d}$$

where $n = 1, 2, \ldots, n_p$, n_p presents the predicted horizontal of the state predictor. It is noted that in a relatively short prediction time, the acceleration values $\dot{v}_x, \dot{v}_y, \dot{\omega}_z$ can be assumed as constant values and the vehicle states estimated by Equation (45) within a small, predicted horizontal can have acceptable prediction performance.

4.2. The Upper-Level Control Mode Supervisor

Based on the predicted vehicle states from the vehicle's future state predictor and the diagram of the vehicle stability region determined in Figures 5 and 6, the upper-level control supervisor can determine the best suitable control mode from cruise control mode, yaw-plane stability control mode and roll-stability control mode.

The cruise control mode only aims to maintain the desired longitudinal velocity. In the yaw-plane stability control mode, the desired yaw rate and body side-slip angle can be achieved. In the roll-stability control mode, the vehicle roll stability can be improved and rollover can be prevented.

The control mode selection rules can be presented as follows, and are also illustrated in Figure 8:

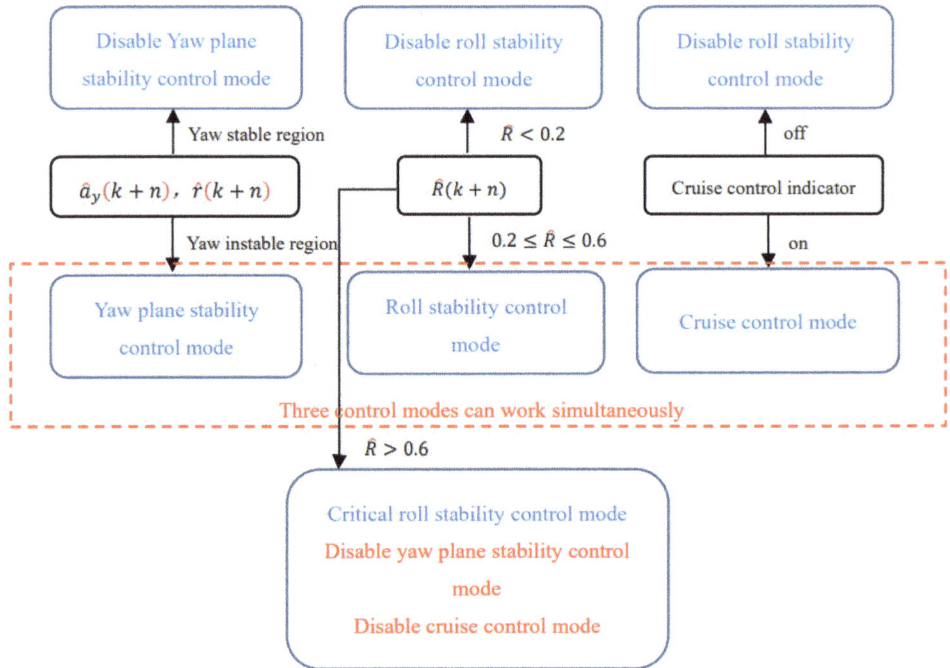

Figure 8. The diagram of the proposed control mode selection rules.

(1)   Determine the predicted state values of lateral velocity $\hat{v}_y(k+n)$, yaw rate $\hat{r}(k+n)$ and the predicted load transfer ratio $\hat{R}(k+n)$. The load transfer ratio can be presented by the following equation, according to [25]:

$$\hat{R}(k+n) = \frac{2K_\phi\phi(k+n) + 2C_\phi\dot{\phi}(k+n)}{cmg} \tag{46}$$

(2)   According to the predicted vehicle lateral velocity $\hat{v}_y(k+n)$, yaw rate $\hat{r}(k+n)$ and diagram of yaw stability region (as in Figures 6 and 7), if the vehicle is moving outside the yaw stability region, the vehicle's yaw-plane stability control mode is selected.

(3)   According to the predicted value of load transfer ratio (LTR) $\hat{R}(k+n)$, if $\hat{R}(k+n) < 0.2$, the rollover is unlikely to happen and the roll-stability control mode is disabled; if $0.2 \leq \hat{R}(k+n) \leq 0.6$, the vehicle is likely to rollover and the roll-stability control mode is selected. These threshold values are determined according to [28].

(4)   If the driver wants to maintain the desired longitudinal velocity, the cruise control mode is selected with full longitudinal velocity control. It is noted the yaw-plane stability control mode, the roll-stability control mode and the cruise control mode could be activated at the same time when their active threshold conditions are satisfied.

(5)   When $\hat{R}(k+n) > 0.6$ and the vehicle is in the critical roll-stability mode, the cruise control and yaw-plane stability control is disabled and the vehicle is in a full brake. According to [28], during the critical roll-stability mode, the inside wheels of the vehicle may have already lifted off and the vehicle may roll over immediately. Rollover prevention is far more important than yaw stability. Therefore, the full brake strategy is selected for critical roll-stability mode by neglecting other control targets.

4.3. The Lower-Level 4 DOF MPC Algorithm

Assumption 2. *It is assumed that the vehicle states $x_{si}(k)$, $x_{ti}(k)$ and $v_{zui}(k)$ are all assumed to be successfully estimated by the proposed state estimator. The vehicle longitudinal velocity, lateral velocity, vertical velocity, yaw angle, roll angle, pitch angle, yaw rate, roll rate and pitch rate are assumed to be easily measured or estimated. In addition, the sideslip angle and slip ratio of the individual wheels are assumed to be known.*

The cost function of the proposed 4 DOF MPC can be presented as the following equation:

$$\min_{T_j(\text{healthy wheels})} J$$

$$\begin{aligned} = &\sum_{i=1}^{N} [a_1(\hat{v}_x(k+i) - v_{xd}(k+i))^2 + a_2(\hat{\beta}(k+i) - \beta_d(k+i))^2 \\ &+ a_3(\hat{\omega}_z(k+i) - \omega_{zd}(k+i))^2 + a_4\hat{R}(k+i)^2] \\ &+ [a_1(\hat{v}_x(k+N+1) - v_{xd}(k+N+1))^2 + a_2(\hat{\beta}(k+N+1) - \beta_d(k+N+1))^2 \\ &+ a_3(\hat{\omega}_z(k+N+1) - \omega_{zd}(k+N+1))^2 + a_4\hat{R}(k+N+1)^2] \end{aligned} \tag{47}$$

where v_{xd} is the desired longitudinal velocity. It is noted that after the tyre deflation, the allocated braking or traction torque on the deflated tyre will further deteriorate the vehicle stability. Therefore, the optimization algorithm (47) only allocates the individual wheel torque T_i to healthy wheels.

β_d and ω_{zd} are desired side-slip angle of C.G. and desired yaw rate, which are determined by a 2 DOF desired vehicle model:

$$\dot{\beta}_d = \left\{ (C_\alpha + C_\alpha)\delta - 2\beta_d(C_\alpha + C_\alpha) - \left[mv_d\omega_{zd} + \frac{2\left(l_fC_\alpha - l_rC_\alpha\right)\omega_{zd}}{v_d} \right] \right\} \frac{1}{mv_d} \tag{48a}$$

$$\dot{\omega}_{zd} = \left\{ 2l_fC_\alpha\delta - 2\left(l_fC_\alpha - l_rC_\alpha\right)\beta_d - \frac{2\omega_{zd}\left(l_f^2C_\alpha + l_r^2C_\alpha\right)}{v_d} \right\} \frac{1}{J_z} \tag{48b}$$

It is noted that the desired yaw rate cannot exceed the maximum yaw rate:

$$\omega_{zd} = \min\left(\omega_{zd}, \frac{\mu g}{v_{xd}}\right) \tag{49}$$

The scaling factors a_1, a_2, a_3, a_4 can be adjusted to reflect different control modes: (1) when it is required to disable the cruise control mode, $a_1 = 0$; (2) when it is required to disable the yaw stability control mode, $a_2 = a_3 = 0$; (3) when it is required to disable the roll-stability control, $a_4 = 0$. It is noted that in order to progressively disable different modes, the function $\tanh x$ is applied.

$$a_1 = \tanh x_1 b_1 \tag{50a}$$

$$a_2 = \tanh x_2 b_2 \tag{50b}$$

$$a_3 = \tanh x_3 b_3 \tag{50c}$$

$$a_4 = \tanh x_4 b_4 \tag{50d}$$

When $x \to +\infty$, $\tanh x \to 1$; When $x \to 0$, $\tanh x \to 0$. x_{1-4} is related to the evaluation criteria of the different mode selections. b_{1-4} is the weighting factors of each individual term.

In optimization cost function (47), the longitudinal velocity v_x, side-slip angle in C.G., yaw rate ω_z and roll angle ϕ can be predicted by the following equations:

$$\hat{v}_x(k+1) = v_x(k) + \dot{v}_x(k)\Delta t \tag{51a}$$

$$\hat{v}_y(k+1) = v_y(k) + \dot{v}_y(k)\Delta t \tag{51b}$$

$$\hat{\omega}_z(k+1) = \omega_z(k) + \dot{\omega}_z(k)\Delta t \tag{51c}$$

$$\hat{\phi}(k+1) = \phi(k) + (\omega_x + \omega_z(k)\theta)\Delta t \tag{51d}$$

$$\hat{\beta}(k+1) = \tan^{-1}\frac{\hat{v}_y(k+1)}{v_x(k+1)} \tag{51e}$$

where $\dot{v}_x(k)$, $\dot{v}_y(k)$, $\dot{\omega}_z(k)$ can be determined based on Equation (1):

$$\dot{v}_x(k) = \omega_y v_z + \omega_z v_y(k) + \frac{\sum_{i=1,2,3,4} F_{xsi}}{m} + g\sin\theta \tag{52a}$$

$$\dot{v}_y(k) = \omega_z(k)v_z - \omega_z(k)v_x(k) + \frac{\sum_{i=1,2,3,4} F_{ysi}}{m} - g\sin\phi\cos\theta \tag{52b}$$

$$\dot{\omega}_z(k) = \frac{l_f(F_{ys1} + F_{ys2}) - l_r(F_{ys3} + F_{ys4})}{J_z} + \frac{c(-F_{xs1} + F_{xs2} - F_{xs3} + F_{xs4})}{2J_z} \tag{52c}$$

where F_{xsi} and F_{ysi} can be determined by Equations (32)–(40) and estimated vehicle states $\hat{x}_{si}(k)$, $\hat{x}_{ti}(k)$ and $\hat{v}_{zui}(k)$.

Therefore, according to Equations (51) and (52), the cost function (47) of MPC can be clearly rewritten as the equation below:

$$\min_{T_j(\text{healthy wheels})} J = \sum_{i=1}^{N} \left\{ a_1 \left[v_x(k+i) + \left(\left(\sum_{i=1,2,3,4} \cos\theta \frac{T_i(k+i-1)}{R_i m} \right) + A_1 \right) \Delta t - v_{xd}(k+i) \right]^2 \right.$$

$$+ a_2 \left[\tan^{-1} \frac{v_y(k+i) + \left(\left(\sum_{i=1,2,3,4} \sin\theta \sin\phi \frac{T_i(k+i-1)}{R_i m} \right) + A_2 \right) \Delta t}{v_x(k+i) + \left(\left(\sum_{i=1,2,3,4} \cos\theta \frac{T_i(k+i-1)}{R_i m} \right) + A_1 \right) \Delta t} - \beta_d(k+i) \right]^2$$

$$+ a_3 \left[\omega_z(k+i) + \left(\left(\frac{c\cos\theta}{2J_z}(T_2 - T_1 + T_4 - T_3) \right) + A_3 \right) \Delta t - \omega_{zd}(k+i) \right]^2$$

$$\left. + a_4 \left[\frac{2k_\phi(\phi(k+i) + (\omega_x + \omega_z(k+i))\Delta t) + c_\phi(\omega_x + \omega_z(k+i))}{cmg} \right]^2 \right\}$$

$$+ \left[a_1 \left[v_x(k+N+1) + \left(\left(\sum_{i=1,2,3,4} \cos\theta \frac{T_i(k+N)}{R_i m} \right) + A_1 \right) \Delta t - v_{xd}(k+N+1) \right]^2 \right.$$

$$+ a_2 \left[\tan^{-1} \frac{v_y(k+N+1) + \left(\left(\sum_{i=1,2,3,4} \sin\theta \sin\phi \frac{T_i(k+N)}{R_i m} \right) + A_2 \right) \Delta t}{v_x(k+N+1) + \left(\left(\sum_{i=1,2,3,4} \cos\theta \frac{T_i(k+N)}{R_i m} \right) + A_1 \right) \Delta t} - \beta_d(k+N+1) \right]^2$$

$$+ a_3 \left[\omega_z(k+N+1) + \left(\left(\frac{c\cos\theta}{2J_z}(T_2 - T_1 + T_4 - T_3) \right) + A_3 \right) \Delta t - \omega_{zd}(k+N+1) \right]^2$$

$$\left. + a_4 \left[\frac{2k_\phi(\phi(k+N+1) + (\omega_x + \omega_z(k+N+1))\Delta t) + c_\phi(\omega_x + \omega_z(k+N+1))}{cmg} \right]^2 \right] \tag{53}$$

where $A_1 = \omega_y v_z + \omega_z(k)v_y(k) + g\sin\theta + \sum_{i=1,2,3,4} \frac{-\sin\theta F_{zgi} + m_u g \sin\theta}{m}$, $A_2 = \omega_z(k)v_z -$

$\omega_z(k)v_x \quad - \quad g\sin\phi\cos\theta \quad + \quad \sum_{i=1,2,3,4} \frac{\cos\phi F_{ygi} + \sin\phi\cos\theta F_{zgi} - m_u g \sin\phi\cos\theta}{m}$,

$A_3 = \frac{l_f(F_{ys1} + F_{ys2}) - l_r(F_{ys3} + F_{ys4})}{J_z} + \frac{c\sin\theta(F_{zg1} - F_{zg2} + F_{zg3} - F_{zg4})}{2J_z}$.

The stability proof of the proposed MPC controller is presented in the Appendix A.

5. Simulation Results

In this section, the proposed 14 DOF MPC is implemented on the simulation platform of Matlab Simulink to present the combined yaw-plane stability and roll-stability control performance. Furthermore, in order to do the comparative study and show the advantages of the proposed 14 DOF MPC, the control performance of the traditional 8 DOF MPC is also presented. This 8 DOF model considers the longitudinal motion, lateral motion, yaw motion, roll motion and rotational motion of four wheels and includes the yaw-plane stability control mode and roll-stability control mode. If $R < 0.6$, the combined yaw-plane stability control mode and roll-stability control mode is enabled; if $R \geq 0.6$, only the roll-stability control mode is enabled, and the vehicle has the full brake.

Three sets of simulation results are presented in the following paragraphs: in the first set of simulations, the proposed MPC is working under the normal driving mode and the yaw-plane stability is the focus; in the second and third sets of simulations, the proposed MPC is under the yaw and roll-stability control mode and the roll stability is the focus. The sampling time of the proposed 14 DOF MPC and 8 DOF MPC was 0.005 s and the prediction horizon was five steps. Due to the large computational effort, the MPC sampling time was chosen as 0.005 s, which is the same as the sampling time constant of vehicle plant dynamics model. The control horizon is also five steps.

In the first set of simulations, the vehicle is assumed to move along the straight line with an initial longitudinal velocity of 40 m/s. The tyre–road friction coefficient is assumed as 0.9. Tyre blow-out happens at the front left wheel after 2 s and the changing of the front-left tyre parameter after tyre blow-out is shown in Figure 9. Figure 10 compares the vehicle dynamics responses when the proposed 14 DOF MPC and traditional 8 DOF MPC are applied. Figure 10f presents the changed real-time scaling factors of the optimization cost function of 14 DOF MPC determined by the upper-level control mode supervisor, which shows that 14 DOF MPC only chooses the cruise control mode 2 s before, when all the tyres are in a healthy condition. After 2 s, the 14 DOF MPC switches into combined cruise control mode and yaw stability control mode. Figure 10a shows the similar control performances of longitudinal velocity for both of the two methods. Figure 10b,c prove that the proposed 14 DOF MPC has a better yaw rate and body side-slip angle control performance than 8 DOF MPC. The 8 DOF MPC has a larger over-shoot of yaw rate and body side-slip angle response after tyre blow-out at 2 s than the no controller applied condition. According to Figure 10d,e, since the LTR is less than the roll-stability control threshold value of 0.2, the roll-stability control is disabled and the proposed 14 DOF MPC cannot control the roll angle and the value of LTR. The motor control torques of the different controllers are shown in Figure 11.

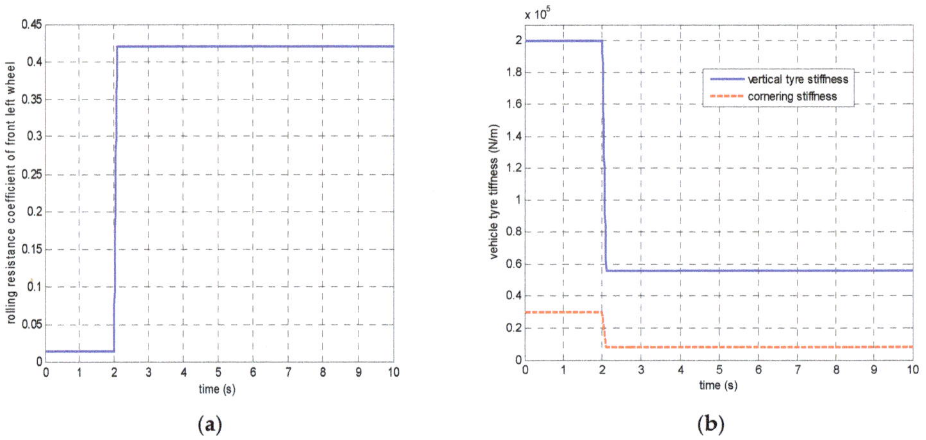

(a)　　　　　　　　　　　　　　(b)

Figure 9. The changing of front-left tyre parameter after tyre blow-out (**a**) rolling resistance coefficient (**b**) tyre stiffness.

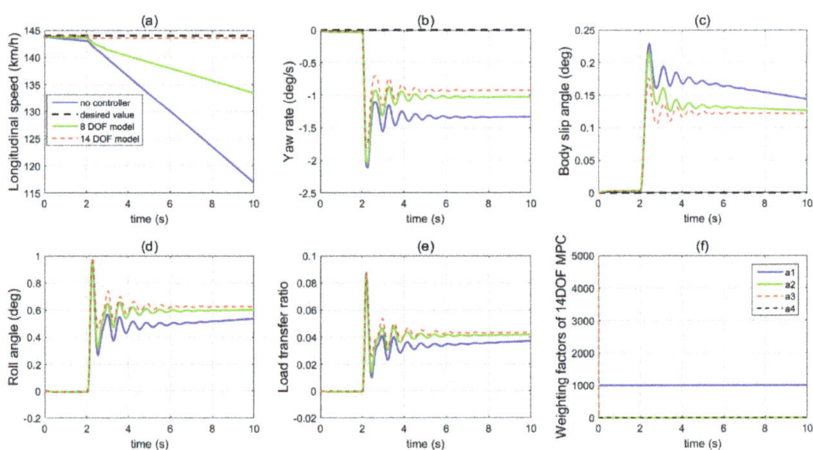

Figure 10. Vehicle dynamics performance when proposed controller applied in the first set of simulations (**a**) longitudinal velocity, (**b**) yaw rate, (**c**) body side-slip angle, (**d**) roll angle, (**e**) LTR, (**f**) scaling factors of 14 DOF MPC.

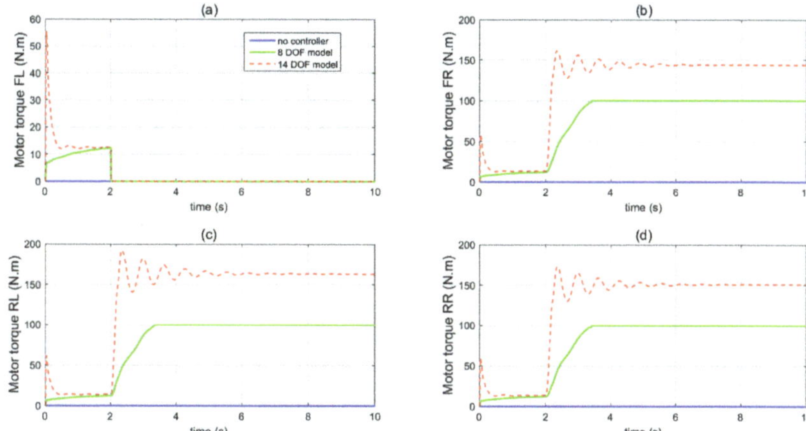

Figure 11. Vehicle motor control inputs in the first set of simulations (**a**) front left wheel, (**b**) front right wheel, (**c**) rear left wheel, (**d**) rear right wheel.

Figure 12 shows the sensitivity analysis of the proposed 14 DOF MPC under different tyre–road friction coefficient conditions and different rolling resistance coefficients after a tyre blow-out. The proposed 14 DOF MPC controller shows a quite robust control performance.

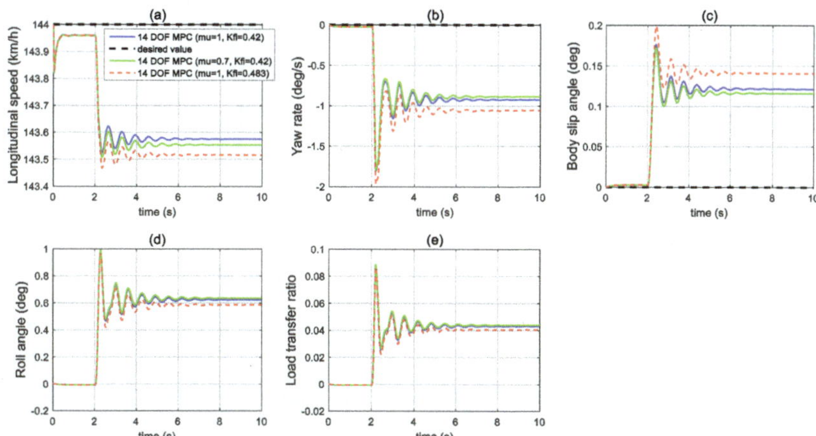

Figure 12. Sensitivity analysis of the dynamics performance of 14 DOF MPC in the first set of simulations (**a**) longitudinal velocity, (**b**) yaw rate, (**c**) body side-slip angle, (**d**) roll angle, (**e**) LTR.

In the second set of simulations, the vehicle started to have the J-turn motion after 2 s (the input steering angle is shown in Figure 13). The initial longitudinal velocity was 40 m/s and tyre–road friction coefficient was 0.9. After 5 s, the front left tyre blows out and the changed tyre parameters are shown in Figure 14. Figure 15 presents and compares the vehicle dynamics responses of the proposed 14 DOF MPC and traditional 8 DOF MPC. Figure 15f shows that 14 DOF MPC switches from the pure cruise control mode into the combined cruise control and yaw stability control mode after the beginning of the J-turn in 2 s, then switches into combined cruise control mode, roll-stability and yaw-stability control mode after 7 s. Figure 15a–c all prove that the proposed 14 DOF MPC can significantly improve the longitudinal velocity, yaw rate and body side-slip angle response after a tyre blow-out compared with 8 DOF MPC. According to Figure 15d,e, after 5 s, the dynamics responses of LTR and roll angle of 14 DOF MPC are improved because the roll-stability control mode is enabled according to Figure 15f. The motor control torques of the different controllers are shown in Figure 16. Figure 17 suggests the control performance of 14 DOF MPC when considering the measurement noise (measured yaw rate with white noise variance of 0.02 rad/s) and shows good robustness on measurement noise.

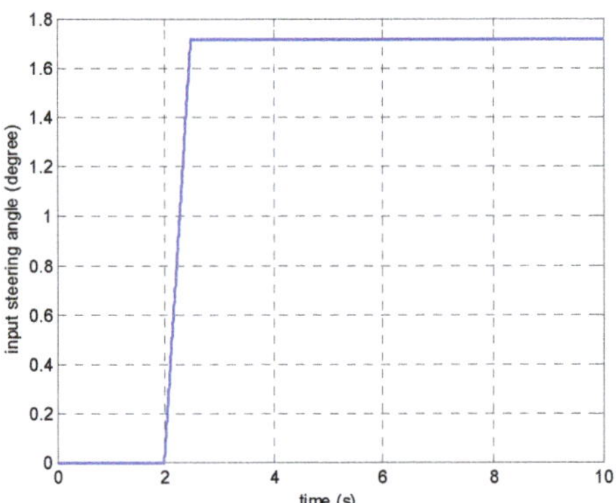

Figure 13. Driver's input steering angle in the second set of simulations.

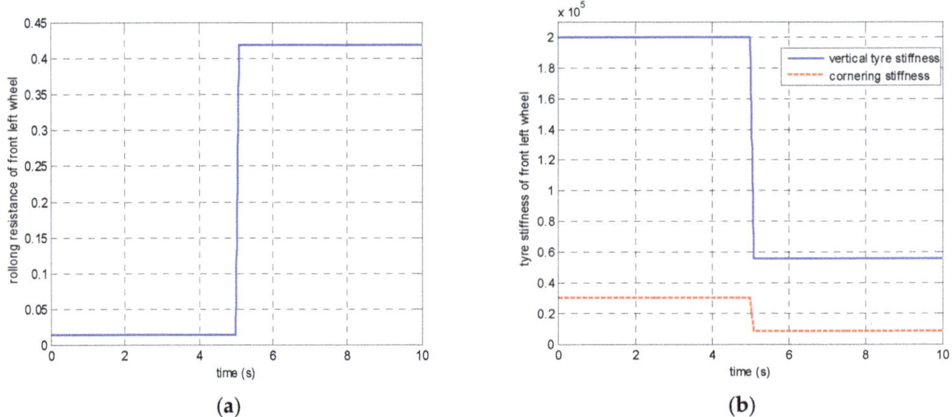

(**a**) (**b**)

Figure 14. The changing of front-left tyre parameters after tyre blow-out in the second set of simulations (**a**) rolling resistance, (**b**) tyre vertical and cornering stiffness.

In the third set of simulations, the fishhook steering manoeuvre (shown in Figure 18) was applied to test the control performance of the proposed method. The initial longitudinal velocity and tyre–road friction coefficient were the same as the second set of simulations. After 5 s, the front left tyre blows out and the changed tyre parameters are the same as in Figure 14. Figure 19 presents and compares the dynamics performance of the proposed 14 DOF MPC and traditional 8 DOF MPC. Figure 19f shows that after 3 s, the control mode of 14 DOF MPC switches from pure cruise control mode into the combined cruise control, yaw stability and roll-stability control mode. Figure 19b,c shows that the proposed 14 DOF MPC and 8 DOF MPC cannot achieve the desired yaw rate and side-slip angle. Figure 19d,e prove that the proposed 14 DOF MPC has much better roll-stability control performance than 8 DOF MPC. In all three sets of simulations, the yaw-stability and roll-stability dynamics control performance of 8 DOF MPC was significantly compromised. This is mainly because the 8 DOF MPC is based on the 8 DOF vehicle dynamics model which cannot accurately present the vehicle dynamics performance during tyre blow-out. The motor control torques of different controllers are shown in Figure 20.

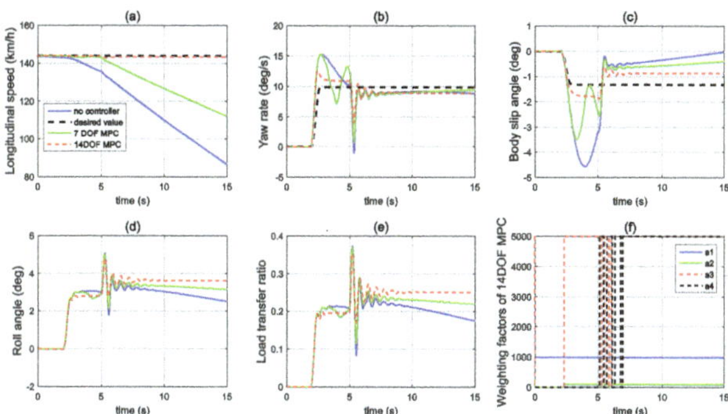

Figure 15. Vehicle dynamics performance when proposed controller applied in the second set of simulations (**a**) longitudinal velocity, (**b**) yaw rate, (**c**) body side-slip angle, (**d**) LTR, (**e**) roll angle, (**f**) scaling factors of 14 DOF MPC.

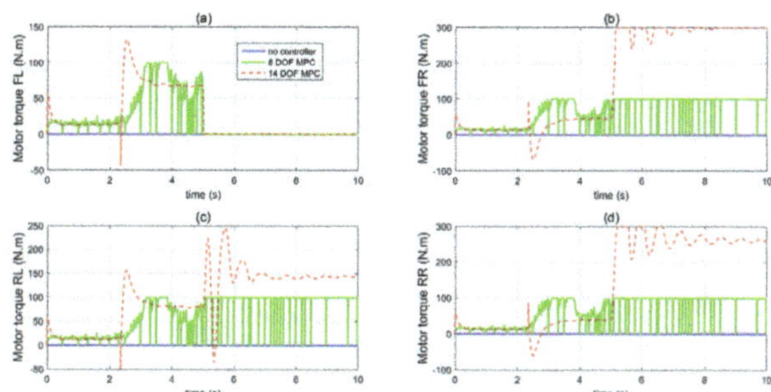

Figure 16. Vehicle motor control inputs in the second set of simulations (**a**) front left wheel, (**b**) front right wheel, (**c**) rear left wheel, (**d**) rear right wheel.

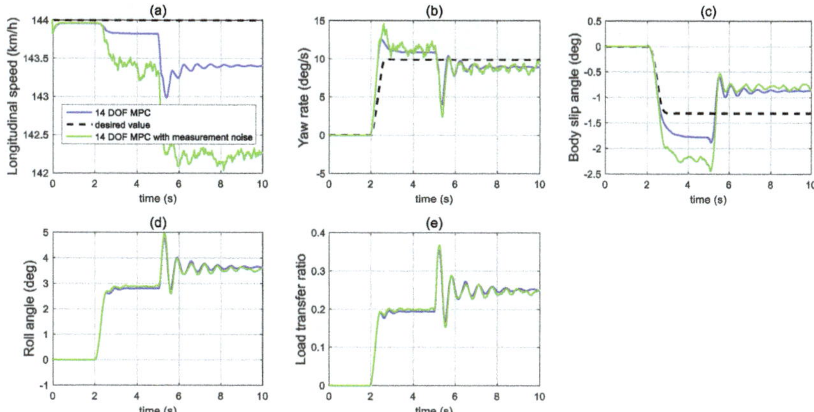

Figure 17. Vehicle dynamics performance when considering the measurement noise in the second set of simulations (**a**) longitudinal velocity, (**b**) yaw rate, (**c**) body side-slip angle, (**d**) LTR, (**e**) roll angle.

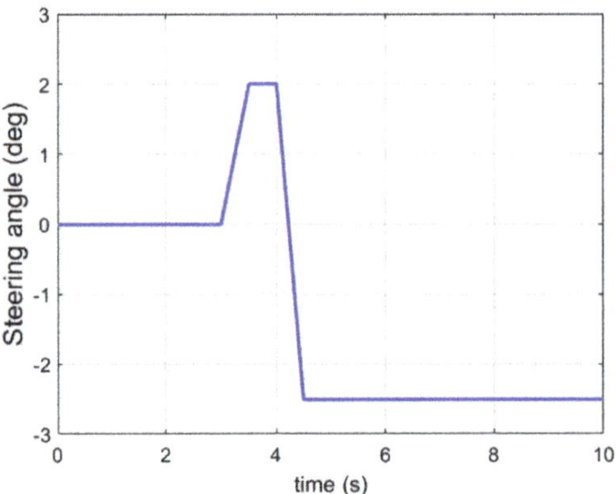

Figure 18. Input steering angle in the third set of simulations.

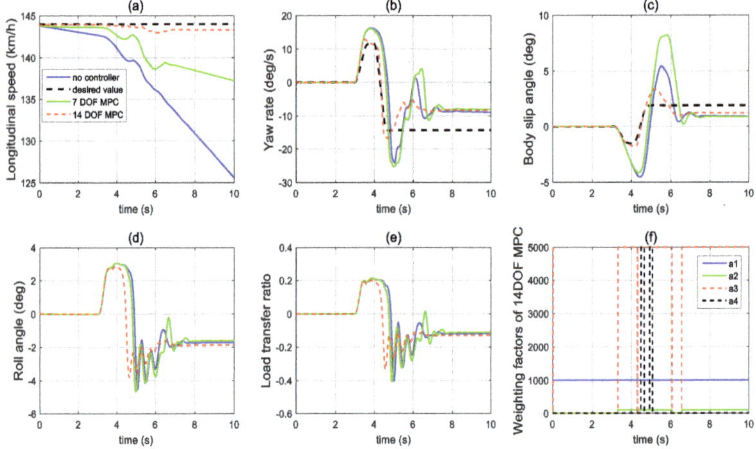

Figure 19. Vehicle dynamics performance when proposed controller applied in the third set of simulations (**a**) longitudinal velocity, (**b**) yaw rate, (**c**) body side-slip angle, (**d**) roll angle, (**e**) LTR, (**f**) scaling factors of 14 DOF MPC.

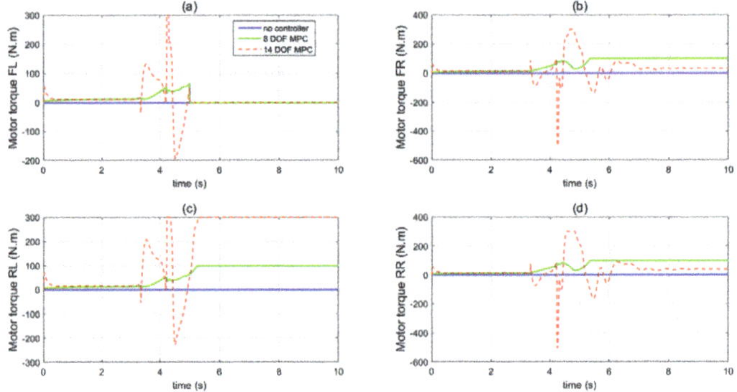

Figure 20. Vehicle motor control inputs in the third set of simulations (**a**) front left wheel, (**b**) front right wheel, (**c**) rear left wheel, (**d**) rear right wheel.

6. Conclusions

This study first proposes a comprehensive 14 DOF vehicle dynamics model to describe the vehicle dynamics performance after a tyre blow-out. Then, based on the proposed 14 DOF model, a non-linear coordinate control framework based on MPC is proposed. The simulation results can be summarised as follows:

(1) The proposed 14 DOF vehicle dynamics model can successfully describe the effect of the changed tyre vertical stiffness, cornering stiffness and rolling resistance after a tyre blow-out on the vehicle dynamics performance.

(2) The proposed vehicle state predictor can successfully predict the vehicle's future states and the proposed upper-level control mode supervisor can use the predicted vehicle states to select the suitable control mode.

(3) The proposed lower-level MPC based on the 14 DOF model can successfully improve the vehicle yaw-dynamics performance including the yaw rate and side-slip angle in the scenarios of tyre blow-out during straight line moving and J-turn manoeuvre.

(4) The proposed lower-level MPC based on the 14 DOF model can successfully improve the roll stability in the challenging scenario of a tyre blow-out during a fishhook manoeuvre when the vehicle has a big load transfer.

(5) The traditional MPC based on the 8 DOF model cannot successfully improve the vehicle yaw stability and roll stability of the vehicle after tyre blow-out.

In the future, the effect of tyre blow-out on the autonomous steering system of autonomous vehicles will be investigated and the design of a fault-tolerant steering control strategy to overcome the issue of tyre blow-out will be focused on.

Author Contributions: Conceptualization, Y.W.; writing—original draft preparation, B.L.; writing—review and editing, C.H.; supervision, B.Z. and H.D. All authors have read and agreed to the published version of the manuscript.

Funding: This research was funded by the Open Research Fund Program of the State Key Laboratory of Advanced Design and Manufacturing for Vehicle Body, Hunan University grant number [31715002] and China Postdoctoral Science Foundation Grant grant number [2018M632958].

Data Availability Statement: Not applicable.

Conflicts of Interest: The authors declare no conflict of interest.

Nomenclature

Vehicle body:

l_f/l_r	Distance of C.G. from front axle/rear axle (m)
c	Track width (m)
h	Height of C.G. (m)
H_{roll}	Distance between and C.G. and sprung mass roll centre (m)
$J_x/J_y/J_z$	Pitch inertia/roll inertia/yaw inertia (kg.m^2)
m_s/m_u	Vehicle sprung mass/unsprung mass (kg)
$v_x/v_y/v_z$	Longitudinal velocity/lateral velocity/vertical velocity of C.G. in fixed-bodycoordinate system (m/s)
$\theta/\phi/\psi$	Pitch angle/roll angle/yaw angle (rad)
$\omega_x/\omega_y/\omega_z$	Pitch rate/roll rate/yaw rate (rad/s)
g	Acceleration of gravity (m/s^2)

Suspension and tyre:

$F_{xti}/F_{yti}/F_{zti}$	Tyre longitudinal force/lateral force/vertical force (N)
$i = 1, 2, 3, 4$	Front left wheel, front right wheel, rear left wheel, rear right wheel
$F_{xgi}/F_{ygi}/F_{zgi}$	Longitudinal force/lateral force/vertical force at tyre contact patch in coordinate system attached to tyre contact patch (N)
$F_{xgsi}/F_{ygsi}/F_{zgsi}$	Longitudinal force/lateral force/vertical force at tyre contact patch in coordinate system attached to C.G. (N)
$F_{xsi}/F_{ysi}/F_{zsi}$	Longitudinal force/lateral force/vertical force transferred to C.G. in the coordinate system attached to C.G. (N)
F_{dzi}	Load transfer of each wheel (N)
M_{xi}/M_{yi}	Roll moment and pitch moment transmitted to the sprung mass (N.m)
I_ω	Wheel moment of inertia (kg.m^2)
k_{si}/k_{ti}	Suspension stiffness/tyre vertical stiffness (N/m)
b_{si}	Suspension damping coefficient (N.s/m)
$k_{ARB,i}$	Stiffness of anti-roll bar
$b_{ARB,i}$	Damping coefficient of anti-roll bar
l_{si}	Instantaneous length of strut (m)
R_i	Instantaneous length of tyre radius (m)
x_{si}/x_{ti}	Suspension spring compression/tyre spring compression (m)
v_{xgi}/v_{ygi}	Longitudinal velocity/lateral velocity at tyre contact patch in coordinate system attached to tyre contact patch (m/s)
$v_{xsi}/v_{ysi}/v_{zsi}$	Longitudinal velocity/lateral velocity/vertical velocity at suspension cor. in coordinate system attached to C.G. (m/s)
$v_{xui}/v_{yui}/v_{zui}$	Unsprung mass longitudinal velocity/lateral velocity/vertical velocity in coordinate system attached to C.G. (m/s)
ω_i	Angular velocity of wheel rotation (rad/s)
α_i	Tyre lateral side-slip angle (rad)
s_i	Tyre longitudinal slip ratio
δ_i	Steering input of each wheel (rad)
T_i	Traction/brake input of each wheel (N.m)

Appendix A

The Lyapunov method is used to prove the stability of the proposed integrated MPC approach. The Lyapunov function can be chosen as the optimal value of the optimisation cost function of MPC according to Equation (53):

$$V^0(k) = \min_{u_j} \sum_{i=1}^{N} \left[a_1(\hat{v}_x(k+i) - v_{xd}(k+i))^2 + a_2(\hat{\beta}(k+i) - \beta_d(k+i))^2 \right.$$
$$\left. + a_3(\hat{\omega}_z(k+i) - \omega_{zd}(k+i))^2 + a_4 \hat{R}(k+i)^2 \right] = \min_{u_j} \sum_{i=1}^{N} L(x(k+i), u(k+i-1)) \tag{A1}$$

It can be noted that $V^0(k)$ is positively defined. Now we need to prove $V^0(k+1) \leq V^0(k)$ to finish the Lyapunov proof.

$$V^0(k+1) = \min_{u_j} \sum_{i=1}^{N} L(x(k+i+1), u(k+i))$$
$$= \min_{u_j} \left\{ \left[\sum_{i=1}^{N} L(x(k+i), u(k+i-1)) \right] - L(x(k+1), u(k)) \right.$$
$$\left. + L(x(k+N+1), u(k+N)) \right\}$$
$$\leq -L(x(k+1), u(k)) + V^0(k) + L(x(k+N+1), u(k+N)) \tag{A2}$$

If the absolute value of terminal term $L(x(k+N+1), u(k+N))$ is smaller than the absolute value of initial term $L(x(k+1), u(k))$, the Lyapunov stability of the MPC can be proved.

Since over the whole prediction horizon, the desired optimisation targets are constant. If the four wheels are all in a healthy condition or only one of the four tyres has a blow-out, the left and right motors can generate different motor torques and an additional yaw moment (torque-vectoring function) to improve the yaw stability performance. Thus, the absolute value of the terminal term $L(x(k+N+1), u(k+N))$ is smaller than the absolute value of the initial term $L(x(k+1), u(k))$ and the stability of the MPC can be proved. On the other hand, if two or more of the tyres have a blow-out, the torque-vectoring function cannot be surely achieved so the absolute value of the terminal term $L(x(k+N+1), u(k+N))$ is not surely smaller than the absolute value of the initial term $L(x(k+1), u(k))$, which cannot prove the stability of the MPC.

References

1. General Estimates System (GES). *User's Manual, National Center for Statistics and Analysis*; NHTSA, Department of Transportation: Washington, DC, USA, 1996.
2. Choi, E.-H. *Tire-Related Factors in the Pre-Crash Phase*; National Highway Traffic Safety Administration: Washington, DC, USA, 2012.
3. Martini, A.; Bonelli, G.P.; Rivola, A. Virtual Testing of Counterbalance Forklift Trucks: Implementation and Experimental Validation of a Numerical Multibody Model. *Machines* **2020**, *8*, 26. [CrossRef]
4. Rebelle, J.; Mistrot, P.; Poirot, R. Development and validation of a numerical model for predicting forklift truck tip-over. *Veh. Syst. Dyn.* **2009**, *47*, 771–804. [CrossRef]
5. Jing, H.; Liu, Z. Gain-scheduling robust control for a tire-blow-out road vehicle. *Proc. Inst. Mech. Eng. Part D J. Automob. Eng.* **2018**, *233*, 344–362. [CrossRef]
6. Lozia, Z. Simulation Tests of Biaxial Vehicle Motion after a "Tire Blow-Out". *Simulation* **2005**, *1*, 0410. [CrossRef]
7. Blythe, W.; Day, T.D.; Grimes, W.D. 3-Dimensional Simulation of Vehicle Response to Tire Blow-outs. *SAE Trans.* **1998**, *1*, 361–375. [CrossRef]
8. Tandy, D.F.; Ault, B.N.; Colborn, J.; Pascarella, R. Objective measurement of vehicle steering and handling performance when a tire lost its air. *SAE Int. J. Passeng. Cars-Mech. Syst.* **2013**, *6*, 741–769. [CrossRef]
9. Wang, F.; Chen, H.; Cao, D. Nonlinear Coordinated Motion Control of Road Vehicles After a Tire Blowout. *IEEE Trans. Control Syst. Technol.* **2015**, *24*, 956–970. [CrossRef]
10. Huang, C.; Lv, C.; Hang, P.; Hu, Z.; Xing, Y. Human-Machine Adaptive Shared Control for Safe Automated Driving Under Automation Degradation. *IEEE Intell. Transp. Syst. Mag.* **2021**, 2–15. [CrossRef]
11. Wang, F.; Chen, H.; Guo, K.; Cao, D. A novel integrated approach for path following and directional stability control of road vehicles after a tire blow-out. *Mech. Syst. Signal Process.* **2017**, *93*, 431–444. [CrossRef]
12. Savaresi, S.; Tanelli, M.; Langthaler, P.; Del Re, L. New Regressors for the Direct Identification of Tire Deformation in Road Vehicles Via "In-Tire" Accelerometers. *IEEE Trans. Control Syst. Technol.* **2008**, *16*, 769–780. [CrossRef]
13. López-Estrada, F.-R.; Rotondo, D.; Valencia-Palomo, G. A Review of Convex Approaches for Control, Observation and Safety of Linear Parameter Varying and Takagi-Sugeno Systems. *Processes* **2019**, *7*, 814. [CrossRef]
14. Patwardhan, S.; Tomizuka, M. Theory and experiments of tire blow-out effects and hazard reduction control for automated vehicle lateral control system. In Proceedings of the American Control Conference, Baltimore, MD, USA, 29 June–1 July 1994.
15. Chen, Q.; Liu, Y.; Li, X. Stability Control of Vehicle Emergency Braking with Tire Blowout. *Int. J. Veh. Technol.* **2014**, *2014*, 436175. [CrossRef]
16. Wielenga, T.J. A Method for Reducing On-Road Rollovers—Anti-Rollover Braking. In Proceedings of the SAE International Congress and Exposition, Detroit, MI, USA, 1–4 March 1999. [CrossRef]
17. Chen, B.; Peng, H. Differential-braking-based rollover prevention for spot utility vehicles with human-in-the-loop evaluations. *Veh. Syst. Dyn.* **2001**, *36*, 359–389. [CrossRef]
18. Schofield, B.; Hagglund, T. Optimal control allocation in vehicle dynamics control for rollover mitigation. In Proceedings of the American Control Conference, Seattle, WA, USA, 11–13 June 2008; pp. 3231–3236. [CrossRef]
19. Odenthal, D.; Bunte, T.; Ackermann, J. Nonlinear steering and braking control for vehicle rollover avoidance. In Proceedings of the European Control Conference, Karlsruhe, Germany, 31 August–3 September 1999. [CrossRef]
20. Solmaz, S.; Corless, M.; Shorten, R. A methodology for the design of robust rollover prevention controllers for automotive vehicles with active steering. *Int. J. Control* **2007**, *80*, 1763–1779. [CrossRef]
21. Sampson, D.J.; Cebon, D. Active Roll Control of Single Unit Heavy Road Vehicles. *Veh. Syst. Dyn.* **2003**, *40*, 229–270. [CrossRef]
22. Lee, A.Y. Coordinated Control of Steering and Anti-Roll Bars to Alter Vehicle Rollover Tendencies. *J. Dyn. Syst. Meas. Control* **2002**, *124*, 127–132. [CrossRef]
23. Gaspar, P.; Szaszi, I.; Bokor, J. Reconfigurable control structure to prevent the rollover of heavy vehicles. *Control Eng. Pract.* **2005**, *13*, 699–711. [CrossRef]
24. Rajamani, R.; Piyabongkarn, D.N. New paradigms for the integration of yaw stability and rollover prevention functions in vehicle stability control. *IEEE Trans. Intell. Transp. Syst.* **2013**, *14*, 249–261. [CrossRef]
25. Alberding, M.; Tjonnas, J.; Johansen, T. Integration of Vehicle Yaw Stabilisation and Rollover Prevention through Nonlonear Hierarchical Control Allocation. *Veh. Syst. Dyn.* **2014**, *52*, 1607–1621. [CrossRef]
26. Yin, C.; Xu, B.; Chen, X.; Qin, Z.; Bian, Y.; Sun, N. Nonlinear Model Predictive Control for Path Tracking Using Discrete Previewed Points. In Proceedings of the 2020 IEEE 23rd International Conference on Intelligent Transportation Systems (ITSC), Rhodes, Greece, 20–23 September 2020; pp. 1–6.
27. Chen, L.; Qin, Z.; Kong, W.; Chen, X. Lateral control using LQR for intelligent vehicles based on the optimal front-tire lateral force. *J. Tsinghua Univ. Sci. Technol.* **2021**, *61*, 906–912.

28. Li, L.; Lu, Y.; Wang, R.; Chen, J. A Three-Dimensional Dynamics Control Framework of Vehicle Lateral Stability and Rollover Prevention via Active Braking With MPC. *IEEE Trans. Ind. Electron.* **2017**, *64*, 3389–3401. [CrossRef]
29. Ataei, M.; Khajepour, A.; Jeon, S. Reconfigurable Integrated Stability Control for Four- and Three-wheeled Urban Vehicles With Flexible Combinations of Actuation Systems. *IEEE/ASME Trans. Mechatron.* **2018**, *23*, 2031–2041. [CrossRef]
30. Li, H.-Q.; Zhao, Y.-Q.; Lin, F.; Xiao, Z. Integrated yaw and rollover control based on differential braking for off-road vehicles with mechanical elastic wheel. *J. Central South Univ.* **2019**, *26*, 2354–2367. [CrossRef]
31. Guo, N.; Zhang, X.; Zou, Y.; Lenzo, B.; Du, G.; Zhang, T. A Supervisory Control Strategy of Distributed Drive Electric Vehicles for Coordinating Handling, Lateral Stability, and Energy Efficiency. *IEEE Trans. Transp. Electrif.* **2021**, *7*, 2488–2504. [CrossRef]
32. Feng, J.; Chen, S.; Qi, Z. Coordinated Chassis Control of 4WD Vehicles Utilizing Differential Braking, Traction Distribution and Active Front Steering. *IEEE Access* **2020**, *8*, 81055–81068. [CrossRef]
33. Shim, T.; Ghike, C. Understanding the limitations of different vehicle models for roll dynamics studies. *Veh. Syst. Dyn.* **2007**, *45*, 191–216. [CrossRef]
34. Boada, B.L.; Boada, M.J.L.; Diaz, V. Fuzzy-logic applied to yaw moment control for vehicle stability. *Veh. Syst. Dyn.* **2005**, *43*, 753–770. [CrossRef]
35. Li, B.; Du, H.; Li, W. A Potential Field Approach-Based Trajectory Control for Autonomous Electric Vehicles With In-Wheel Motors. *IEEE Trans. Intell. Transp. Syst.* **2016**, *18*, 2044–2055. [CrossRef]
36. Steyn, W.J.; Lombard, D.; Mashabela, G.; Rudolph, J.S.; Francois, W.; Singh, D.; Hu, C.; Valentin, J.; Liu, Z. Vehicle Rolling Resistance as Affected by Tire and Road Conditions. In Proceedings of the 4th Geo-China International Conference, Shandong, China, 25–27 July 2016; pp. 129–136. [CrossRef]
37. Li, B.; Du, H.; Li, W. A Novel Method for Side Slip Angle Estimation of Omni-Directional Vehicles. *SAE Int. J. Passeng. Cars-Electron. Electr. Syst.* **2014**, *7*, 471–480. [CrossRef]
38. Wang, J.; Alexander, L.; Rajamani, R. Friction estimation on high-way vehicles using longitudinal measurements. *ASME J. Dyn. Syst. Meas. Control* **2004**, *126*, 265–275. [CrossRef]

Article

A New Trajectory Tracking Algorithm for Autonomous Vehicles Based on Model Predictive Control

Zhejun Huang [1,2,3], Huiyun Li [1,2,3,*], Wenfei Li [1,2,3], Jia Liu [1,2,3], Chao Huang [4], Zhiheng Yang [1,2,3] and Wenqi Fang [5]

[1] Shenzhen Institutes of Advanced Technology, Chinese Academy of Sciences, Shenzhen 518055, China; zj.huang@siat.ac.cn (Z.H.); wf.li1@siat.ac.cn (W.L.); jia.liu1@siat.ac.cn (J.L.); zh.yang@siat.ac.cn (Z.Y.)
[2] CAS Key Laboratory of Human-Machine Intelligence-Synergy Systems, Shenzhen Institutes of Advanced Technology, Shenzhen 518055, China
[3] Guangdong-Hong Kong-Macao Joint Laboratory of Human-Machine Intelligence-Synergy Systems, Shenzhen 518055, China
[4] Department of Industrial and Systems Engineering, The Hong Kong Polytechnic University, Hong Kong 999077, China; hchao.huang@polyu.edu.hk
[5] Research Center of Digital Intelligence Technology, Nanhu Lab, Jiaxing 314033, China; wqfang@nanhulab.ac.cn
* Correspondence: hy.li@siat.ac.cn; Tel.: +86-132-6580-5460

Abstract: Trajectory tracking is a key technology for precisely controlling autonomous vehicles. In this paper, we propose a trajectory-tracking method based on model predictive control. Instead of using the forward Euler integration method, the backward Euler integration method is used to establish the predictive model. To meet the real-time requirement, a constraint is imposed on the control law and the warm-start technique is employed. The MPC-based controller is proved to be stable. The simulation results demonstrate that, at the cost of no or a little increase in computational time, the tracking performance of the controller is much better than that of controllers using the forward Euler method. The maximum lateral errors are reduced by 69.09%, 47.89% and 78.66%. The real-time performance of the MPC controller is good. The calculation time is below 0.0203 s, which is shorter than the control period.

Keywords: autonomous driving; trajectory tracking; real-time control; model predictive control

Citation: Huang, Z.; Li, H.; Li, W.; Liu, J.; Huang, C.; Yang, Z.; Fang, W. A New Trajectory Tracking Algorithm for Autonomous Vehicles Based on Model Predictive Control. *Sensors* **2021**, *21*, 7165. https://doi.org/10.3390/s21217165

Academic Editor: Soufiene Djahel

Received: 3 September 2021
Accepted: 26 October 2021
Published: 28 October 2021

Publisher's Note: MDPI stays neutral with regard to jurisdictional claims in published maps and institutional affiliations.

Copyright: © 2021 by the authors. Licensee MDPI, Basel, Switzerland. This article is an open access article distributed under the terms and conditions of the Creative Commons Attribution (CC BY) license (https://creativecommons.org/licenses/by/4.0/).

1. Introduction

Research in autonomous driving has aroused increasingly more attention of late [1,2]. The most basic and important goal of an autonomous passenger vehicle is to free people from driving and safely take passengers from an initial state to a final state in a desired interval of time. The architecture of contemporary autonomous driving systems is typically organized into the perception system and the decision-making system [3]. The perception system takes charge of estimating the vehicle states and representing the surrounding environment using data from sensors, including Light Detection and Ranging (LIDAR), Radio Detection and Ranging (RADAR), cameras, a Global Positioning System (GPS), and an Inertial Measurement Unit (IMU). In particular, camera data is of vital importance. Tesla released its fully self-driving version 9 Beta software on 10 July 2021, which relies on camera vision and neural net processing to deliver autopilot. The Lane Support System (LSS) uses cameras to identify the road lines and alert drivers to potential hazards. However, there is still much uncertainty regarding the vision needs of LSS and the results of the experimental tests for LSS are quite limited [4,5]. Cafiso and Pappalardo [4] developed logit models to investigate road characteristics and conditions that affects LSS performance and employed the Firth's penalized maximum-likelihood method to estimate the logistic regression coefficients and standard errors to describe the rareness of the events. They gave threshold values for the luminance coefficient in diffuse lighting conditions and horizontal curvature radius, and presented remarks on road maintenance and design standards. Pappalardo et al. [5] experimentally tested LSS performance in two lane rural roads with distinct geometric alignments and road marking conditions. They proposed a decision tree method to analyze the cause of the LSS faults and the effects of the variables involved. On the other hand, the decision-making system takes charge of navigating a car from the current position to a goal position safely, feasibly, timely, and comfortably [6]. The decision-making system can be further divided into three subsystems: a decision and planning system, a control system, and an actuation system. Of them, motion planning is a key autonomous driving technique. Li and Shao [7] proposed a motion planner for autonomous parking, and the time-optimal dynamic optimization problem with vehicle kinematics, collision-avoidance conditions and mechanical constraints was solved using a simultaneous approach using the interior-point method. Zhang [8] proposed a hierarchical three-layer trajectory planning framework to realize real-time collision avoidance on highways under complex driving conditions. Therefore, a general framework of an autonomous driving system is shown in Figure 1. Besides a perception and decision-making system, advanced X-by-wire chassis, including drive-by-wire, steer-by-wire, brake-by-wire and active/semi-active suspension subsystems are of vital importance to improving the performance and safety of connected and autonomous vehicles. Zhang et al. [9] proposed a fault-tolerant control method for steer-by-wire systems to mitigate the undesirable influence of front wheel steering angle sensor faults via the use of the Kalman filtering technique. A complete and systematic survey on chassis coordinated control methods for full X-by-wire vehicles can be found in [10]. Here we focus on trajectory tracking, which is a key technology for precisely controlling autonomous vehicles.

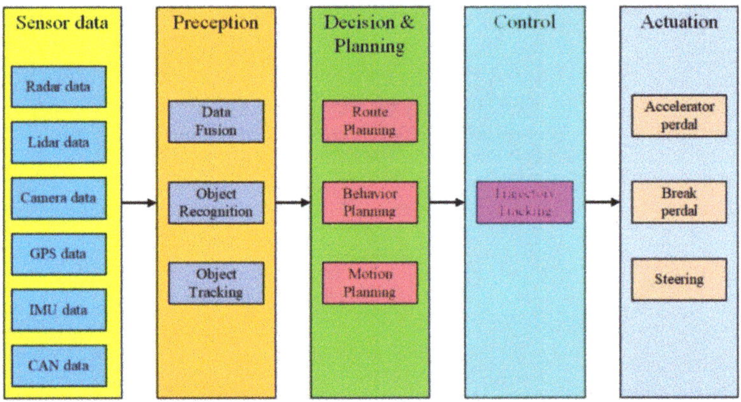

Figure 1. General framework of an autonomous driving system comprising perception, decision and planning, control and actuation.

The trajectory tracking algorithms are designed to ensure that a vehicle follows a predetermined trajectory generated either offline using navigation systems or online using the motion planning module. The performance of trajectory tracking directly determines the performance of autonomous vehicles, which involves driving safety, passenger comfort, travel efficiency, and energy consumption [11]. The trajectory tracking control of autonomous vehicles is a challenging research area because these systems typically are nonlinear systems with non-holonomic constraints.

Pure-pursuit [12] and the Stanley method [13] are two prevalent geometric controllers. The main advantage of these methods is that they use simple geometric models with few parameters, and therefore can give timely feedback on the current state and constraints to meet the real-time requirement of an autonomous vehicle. The pure-pursuit method and its variants are one of the most commonly used methods to solve the path-tracking problem for mobile robots [14]. The Stanley method is the path-tracking approach used by

Stanley, Stanford University's autonomous car; Stanley won the DARPA Grand Challenge in 2005 [13]. However, these methods have their limitations. Pure-pursuit control works as a proportional controller of the steering angle operating on the cross-track error by calculating the curvature from the current position to some goal position. When the look-ahead distance is too large, its performance is poor, and the vehicle may cut corners when changing direction or making a U-turn. The Stanley method considers both the heading and cross-track errors and therefore it is more effective and steady than the pure-pursuit method. But it does not perform well on discontinuous paths. To sum up, geometric-based tracking controllers (pure pursuit, Stanley, etc.) have a simple structure and are easy to implement. However, they are not suitable for applications that need to consider vehicle dynamics (e.g., high-speed trajectory tracking, extreme path curvature, etc.). It is also difficult to achieve a trade-off between stability and tracking performance [15].

Proportional-Integral-Derivative controllers (PID) [16] and sliding model controllers (SMC) [17] are two prevalent classical control algorithms. Although PID controllers have good tracking performance, there is a major challenge in the tuning of the parameters because of the vehicle and tire nonlinearities. SMC is a well-developed nonlinear state-feedback controller and has been used to design vehicle trajectory tracking controllers. Because of the nonlinear control law, SMC shows good tracking accuracy. However, there are several drawbacks: first, its performance is sensitive to the sampling rate of the controller; second, chattering problems exist under certain conditions [18]; third, robustness is only guaranteed on the sliding surface; and lastly, it needs prior knowledge [19]. To sum up, compared with geometric-based tracking controllers, model-based tracking methods are more feasible and reliable in real driving scenarios at the cost of the increase in computational burden and complexity.

Reinforcement learning (RL) has shown an ability to achieve super-human results at turn-based games like Go [20] and chess [21]. Deep RL has been applied to the decision-making system of autonomous driving in several simulated environments [22]. Mohammadi et al. [23] proposed an optimal tracking controller for nonlinear continuous-time systems with time-delay, mismatched external disturbances, and input constraints, using the technique of integral reinforcement learning and a Hamilton-Jacobi-Bellman equation. However, there are two main limitations for RL-based methods. First, they require large amounts of data to build up a feasible model; specifically, data is sometimes expensive and hard to obtain. Second, they require a sufficiently long time to train the model to complete the specific tasks due to significant data manipulation. The performance of the controllers using machine-learning methods relies on the learning capability of the model and the quality of the data.

Model Predictive Control (MPC) has been applied to trajectory planning and tracking of an autonomous vehicle due to its flexibility and ability to compute optimal solutions with hard and soft constraints [24,25]. Shen et al. [24] proposed a unified receding-horizon optimization scheme for the integrated path-planning and tracking control of an autonomous underwater vehicle using nonlinear MPC techniques. Borrelli et al. [25] proposed a novel approach to autonomous steering systems based on an MPC scheme. The general framework of an MPC structure is shown in Figure 2. However, these MPC-based tracking controllers are feasible only in low-speed scenarios.

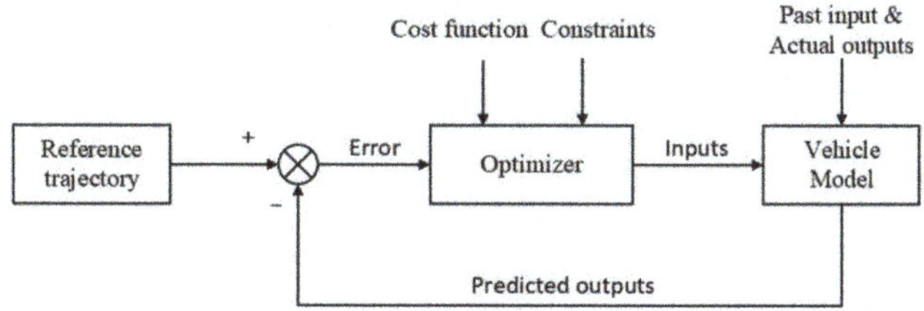

Figure 2. General framework of an MPC structure.

The accuracy of the trajectory tracking control can be greatly improved by improving the accuracy of the predictive model. Most researchers have attempted to improve the accuracy of the kinematic or dynamic model to improve the accuracy of the controller. Few people paid attention to the computation errors during the integration process. With the accumulation of the computation errors, the controller could lose its stability or an accident might even result.

In this paper, we propose a new trajectory-tracking algorithm based on MPC. Instead of using the forward Euler integration method, the backward Euler integration method is used to establish the predictive model.

The contributions here can be summarized as follows:

- A trajectory tracking method is proposed based on MPC. Instead of using the forward Euler method, the backward Euler method is used to establish the predictive model. The proposed method is designed to meet the real-time requirement of autonomous vehicles by structuralizing the control law and employing the warm-start strategy.
- Unlike conventional MPC-based controllers, both the acceleration and steer angle are control inputs. The proposed MPC-based controller can automatically adjust the velocity according to the information of the reference trajectory.
- The dynamic regret of the proposed controller is tightly bounded, and the closed-loop controller is proved to be stable.
- The MPC controller using the backward Euler method has a better tracking accuracy in the lateral error, and it is more robust.

This paper is organized as follows. The MPC-based controller of autonomous vehicles is described in Section 2. After that, the stabilizability of the controller is discussed in Section 3. Simulation results are shown in Section 4. Section 5 concludes this paper by summarizing all of the main results.

2. Control Design

Establishing a prediction model and designing a rolling optimization function are the kernels of designing a path tracking controller. Due to the strongly nonlinearity of vehicle dynamics, it is very hard to establish a model to describe the actual vehicle dynamics. Researchers generally use Ackermann steering geometry and its simplified bicycle models to describe the vehicle kinematics and dynamics. MPC schemes using dynamic vehicle models and various tire models are generally computationally expensive, and tire models may become singular at low speeds [26]. Kong et al. [26] compared a kinematic and a dynamic bicycle model, and showed that both models could correctly predict a vehicle's future states, and combining MPC schemes with a simple kinematic bicycle model is less computationally expensive. Polack et al. [27] compared a 3-DOF kinematic bicycle model with a 9-DOF model, and showed that the 3-DOF model could capture enough of the non-holonomic constraints of the actual vehicle dynamics. When the maximum-allowed lateral acceleration of a vehicle was no greater than $0.5\,g$ m/s^2, where g is the acceleration due to gravity, using a 3-DOF kinematic bicycle model produces acceptable results and could generate a feasible track. Chen et al. [28] implemented an MPC-based controller for path-tracking using three vehicle dynamics models: a bicycle model, an 8-DOF model and a 14-DOF model. They showed that the bicycle controller could successfully navigate a vehicle along the given path and calculate the optimal steering sequences faster than the controllers with the 8-DOF and 14-DOF vehicle. They concluded that the bicycle controller is suitable for a possible physical implementation with real-time requirements. Therefore, in this paper, the kinematic model of autonomous vehicles is used [26,29]

$$
\begin{aligned}
\dot{x} &= v\cos(\varphi + \beta), \\
\dot{y} &= v\sin(\varphi + \beta), \\
\dot{\varphi} &= \frac{v\sin(\beta)}{l_r}, \\
\dot{v} &= a, \\
\beta &= \tan^{-1}\left(\frac{l_r}{l_f + l_r}\tan(\delta)\right),
\end{aligned}
\tag{1}
$$

where x and y are the coordinates of the center of mass in an inertial frame (X,Y). φ is the inertial heading, and v is the longitudinal speed of the vehicle. The parameters l_f and l_r are the distance from the center to the front and rear axles, and δ is the front steering angle. Two front and two rear wheels of the vehicle are combined into single wheels located at the center of the front and rear axle, respectively, as illustrated in Figure 3. β is the slip angle at the center of gravity.

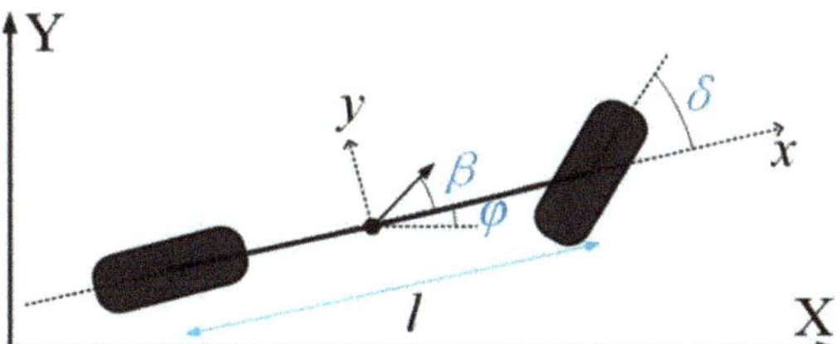

Figure 3. Kinematic rear axle bicycle model of the vehicle.

In our problem, $X = [x, y, \varphi, v]$ is the vehicle state, $U = [a, \delta]$ is the control state. The model is established based on the following assumptions.

- The vehicle is traveling on a flat surface, with the vehicle's movement perpendicular to the road surface ignored.
- Only the front wheel can be steered.
- The wind resistance and ground-side friction that the wheels are subjected to while driving are ignored.
- The wheels always maintain good rolling contact with the ground.
- The impact of the vehicle suspension is not taken into account.
- Load transfer is not considered.

The state-space equations of the vehicle system (1) are continuous in time and cannot be used for the design of the MPC algorithm directly. Therefore, the model of the system was converted to discrete state-space equations by discretizing the state-space equations. We assume that the model can be rewritten as

$$\dot{X} = f(X, U). \tag{2}$$

Generally, the state at $k + 1$ instant at time t is computed using the forward Euler integration method

$$X(k+1|t) = X(k|t) + T_s \dot{X}(k|t) = X(k|t) + T_s f(X(k|t), U(k|t)), \tag{3}$$

where T_s is the sampling time.

In this paper, instead of the forward Euler method, the backward Euler method is used to establish the predictive model. Although it requires an extra computation at each iteration, the backward Euler method has great stability properties and its local truncation error is of order $O(T_s^3)$, which is much smaller than $O(T_s^2)$ using the forward Euler method. Hence, the backward Euler method's error generally decreases faster as $T_s \to 0$.

The state at $k + 1$ instant at time t is computed using the backward Euler method

$$\begin{aligned} \widetilde{X}(k+1|t) &= X(k|t) + T_s f(X(k|t), U(k|t)), \\ X(k+1|t) &= X(k|t) + T_s f(\widetilde{X}(k+1|t), U(k|t)), \end{aligned} \tag{4}$$

Equation system (8) can be rewritten as

$$\begin{aligned} X(k+1|t) &= X(k|t) + T_s \widetilde{f}(X(k|t), U(k|t)), \\ \widetilde{f}(X(k|t), U(k|t)) &= f(X(k|t) + T_s f(X(k|t), U(k|t)), U(k|t)). \end{aligned} \tag{5}$$

Therefore, the state information of vehicles in the prediction horizon N_P can be obtained

$$X(k+1|t) = X(k|t) + T_s\widetilde{f}(X(k|t), U(k|t)),$$
$$\vdots$$
$$X(k+i|t) = X(k+i-1|t) + T_s\widetilde{f}(X(k+i-1|t), U(k+i-1|t)),$$
$$\vdots \tag{6}$$
$$X(k+N_c+1|t) = X(k+N_c|t) + T_s\widetilde{f}(X(k+N_c|t), U(k+N_c|t)),$$
$$\vdots$$
$$X(k+N_P|t) = X(k+N_P-1|t) + T_s\widetilde{f}(X(k+N_P-1|t), U(k+N_c|t)),$$

where N_c is the control horizon and $1 \leq N_c \leq N_P$, which denotes component-wise inequality.

The differences between the predictive states and the reference trajectory X_{ref} are defined as follows

$$e(k+1|t) = X(k+1|t) - X_{ref}(k+1|t),$$
$$\vdots \tag{7}$$
$$e(k+N_P|t) = X(k+N_P|t) - X_{ref}(k+N_P|t).$$

To ensure the passenger comfort and feasibility of the vehicle, the output control should be varied as smoothly as possible. Therefore, the optimization objective function is defined as

$$J(e(t), U(t)) = \sum_{i=1}^{N_P} \|e(k+i|t)\|_Q^2 + \sum_{i=1}^{N_c} \|U(k+i|t) - U(k+i-1|t)\|_R^2, \tag{8}$$

where Q and R are the weight matrices for the vehicle states and control states, respectively. Consequently, the rolling optimization can be obtained by solving the constrained optimization problem in every sampling period

$$\min_{U(t)} J(e(t), U(t))$$
$$\text{s.t.}$$
$$a_{\min} \leq a(k+i|t) \leq a_{\max}, \quad i = 1, 2, \cdots, N_c,$$
$$\delta_{\min} \leq \delta(k+i|t) \leq \delta_{\max}, \quad i = 1, 2, \cdots, N_c, \tag{9}$$
$$e_{\min} \leq e(t) \leq e_{\max}, \quad\quad t = k + T_s, \cdots, k + N_P T_s,$$

where $(a_{\min}, a_{\max})$ and $(\delta_{\min}, \delta_{\max})$ are the hard constraints of the vehicle. The last constraints are added to ensure safety driving.

The control inputs are obtained by solving the optimization problem (9). The first element in the control inputs is taken as the optimal control at the current time. After the prediction and control of the current time step are completed, the states are updated with the actual ones, which are then used as the initial states for the optimization problem in the next predictive horizon. The process is repeated until the vehicle reaches the final state.

The problem (9) is a quadratic programming (QP) one which is a traditional optimization problem for trajectory tracking. The first term in the cost function requires that the actual trajectory be as close as possible to the reference trajectory to ensure the safety and feasibility of the trajectory. The second term requires that the control input be varied smoothly to ensure the feasibility of the vehicle and the comfort of passengers. The difference between the reference and the actual trajectory must be sufficiently small. Otherwise, it may lead to a crash, and the trajectory is no longer feasible.

To meet the real-time requirement, instead of directly calculating a control sequence by solving (9), we solve an approximate optimization problem by imposing the constraints $u_{k+1} = u_{k+2} = \ldots = u_{k+Nc}$ on the control law. Therefore, we only need to calculate a 'mediocre' control to follow the given trajectory. This significantly reduces the complexity of the primal problem as it dramatically reduces the number of the variables. It is worth noting that imposing the constraint conditions $u_{k+1} = \ldots = u_{k+Nc}$ on (9) is equivalent to setting N_c to 1. Besides greatly reducing the computational burden, one of the most telling advantages of structuralizing the control is to produce an improvement in the robustness and in the general behavior of the system, because allowing the free evolution of the manipulated variables could lead to undesirable high-frequency control signals and even to instability as noted in [30]. We note that, if the coefficient matrices Q and R are positive semi-definite, the primal problem is tightly bounded and the approximate problem is also tightly bounded. Since Q and R are positive semi-definite, $x_1{}^T Q x_1 \geq 0$, $x_2{}^T R x_2 \geq 0$. Hence, the cost function in (9) is convex. The feasible region subjected to the constraint conditions (linear equations and inequalities) in (9) is also convex. Thus, the optimal solution of (9) is located in either the interior or the boundary of the feasible region. Therefore, the value of the cost function does not go to infinity, and the primal problem is tightly bounded. When imposing the constraint condition $u_{k+1} = \ldots = u_{k+Nc}$, the corresponding feasible region is still convex since the intersection of convex sets is still a convex set. Similarly, the approximate problem is tightly bounded. In the next section, we prove that the proposed close-loop MPC controller is stable if $N_c = N_1 = 1$, $\lambda = 0$ and N_P is large.

3. Stabilizability of Controller

Combining (1) and (5) leads to

$$
\begin{aligned}
x_{k+1} &= x_k + T_s(v_k + aT_s)\cos(\varphi_k + T_s v_k \sin(\beta)/l_r + \beta), \\
y_{k+1} &= y_k + T_s(v_k + aT_s)\sin(\varphi_k + T_s v_k \sin(\beta)/l_r + \beta), \\
\varphi_{k+1} &= \varphi_k + T_s(v_k + aT_s)\sin(\beta)/l_r, \\
v_{k+1} &= v_k + aT_s.
\end{aligned}
\tag{10}
$$

Equation system (10) can be rewritten as

$$
X_{k+1} = (I + A_k T_s)X_k + B_k T_s U_k,
$$

$$
A_k = \begin{bmatrix}
0 & 0 & -(v_k + aT_s)\sin(\gamma) & \cos(\gamma) - (v_k + aT_s)T_s \sin(\gamma)\sin(\beta)/l_r \\
0 & 0 & (v_k + aT_s)\cos(\gamma) & \sin(\gamma) + (v_k + aT_s)T_s \cos(\gamma)\sin(\beta)/l_r \\
0 & 0 & 0 & \sin(\beta)/l_r \\
0 & 0 & 0 & 0
\end{bmatrix},
$$

$$
B_k = \begin{bmatrix}
T_s \cos(\gamma) & -(v_k + aT_s)\sin(\gamma)(1 + v_k T_s \cos(\beta)/l_r)\beta_\delta \\
T_s \sin(\gamma) & (v_k + aT_s)\cos(\gamma)(1 + v_k T_s \cos(\beta)/l_r)\beta_\delta \\
T_s \sin(\beta)/l_r & (v_k + aT_s)\cos(\beta)\beta_\delta/l_r \\
1 & 0
\end{bmatrix},
$$

$$
\gamma = \varphi_k + T_s v_k \sin(\beta)/l_r + \beta, \quad \beta_\delta = \frac{l_r l}{l^2 \cos^2(\delta) + l_f^2 \sin^2(\delta)}, \quad l = l_f + l_r.
$$

$$
\tag{11}
$$

For the sake of convenience, we omit the subscript k in the remainder of this paper.

Theorem 1. *System (11) is controllable.*

Proof of Theorem 1. First, we seek the eigenvalues λ of A. By solving the characteristic polynomial $\det(\lambda I - A) = 0$, we have

$$
\lambda_1 = \lambda_2 = \lambda_3 = \lambda_4 = 0.
\tag{12}
$$

According to the definition of controllability proposed by Hautus [31], system (4) is controllable if and only if, for all λ_i, $i = 1, 2, 3, 4$ Rank($[\lambda_i I\text{-}A,B]$) = 4. Here we only need to consider Rank($[\lambda_1 I\text{-}A,B]$) due to (12)

$$[\lambda_1 I - A, B] = \begin{bmatrix} 0 & 0 & -A_{13} & -A_{14} & B_{11} & B_{12} \\ 0 & 0 & -A_{23} & -A_{24} & B_{21} & B_{22} \\ 0 & 0 & -A_{33} & -A_{34} & B_{31} & B_{32} \\ 0 & 0 & -A_{43} & -A_{44} & B_{41} & B_{42} \end{bmatrix} = [0_{4\times2}, \Omega_{4\times4}]. \tag{13}$$

The determinant of Ω is

$$\det(\Omega) = \frac{(v_k + aT_s)^2 l \cos(\beta)}{l^2 \cos^2(\delta) + l_r^2 \sin^2(\delta)} \neq 0. \tag{14}$$

Hence we have Rank(Ω) = 4 and $4 \geq$ Rank($[\lambda_1 I\text{-}A,B]$) $\geq$ Rank(Ω) = 4. $\square$

Theorem 2. *System (11) is observable.*

Proof of Theorem 2. According to the definition of observability proposed by Hautus [31], system (4) is observable if and only if, for all λ_i, $i = 1, 2, 3, 4$ Rank($[\lambda_i I\text{-}A;C]$) = 4.

In our problem, the output function is $Y = X = CX$, and thus $C = I_{4\times4}$ and Rank(C) = 4. Therefore, $4 \geq$ Rank($[\lambda_1 I\text{-}A;C]$) $\geq$ Rank(C) = 4. $\square$

Theorem 3. *System (11) is stabilizable.*

Proof of Theorem 3. According to the definition of stabilizability proposed by Hautus [31], system (11) is stabilizable if and only if $\lambda_i \geq 0$, $i = 1, 2, 3, 4$, and the system is controllable. Combining Theorem 1 and (12) proves that Theorem 3 holds. $\square$

Theorem 4. *The closed-loop MPC controller is stable for $N_c = 1$, $\lambda = 0$ and large N_P.*

Proof of Theorem 4. This proof is similar as that for Theorem 4 in [32] for generalized predictive control. When N_P is sufficiently large, we have

$$G^T G > 0, \tag{15}$$

where

$$G = \left[B; AB; \cdots ; A^{N_P - 1} B \right]_{N_P \times N_c}. \tag{16}$$

Therefore, $G^T G$ is a positive scalar, which is always invertible. Therefore, the matrix $G^T G + \lambda I$ is invertible, and a feasible control can be obtained using the expression in [32]

$$u_{opt} = (G^T G + \lambda I)^{-1} G^T (X_r - X). \tag{17}$$

Since our optimization problem is convex, there is only one optimal solution and thus our controller will asymptotically converge to (17). $\square$

4. Simulation

The simulation environment is MATLAB/Simulink R2020a, and (9) is solved using 'fmincon', a built-in function in MATLAB. Sequential quadratic programming is used as the nonlinear solver. The warm-start technique is employed by using the result of the previous optimization problem as a guess for the current optimization problem to further speed up the efficiency of the nonlinear solver. The accuracy of 'fmincon' is set to 10^{-6}. The processor used in the simulation is Intel(R) Core(TM) i7-4510U @ 2.00 GHz 2.6 GHz.

Real-Time Synchronization is enabled to test the real-time performance of the controllers. The simulation system consists of a kinematic model of autonomous vehicles and the trajectory tracking controller proposed in this paper. The parameters of the vehicle model and the controller are shown in Table 1 and can be found in [33]. The road conditions are assumed to be dry and clean and they can support the forces required for braking, accelerating and steering.

Table 1. Parameters of the vehicle and the controller.

Parameter	Value
l_f	1.232 m
l_r	1.468 m
Range of a	$[-1 \text{ m/s}^2, 1 \text{ m/s}^2]$
Range of δ	$[-0.44 \text{ rad}, 0.44 \text{ rad}]$
Range of lateral error	$[-0.5 \text{ m}, 0.5 \text{ m}]$
N_p	15
N_c	1
Q	$100I$ [1]
R	I

[1] I is the unit matrix.

4.1. Sinusodial Path Following

First, we present the tracking results of a sinusoidal trajectory with an amplitude of 4 m and a wavelength of 100 m in [20]. The reference speed along x-axis $Vref$ is set to be a constant. The open-loop reference trajectory is given by

$$Yref = 4\sin(2\pi Xref/100). \tag{18}$$

The tracking result of the sinusoidal trajectory is shown in Figure 4. The sampling time is set to $T_s = 0.05$ s. The reference trajectory was indicated by the black solid line. The obtained trajectories using the forward and backward Euler method were represented by a blue dotted line and a red dashed line, respectively. When the reference velocity is set to $Vref = 40$ km/h, the maximum lateral error using the backward Euler method was 0.0767 m, in contrast to 0.2481 m using the forward Euler method. The maximum longitudinal errors using the forward and backward Euler method were 0.07 m and 0.0703 m, respectively. The maximum calculation time using the backward Euler method was 0.0203 s, and the average calculation time was 0.0081 s, in contrast to 0.0197 s and 0.01 s using the forward Euler method. The maximum heading errors were 0.0277 rad using the backward Euler method and 0.019 rad using the forward Euler method. When the reference velocity is set to $Vref = 60$ km/h, the maximum lateral error using the Euler method was 0.4191 m; whereas, it was 0.2184 m using the backward Euler method. The maximum calculation times using the forward and backward Euler method were 0.0183 s and 0.0143 s, respectively; the average computation times were 0.0086 s and 0.0084 s; the maximum heading errors were 0.0293 rad and 0.0355 rad. The maximum longitudinal errors are 0.1059 m and 0.1085 m. To sum up, the lateral error using the backward Euler method was much smaller than that using the forward Euler method. However, the longitudinal error and heading error using the backward Euler method were slightly larger than that using the forward Euler method. Besides that, the backward Euler method required a little more calculation time. The state errors, including the lateral, longitudinal and heading errors, increased with the reference velocity.

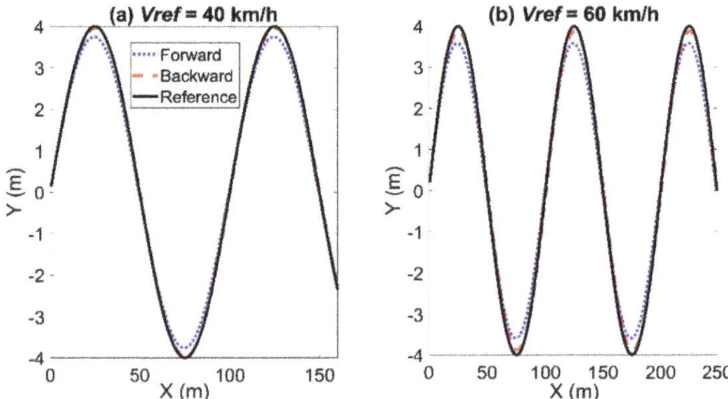

Figure 4. Results for tracking the sinusoidal trajectory with $T_s = 0.05$ s: left for Case (**a**) with *Vref* = 40 km/h and right for Case (**b**) with *Vref* = 60 km/h.

The comparison of the calculation time between the two controllers is shown in Figure 5. The calculation time of the MPC controller using the backward Euler method at each control period was almost the same or slightly larger than that of the MPC-based controller using the forward Euler method.

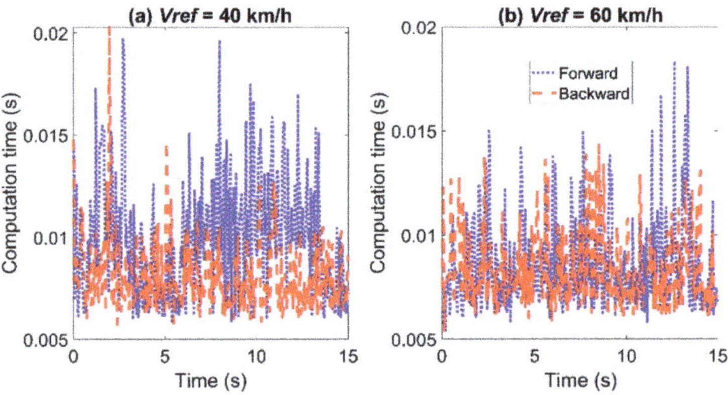

Figure 5. Comparison of the computation time for tracking the sinusoidal trajectory with $T_s = 0.05$ s: left for Case (**a**) with *Vref* = 40 km/h and right for Case (**b**) with *Vref* = 60 km/h.

Figures 6 and 7 show the articulated acceleration and steer angle, respectively. The Forward Euler method was more sensitive to the longitudinal velocity, whereas the backward Euler method was more sensitive to the steer angle.

We noted that, as mentioned before, the differences between the reference and actual trajectories increase with the vehicle velocity. There exists a threshold value of velocity to determine the existence of the solution of the optimization problem for trajectory tracking. In other words, when the reference velocity is greater than some value, no feasible solution exists. When $T_s = 0.05$ s, the threshold value of the reference velocity using the forward Euler method was 67.7 km/h (when *Vref* = 67.8 km/h, the maximum lateral error was 0.5009 m), whereas it was 83 km/h using the backward Euler method (when *Vref* = 83.1 km/h, the maximum lateral error was 0.5002 m). Hence, the MPC using the backward Euler method was more robust than that using the forward Euler method.

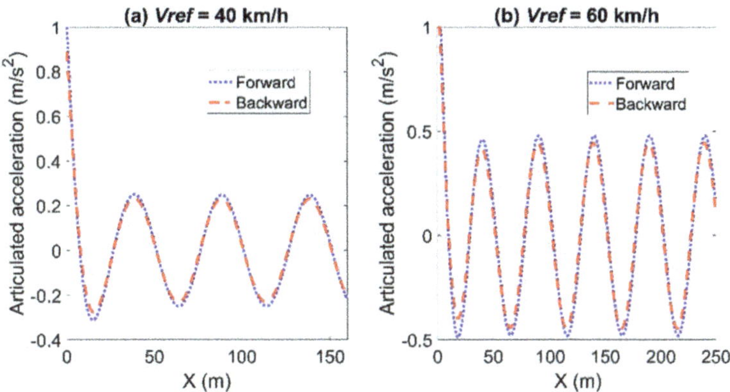

Figure 6. Articulated acceleration for the comparison between the forward and backward Euler method: left for Case (**a**) with *Vref* = 40 km/h and right for Case (**b**) with *Vref* = 60 km/h.

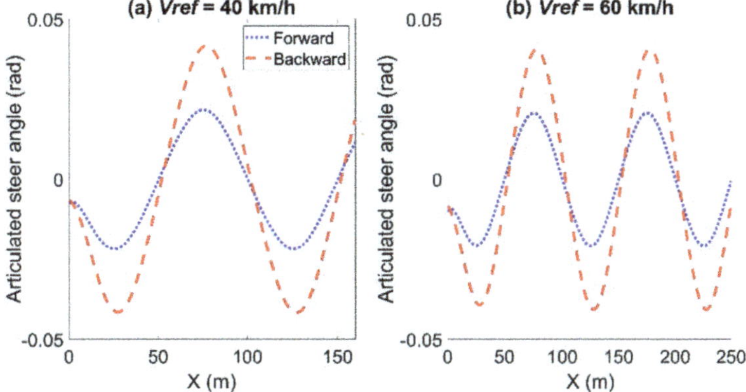

Figure 7. Articulated steer angle for the comparison between the forward and backward Euler method: left for Case (**a**) with *Vref* = 40 km/h and right for Case (**b**) with *Vref* = 60 km/h.

4.2. Circular Path

In the second scenario, the vehicle was required to track a circle with a radius of 40 m. The parametric equations for the circle were

$$\begin{cases} X(t) = Rd\cos(\varphi(t) - \pi/2), \\ Y(t) = Rd + Rd\sin(\varphi(t) - \pi/2), \\ \varphi(t) = t\, Vref/Rd, \end{cases} \tag{19}$$

where Rd is the radius of the reference circle. The initial configuration and constraint conditions were chosen to be same as previously to be consistent. The sampling time and the reference velocity are set to T_s = 0.05 s and *Vref* = 10 m/s, respectively.

The tracking result of the circular path is shown in Figure 8. The maximum lateral and longitudinal errors using the backward Euler method were 0.0596 m and 0.0091 m, in contrast to 0.3664 m and 0.3664 m using the forward Euler method. The maximum calculation time using the backward Euler method was 0.016 s, and the average calculation time was 0.0084 s, in contrast to 0.0199 s and 0.0085 s using the forward Euler method. The maximum heading errors were 0.0411 rad using the backward Euler method and 0.0194 rad using the forward Euler method. In sum, the MPC controller using the backward Euler method had a better tracking accuracy in the circular path than that using the forward Euler method.

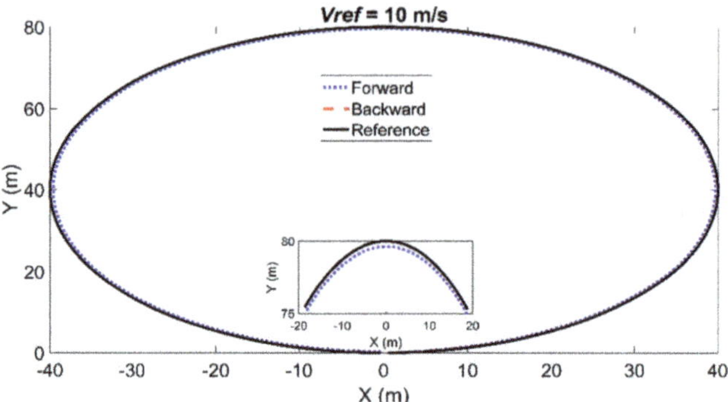

Figure 8. Simulation results for the circular path with radius *Rd* = 40 m. Notations the same as in Figure 4.

Figures 9 and 10 show the articulated acceleration and steer angle, respectively. The articulated acceleration and steer angle using the backward Euler method were quite different from those using the forward Euler method. As can be seen from Figure 8, the backward Euler method was more accurate than the forward Euler method, and the calculation times were almost the same as shown in Figure 11.

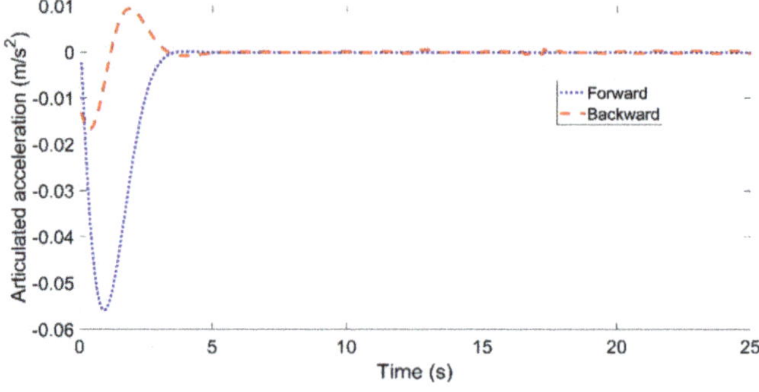

Figure 9. Articulated acceleration for the comparison between the forward and backward Euler method for the circular trajectory.

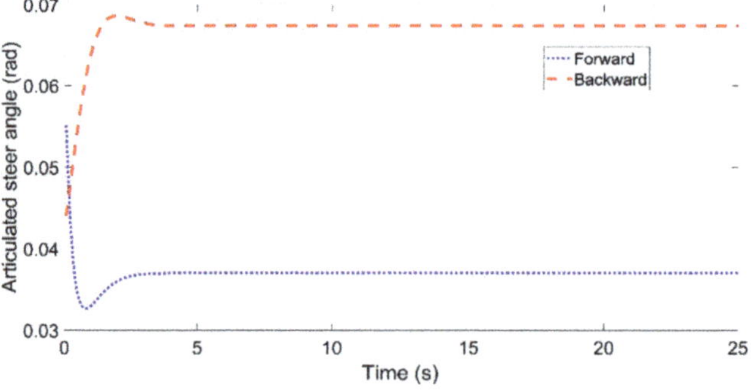

Figure 10. Articulated steer angle for the comparison between the forward and backward Euler method for the circular trajectory.

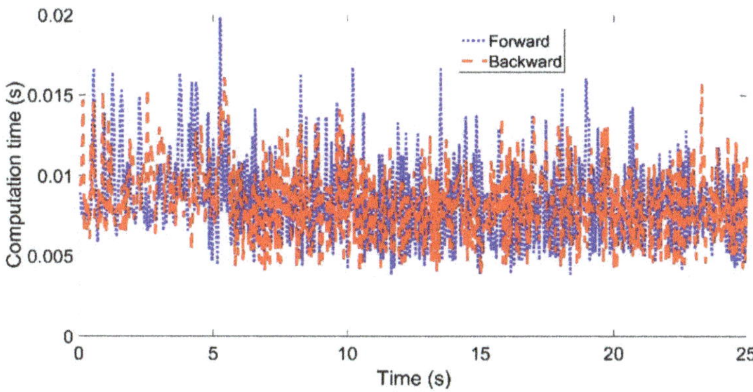

Figure 11. Comparison of the computation time for the circular path. Red dashed curve: MPC-based controller using the backward Euler method; blue dotted curve: MPC-based controller using the forward Euler method.

4.3. Double Line Change Path

In this scenario, the vehicle was required to track a double line change path. The reference trajectory of the double line change path can be found in [33]. The tracking result of the double line change path is shown in Figure 12. When the reference velocity is set to *Vref* = 40 km/h, the maximum lateral and longitudinal errors using the backward Euler method were 0.3034 m and 0.0203 m, in contrast to 0.3827 m and 0.0412 m using the forward Euler method. The maximum heading errors were 0.0673 rad using the backward Euler method and 0.0648 rad using the forward Euler method. When the reference velocity is set to *Vref* = 60 km/h, the maximum lateral and longitudinal errors using the backward Euler method were 0.587 m and 0.0504 m, in contrast to 0.6187 m and 0.0311 m using the forward Euler method. The maximum heading errors were 0.1035 rad using the backward Euler method and 0.0967 rad using the forward Euler method. In sum, the MPC controller using the backward Euler method had a better tracking accuracy in the circular path than that using the forward Euler method. Figures 13 and 14 show the articulated acceleration and steer angle, respectively.

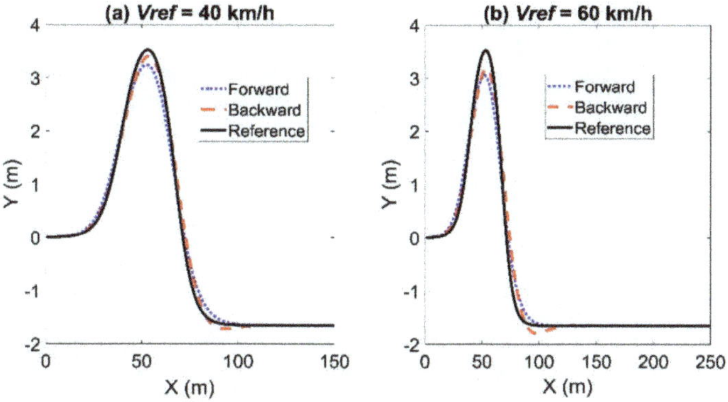

Figure 12. Simulation results for the double line change path: left for Case (**a**) with *Vref* = 40 km/h and right for Case (**b**) with *Vref* = 60 km/h.

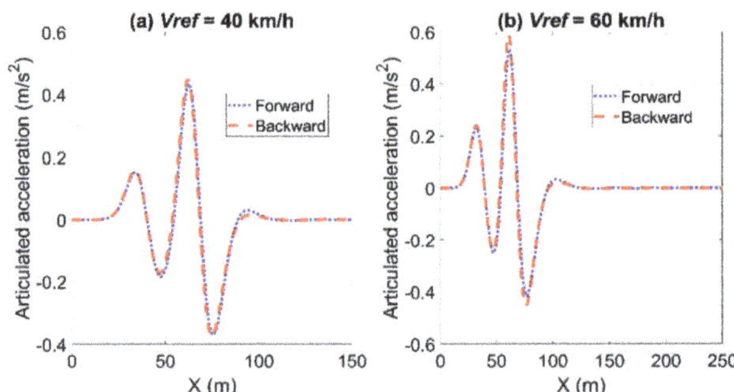

Figure 13. Articulated acceleration for the comparison between the forward and backward Euler method for the circular trajectory: left for Case (**a**) with *Vref* = 40 km/h and right for Case (**b**) with *Vref* = 60 km/h.

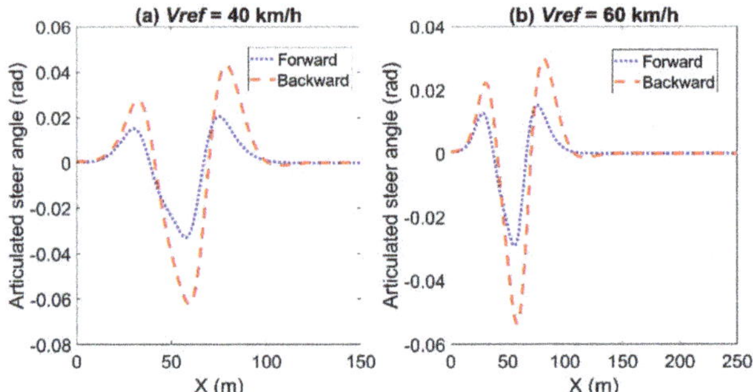

Figure 14. Articulated steer angle for the comparison between the forward and backward Euler method for the circular trajectory: left for Case (**a**) with *Vref* = 40 km/h and right for Case (**b**) with *Vref* = 60 km/h.

The comparison of the calculation time between the two controllers is shown in Figure 15. When the reference velocity is set to *Vref* = 40 km/h, the maximum calculation time using the backward Euler method was 0.0178 s, and the average calculation time was 0.0084 s, in contrast to 0.0157 s and 0.0084 s using the forward Euler method. When the reference velocity is set to *Vref* = 60 km/h, the maximum calculation time using the backward Euler method was 0.0152 s and the average calculation time was 0.008 s, in contrast to 0.0169 s and 0.0083 s using the forward Euler method.

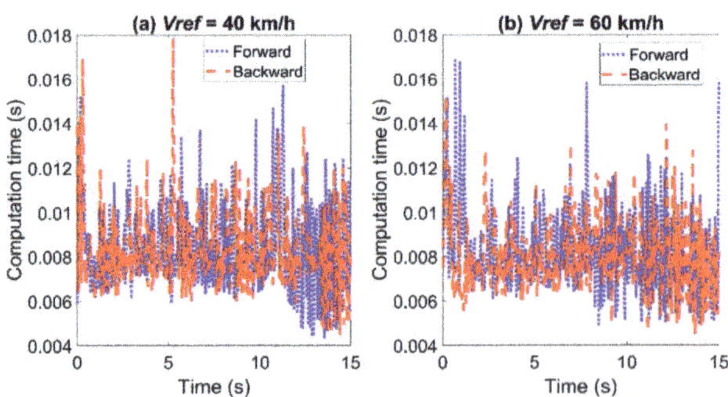

Figure 15. Comparison of the computation time for tracking the double line change path with T_s = 0.05 s: left for Case (**a**) with *Vref* = 40 km/h and right for Case (**b**) with *Vref* = 60 km/h.

Through three sets of comparisons, we can draw some conclusions. First, the lateral tracking errors using the backward Euler method were much smaller than those when using the forward Euler method. Second, the lateral tracking errors using either the forward or backward Euler method increased with the reference velocity. Third, the calculation times using the backward Euler method were almost the same with that using the forward Euler method. Lastly, compared with the articulated acceleration, there was a clear discrepancy in the articulated steer angle.

5. Conclusions

An effective and efficient method for generating a feasible trajectory is of vital importance to meet the requirement of instantaneous control for autonomous driving. In this paper, we have proposed a trajectory tracking controller based on MPC. Most MPC-based and other methods either set the velocity to a constant or cannot actively adjust the longitudinal velocity according to the information of the reference trajectory. To solve this problem, both the acceleration and steer angle are set to control inputs. Hence, the proposed controller can automatically adjust the velocity according to the information of the reference trajectory. Moreover, instead of the forward Euler integration method, the backward Euler integration method is used to establish the predictive model. To meet the real-time requirement, we impose the constraints $u_{k+1} = \ldots = u_{k+Nc}$ on the control law. This significantly reduced the problem complexity. The warm-start technique was used to further accelerate the convergence of the optimization solver of the controller by using the previous results as a guess for the current optimization problem.

The proposed closed-loop MPC controller was stable and validated by simulation experiments. Compared with the MPC controller using the forward Euler method, the MPC controller using the backward Euler method had a much better accuracy in the lateral error, which is an important indicator to ensure driving safety. The lateral error could be reduced by up to 78%. There is little difference in the longitudinal error between the two controllers. However, the heading error of the MPC controller using the backward Euler method was larger than that of the MPC controller using the forward Euler method. The maximum and average computation times using the backward Euler method were almost the same or slightly larger than those using the forward Euler method. Moreover, the MPC controller using backward Euler method was more robust than that using the forward Euler method. The threshold value of the velocity for the MPC controller using the backward Euler method was larger than that using the forward Euler method (83 km/h versus 67.7 km/h for the sinusoidal trajectory). Overall, the MPC controller using the backward Euler method had a better tracking accuracy at the cost of no or little computation time.

The existence of the discrepancy between the actual trajectory and the reference trajectory is mainly due to the modelling errors, computation errors and disturbance errors. Recent studies on MPC-based controllers mainly focus on the modelling errors and disturbance errors. Few authors investigated the computation errors during the discretization for nonlinear systems. We hope that this paper is instructive and allows researchers new insight into creating MPC-based controllers.

From the perspective of science, our contribution is to give a new way to establish a predictive model, which is a cornerstone of designing a path tracking controller. Besides the forward and backward Euler methods, there are several integration methods, such as the midpoint method and Runge-Kutta methods. Improving the accuracy of the prediction model using other integration methods could be a promising way to get an effective control to maintain a good tracking accuracy.

Author Contributions: Conceptualization, Z.H. and H.L.; methodology, Z.H.; software, Z.Y.; validation, W.L., J.L., C.H. and W.F.; formal analysis, Z.H., W.L., C.H., J.L. and W.F.; investigation, Z.H.; writing—original draft preparation, Z.H.; writing—review and editing, H.L.; supervision, H.L.; project administration, H.L.; funding acquisition, J.L. All authors have read and agreed to the published version of the manuscript.

Funding: This research was funded by the National Natural Science Foundation of China, Grant number No. 62003328 and the China Postdoctoral Science Foundation, Grant No. 2020M682985.

Institutional Review Board Statement: Not applicable.

Informed Consent Statement: Not applicable.

Data Availability Statement: Not applicable.

Acknowledgments: The authors would like to thank for support from National Natural Science Foundation of China and the China Postdoctoral Science Foundation.

Conflicts of Interest: The authors declare no conflict of interest.

References

1. Li, L.; Song, J.; Wang, F.Y.; Niehsen, W.; Zheng, N.N. IVS 05: New developments and research trends for intelligent vehicles. *IEEE Intell. Syst.* **2005**, *20*, 10–14. [CrossRef]
2. Ma, Y.; Wang, Z.; Yang, H.; Yang, L. Artificial intelligence applications in the development of autonomous vehicles: A survey. *IEEE/CAA J. Autom. Sin.* **2020**, *7*, 315–329. [CrossRef]
3. Paden, B.; Cap, M.; Yong, S.Z.; Yershov, D.; Frazzoli, E. A survey of motion planning and control techniques for self-driving urban vehicles. *IEEE Trans. Intell. Veh.* **2016**, *1*, 33–55. [CrossRef]
4. Cafiso, S.; Pappalardo, G. Safety effectiveness and performance of lane support systems for driving assistance and automation— Experimental test and logistic regression for rare events. *Accid. Anal. Prev.* **2020**, *148*, 105791. [CrossRef]
5. Pappalardo, G.; Cafiso, S.; Di Graziano, A.; Severino, A. Decision Tree Method to Analyze the Performance of Lane Support Systems. *Sustainability* **2021**, *13*, 846. [CrossRef]
6. Badue, C.; Guidolini, R.; Carneiro, R.V.; Azevedo, P.; Cardoso, V.B.; Forechi, A.; Jesus, L.; Berriel, R.; Paixão, T.; Mutz, F.; et al. Self-driving cars: A survey. *Expert Syst. Appl.* **2021**, *165*, 113816. [CrossRef]
7. Li, B.; Shao, Z. A unified motion planning method for parking an autonomous vehicle in the presence of irregularly placed obstacles. *Knowl.-Based Syst.* **2015**, *86*, 11–20. [CrossRef]
8. Zhang, Z.; Zhang, L.; Deng, J.; Wang, M.; Wang, Z.; Cao, D. An Enabling Trajectory Planning Scheme for Lane Change Collision Avoidance on Highways. *IEEE Trans. Intell. Veh.* **2021**. [CrossRef]
9. Zhang, L.; Wang, Z.; Ding, X.; Li, S.; Wang, Z. Fault-Tolerant Control for Intelligent Electrified Vehicles Against Front Wheel Steering Angle Sensor Faults During Trajectory Tracking. *IEEE Access* **2021**, *9*, 65174–65186. [CrossRef]
10. Zhang, L.; Zhang, Z.; Wang, Z.; Deng, J.; Dorrel, D.G. Chassis Coordinated Control for Full X-by-Wire Vehicles—A Review. *Chin. J. Mech. Eng.* **2021**, *34*, 42. [CrossRef]
11. Fabiani, F.; Grammatico, S. Multi-vehicle automated driving as a generalized mixed-integer potential game. *IEEE Trans. Intell. Transp. Syst.* **2019**, *21*, 1064–1073. [CrossRef]
12. Coulter, R.C. *Implementation of the Pure Pursuit Path Tracking Algorithm*; Technical Report; Carnegie-Mellon UNIV Pittsburgh PA Robotics INST: Pittsburgh, PA, USA, 1992.
13. Thrun, S.; Montemerlo, M.; Dahlkamp, H.; Stavens, D.; Aron, A.; Diebel, J.; Fong, P.; Gale, J.; Halpenny, M.; Hoffmann, G.; et al. Stanley: The Robot That Won the DARPA Grand Challenge. *J. Field Robot.* **2006**, *23*, 661–692. [CrossRef]
14. Amidi, O.; Thorpe, C.E. Integrated mobile robot control. *Mob. Robot. V* **1991**, *1388*, 504–523.
15. Dixit, S.; Fallah, S.; Montanaro, U.; Dianati, M.; Stevens, A.; Mccullough, F.; Mouzakitis, A. Trajectory planning and tracking for autonomous overtaking: State-of-the-art and future prospects. *Annu. Rev. Control* **2018**, *45*, 76–86. [CrossRef]
16. Araki, M. Control Systems, Robotics, and Automation—Vol. II—PID Control. In *Encyclopedia of Life Support Systems*; EOLSS Publishers Ltd.: London, UK, 2009.
17. Young, K.D.; Utkin, V.I.; Ozguner, U. A control engineer's guide to sliding mode control. *IEEE Trans. Control Syst. Technol.* **1999**, *7*, 328–342. [CrossRef]
18. Utkin, V.; Lee, H. Chattering problem in sliding model control systems. In Proceedings of the 2nd IFAC Conference on Analysis and Design of Hybrid Systems, Alghero, Italy, 7–9 June 2006.
19. Amer, N.; Zamzuri, H.; Hudha, K.; Hudha, K.; Kadir, Z. Modelling and control strategies in path tracking control for autonomous ground vehicles: A review of state of the art and challenges. *J. Intell. Robot. Syst.* **2017**, *86*, 225–254. [CrossRef]
20. Silver, D.; Huang, A.; Maddison, C.J.; Guez, A.; Sifre, L.; van den Driessche, G.; Schrittwieser, J.; Antonoglou, I.; Panneershelvam, V.; Lanctot, M.; et al. Mastering the game of go with deep neural networks and tree search. *Nature* **2016**, *529*, 484–489. [CrossRef]
21. Silver, D.; Hubert, T.; Schrittwieser, J.; Antonoglou, I.; Matthew, L.; Guez, A.; Lanctot, M.; Sifre, L.; Kumaran, D.; Graepel, T.; et al. Mastering chess and shogi by self-play with a general reinforcement learning algorithm. *Science* **2018**, *362*, 1140–1144. [CrossRef]
22. Kendall, A.; Hawke, J.; Janz, D.; Mazur, P.; Reda, D.; Allen, J.M.; Lam, V.D.; Bewley, A.; Shah, A. Learning to drive in a day. In Proceedings of the 2019 International Conference on Robotics and Automation (ICRA), Montreal, QC, Canada, 2–6 May 2019; IEEE: Piscataway, NJ, USA, 2019; pp. 8248–8254.

23. Mohammadi, M.; Arefi, M.M.; Setoodeh, P.; Kaynak, O. Optimal tracking control based on reinforcement learning value iteration algorithm for time-delayed nonlinear systems with external disturbances and input constraints. *Inf. Sci.* **2021**, *554*, 84–98. [CrossRef]
24. Shen, C.; Shi, Y.; Buckham, B. Integrated path planning and tracking control of an AUV: A unified receding horizon optimization approach. *IEEE/ASME Trans. Mechatron.* **2016**, *22*, 1163–1173. [CrossRef]
25. Borrelli, F.; Falcone, P.; Keviczky, T.; Asgari, J.; Hrovat, D. MPC-based approach to active steering for autonomous vehicle systems. *Int. J. Veh. Auton. Syst.* **2005**, *3*, 265–291. [CrossRef]
26. Kong, J.; Pfeiffer, M.; Schildbach, G.; Borrelli, F. Kinematic and dynamic vehicle models for autonomous driving control design. In Proceedings of the 2015 IEEE Intelligent Vehicles Symposium (IV), Seoul, Korea, 28 June–1 July 2015; IEEE: Piscataway, NJ, USA, 2016; pp. 1094–1099.
27. Polack, P.; Altché, F.; d'Andréa-Novel, B.; de La Fortelle, A. The kinematic bicycle model: A consistent model for planning feasible trajectories for autonomous vehicles? In Proceedings of the 2017 IEEE intelligent vehicles symposium (IV), Redondo Beach, CA, USA, 11–14 June 2017; IEEE: Piscataway, NJ, USA, 2017; pp. 812–818.
28. Chen, S.; Chen, H.; Negrut, D. Implementation of MPC-Based Path Tracking for Autonomous Vehicles Considering Three Vehicle Dynamics Models with Different Fidelities. *Automot. Innov.* **2020**, *3*, 386–399. [CrossRef]
29. Rajamani, R. *Vehicle Dynamics and Control*; Springer Science & Business Media: Berlin/Heidelberg, Germany, 2011.
30. Camacho, E.F.; Brodons, C. *Model Predictive Control*; Springer: London, UK, 1999.
31. Hautus, M. Stabilization controllability and observability of linear autonomous systems. *Indag. Math. (Proc.)* **1970**, *73*, 448–455. [CrossRef]
32. Clarke, D.W.; Mohtadi, C. Properties of generalized predictive control. *Automatica* **1989**, *25*, 859–875. [CrossRef]
33. Gong, J.W.; Jiang, Y.; Xu, W. *Model Predictive Control for Self-Driving Vehicles*; Beijing Institute of Technology Press: Beijing, China, 2014.

Article

Unified Chassis Control of Electric Vehicles Considering Wheel Vertical Vibrations

Xinbo Chen, Mingyang Wang and Wei Wang *

Institute of Intelligent Vehicles, School of Automotive Studies, Tongji University, No. 4800 Cao'an Highway, Shanghai 201804, China; austin_1@163.com (X.C.); wangmingyang@sina.cn (M.W.)
* Correspondence: lazguronwang@gmail.com

Abstract: In the process of vehicle chassis electrification, different active actuators and systems have been developed and commercialized for improved vehicle dynamic performances. For a vehicle system with actuation redundancy, the integration of individual chassis control systems can provide additional benefits compared to a single ABS/ESC system. This paper describes a Unified Chassis Control (UCC) strategy for enhancing vehicle stability and ride comfort by the coordination of four In-Wheel Drive (IWD), 4-Wheel Independent Steering (4WIS), and Active Suspension Systems (ASS). Desired chassis motion is determined by generalized forces/moment calculated through a high-level sliding mode controller. Based on tire force constraints subject to allocated normal forces, the generalized forces/moment are distributed to the slip and slip angle of each tire by a fixed-point control allocation algorithm. Regarding the uneven road, H∞ robust controllers are proposed based on a modified quarter-car model. Evaluation of the overall system was accomplished by simulation testing with a full-vehicle CarSim model under different scenarios. The conclusion shows that the vertical vibration of the four wheels plays a detrimental role in vehicle stability, and the proposed method can effectively realize the tire force distribution to control the vehicle body attitude and driving stability even in high-demanding scenarios.

Keywords: electric vehicle; unified chassis control; unsprung mass

Citation: Chen, X.; Wang, M.; Wang, W. Unified Chassis Control of Electric Vehicles Considering Wheel Vertical Vibrations. *Sensors* **2021**, *21*, 3931. https://doi.org/10.3390/s21113931

Academic Editors: Chao Huang, Chen Lv, Fuwu Yan, Haiping Du, Wanzhong Zhao and Yifan Zhao

Received: 11 April 2021
Accepted: 3 June 2021
Published: 7 June 2021

Publisher's Note: MDPI stays neutral with regard to jurisdictional claims in published maps and institutional affiliations.

Copyright: © 2021 by the authors. Licensee MDPI, Basel, Switzerland. This article is an open access article distributed under the terms and conditions of the Creative Commons Attribution (CC BY) license (https:// creativecommons.org/licenses/by/ 4.0/).

1. Introduction

With the growing concern about pollution, energy shortage, and also fast development of electric propulsion technologies, modern vehicles are increasingly electrified. From the driver's point of view, an adequate response in critical driving conditions is still a challenging task for non-professional drivers. Therefore, electric or electromechanical systems, featuring energy regeneration capability and fast response, are readily developed and applicated to improve the vehicle dynamics, from the aspects of comfort, stability, safety, maneuverability, and driver's feeling, especially in adverse driving situations.

In order to enhance the driving stability and the overall dynamic performance, a vehicle may be equipped with multi-actuators, which can be classified into three categories as active torque distribution [1], active steering [2], and active suspension control [3]. Among all possible actuators, ABS-based differential braking has received the most attention since it can be executed on almost all vehicles regardless of powertrain configuration, known as the traditional ESC system. In 2003, the active front steering technology was developed and recognized as a supplemental approach to generate desired yaw moment without braking, ensuring enhanced vehicle stability even in high-speed conditions. However, the differential braking systems reduce the driving speed, which may conflict with the driver's intention during acceleration scenarios. Concerning this defect, active torque distribution was implemented by two actuation methods: torque vectoring [4] and individual motors. The corresponding vehicle stability control method was called direct yaw control.

In most vehicle control approaches, different control logic based on various actuators are always separately synthesized and locally tuned without considering the interaction among them, which may lead to sub-optimal or conflicting control efforts. Nowadays, extensive research has been carried out on the coordination of active steering and independent torque control. Furthermore, the integration with active suspension systems provides a new research field under the name of global chassis control [5–7] or UCC (Unified Chassis Control) [8–10]. In general, most integrated control algorithms were developed using two different approaches concerning the actuators hierarchy: (1) In the first method, a supervisor with a higher command hierarchy was used to monitor vehicle states and coordinate different sub-controllers. As illustrated, sometimes the active steering was only considered if the differential braking system exceeded its limits or before the ESC was activated [11]. In [12], the coordination of active front steering and ESC was investigated using a rule-based method according to the different value ranges of lateral acceleration. (2) The second UCC category treated the overall control structure as two levels. In the upper-level, desired yaw moment was computed, then in the lower-level, the moment was distributed into tire forces [8,13].

From a control point of view, a variety of problems arise from the UCC system synthesis, such as multiple input–multiple output control design, system robustness, and non-linearity. Many researchers have tried to solve these challenges from the standpoints of reference track following and control optimization. Fuzzy logic was applied for an intermediate layer of a UCC method [14]. The sliding mode control technique, which possesses good robustness, was used to cope with system uncertainties [7,15]. In [11,16], control objectives were achieved in a linear parameter varying robust framework by providing a solution to the linear matrix inequality problem. The H∞-based observer is also designed for fault estimation and fault-tolerant control [17,18]. Model prediction control becomes a hotspot due to significant development on online computational devices [19–21]. Another challenge stemmed from how to achieve the constrained optimal allocation problems in systems with redundancy. In [9], the desired yaw moment distribution was algebraically solved by Karush–Kuhn–Tucker Conditions. Combined with the desired target following, energy minimization, and tire force saturation, the Holistic Cornering Control architecture was introduced [22,23]. Under this framework, the gain optimization was designed based on linear matrix inequality and genetic algorithm techniques. In [24], the redundant actuator allocation problem was investigated with pseudo-inverse and accelerated fix-point iterative algorithms. Since the tire normal load restraints the boundary of longitudinal and lateral force, the vertical force control indirectly influences the vehicle dynamics in the motion plane. The research in [25] has shown the possible benefits to be attained by modulating normal force through active suspension control during cornering maneuvers. Regarding the multiple targets of vehicle yaw performance and attitude, the integration control method involving suspensions and braking actuators was investigated in [26,27]. However, the integrated suspension control does not consider the tire contact stability.

Different from traditional vehicle chassis structures, the layout with four in-wheel motors is especially suitable for high-performance and off-road vehicles. However, this layout inevitably introduces a larger unsprung mass. With the development of in-wheel motor technology, it becomes a common judgment that the introduced unsprung mass aggravates the wheel vibration and vehicle ride comfort [28]. Especially when the vehicle is running over an uneven road, one or more of the wheels might jump, and the tires will lose adhesion stochastically, which leads to the unbalance of driving/braking torques among four wheels and even makes the vehicle encounter instability situations. This effect occurs not only for in-wheel drive electric vehicles but also for common vehicles. Till now, little research can be found in this area. Vos et al. [29] verified that the vertical accelerations caused by bad roads could increase the roll and pitch movement ranges. Zhang et al. [30] proposed the inference that the unsprung mass plays a different role in the rollover motion between a flat road and an uneven road, and verified it with typical situations. Tan et al. [31] investigated the negative effect of wheel motor vibration on vehicle anti-rollover performance. In addition, the tire contact stability has been given little attention in existing UCC designs.

The purpose of this work is to present a unified control framework with a particular focus on wheel vertical vibrations for enhanced tire contact stability. More specifically, we treat the multi-objective active suspension design problem as a robust tracking problem to achieve desired body attitude and normal forces simultaneously. Besides, a reconfigurable control allocation (CA) method is used to deal with the problematic condition of tire contact force loss due to an uneven road or tire blow-out emergency. Evaluation of the overall system was accomplished by simulation testing with a full-vehicle CarSim model under different test scenarios.

The remaining sections of this paper are organized as follows: Section 2 briefly presents the overall system modeling. In Section 3, nonlinear controller design and reconfigurable CA approach are presented, and the synthesis of the normal force controller considering different targets is described, followed by Section 4, which examines the closed-loop vehicle dynamics performance under different test scenarios through simulation results. Section 5 provides some concluding remarks.

2. System Modeling

2.1. Nonlinear Vehicle Model

In this paper, both the chassis planar motion (longitudinal, lateral, and yaw) and the spatial motion (vertical, roll, and pitch) are considered. Thus, the vehicle body motion is treated as a rigid body with six degrees of freedom. Figure 1 shows the vehicle planar motion model. The dynamics equations of planar motions can be written as

$$
\begin{cases}
m(\dot{v}_x - v_y\omega_z) + m_s e_r \dot{\omega}_z \varphi = F_X - F_r \\
m(\dot{v}_y + v_x\omega_z) - m_s e_r \ddot{\varphi} = F_Y \\
I_{zz}\dot{\omega}_z = M_z
\end{cases}, \tag{1}
$$

with m as the vehicle mass (including the sprung mass m_s and unsprung mass m_u) and I_{zz} as the moment of inertia along Z-axis. The vehicle states are defined as the body motion velocity in three directions, namely ($v_x\ v_y\ \omega_z$), with the vehicle coordinate system fixed at the vehicle Center of Gravity (CG). e_r denotes the distance of CG from the rolling center and φ is the rolling angle. The resultant forces exerted on the vehicle are defined as F_X, F_Y, and M_Z, with the following expressions

$$
\begin{cases}
F_X = \sum F_{Xi}, \ i \in Q := fl, fr, rl, rr \\
F_Y = \sum F_{Yi}, \ i \in Q := fl, fr, rl, rr
\end{cases}, \tag{2}
$$

$$
M_Z = a\left(F_{Yfl} + F_{Yfr}\right) - b(F_{Yrl} + F_{Yrr}) + d\left(F_{Xfr} + F_{Xrr} - F_{Xfl} - F_{Xrl}\right) \tag{3}
$$

F_r stands for the total longitudinal resistance. F_{Xi} and F_{Yi} are the tire forces under the vehicle coordinate system, whereas the longitudinal and lateral forces of a single tire under the tire coordinate system are expressed as F_{xi} and F_{yi}. Their relationship is described as follows.

$$
\begin{bmatrix} F_{Xi} \\ F_{Yi} \end{bmatrix} = \begin{bmatrix} \cos\delta_i & -\sin\delta_i \\ \sin\delta_i & \cos\delta_i \end{bmatrix} \begin{bmatrix} F_{xi} \\ F_{yi} \end{bmatrix}, i \in Q := fl, fr, rl, rr \tag{4}
$$

Considering the assumption that the vehicle has a four-wheel independent steering system, δ_i stands for the steering angle of a given wheel with the subscript representing the position which is controlled by every steering actuator. In Figure 2, the vehicle spatial motion models are illustrated. Concerning the unsprung mass dynamics, we treat the vehicle as a whole system under the influence of lateral and longitudinal accelerations.

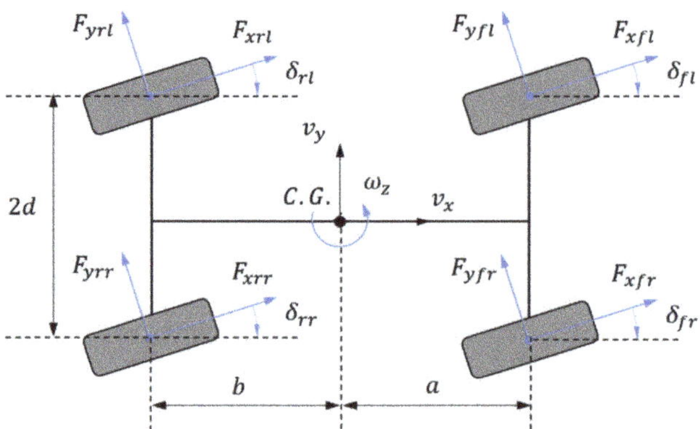

Figure 1. Vehicle planar motion model.

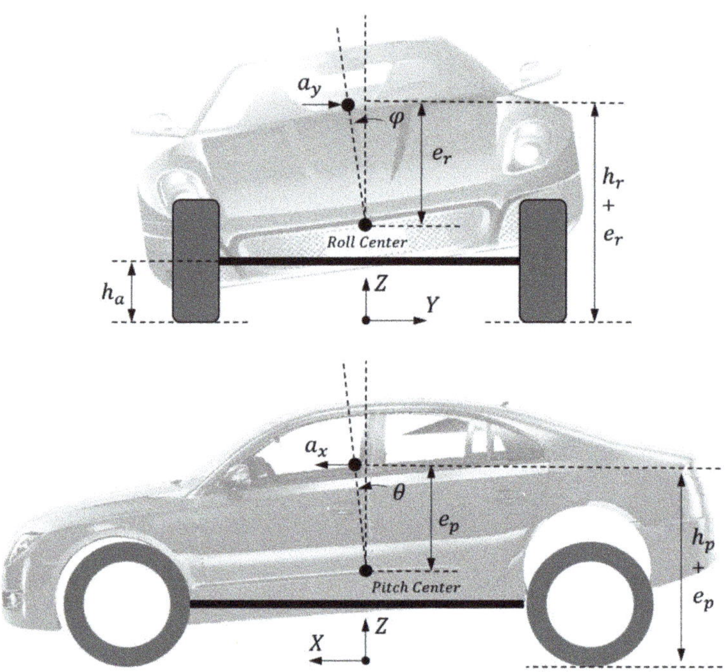

Figure 2. Vehicle roll and pitch motion.

Based on the reference direction in Figure 2, if we treat the sprung/unsprung mass as a whole system, the roll and pitch dynamic equations can be respectively expressed as

$$I_{xx}\ddot{\varphi} = M_\varphi + a_y m_s (h_r + e_r) + m_s g \varphi e_r + a_y m_u h_a, \tag{5}$$

$$I_{yy}\ddot{\theta} + a_x m_s (h_p + e_p) + a_x m_u h_a = M_\theta + m_s g \theta e_p, \tag{6}$$

where, I_{xx} and I_{yy} are the principal moments of inertia of the vehicle body part along X and Y axis, respectively. $(h_r + e_r)$ and $(h_p + e_p)$ represent the heights of the rolling center and pitching center, respectively, with e_r as the distance between vehicle CG and rolling center and h_p as the distance between vehicle CG and pitching center. h_a represents the height of the unsprung mass center. The vehicle longitudinal/lateral accelerations can be calculated as,

$$a_x = \dot{v}_x - v_y \omega_z, \; a_y = \dot{v}_y + v_x \omega_z$$

These values are assumed to be available through a sensing system. Here, we ignore the mass of suspension links and assume that the unsprung mass CG is located in the wheel center. M_θ and M_φ denote the moments generated by the load transfer among vertical tire forces, and we describe this relationship in the form of

$$\begin{cases} F_{Zfl} = \overline{F}_{Zfl} - \frac{M_\theta}{2(a+b)} + \frac{M_\varphi}{4d}, F_{Zfr} = \overline{F}_{Zfr} - \frac{M_\theta}{2(a+b)} - \frac{M_\varphi}{4d}, \\ F_{Zrl} = \overline{F}_{Zrl} + \frac{M_\theta}{2(a+b)} + \frac{M_\varphi}{4d}, F_{Zrr} = \overline{F}_{Zrr} + \frac{M_\theta}{2(a+b)} - \frac{M_\varphi}{4d}, \end{cases} \tag{7}$$

where, $\overline{F}_{Zi}$ are static normal tire forces and F_{Zi} are desired tire normal forces, which can be achieved using four independent active suspension actuators.

2.2. Nonlinear Tire Model

We applied Pacejka's Magic Formula tire model [32] to describe the nonlinear characteristics of tires, which is an empirical approach and can be effectively matched with experimental data. The longitudinal force, lateral force, and self-aligning moment of tire can be expressed by the following unified form with different parameter sets.

$$\begin{cases} y(x) = D\sin\{C\arctan[Bx - E(Bx - \arctan Bx)]\} \\ Y(X) = y(x) + S_v \\ x = X + S_h \end{cases} \tag{8}$$

where Y represents the tire force that could be the longitudinal/lateral tire force or self-aligning moment. X denotes the model input, which corresponds to the tire slip or the slip angle. S_v and S_h correspond to the vertical and horizontal bias, respectively. Coefficients B, C, D, and E are the stiffness, shape, peak, and curvature factors, respectively. When considering the coupling of lateral and longitudinal force, we usually neglect the curvature factor for simplicity, and then the following relations hold approximately.

$$F_{xi} = \frac{\sigma_{xi}}{\sigma_i} F_{x0i}, \; F_{yi} = \frac{\sigma_{yi}}{\sigma_i} F_{y0i} \tag{9}$$

F_{x0i} and F_{y0i} can be calculated by applying their corresponding magic formulas, with σ_{xi} as the longitudinal tire slip expressed below and σ_{yi} as the lateral tire slip.

$$\begin{cases} \sigma_{xi} = \frac{R_i \omega_i - v_{wxi}}{v_{wxi}}, \text{ during braking} \\ \sigma_{xi} = \frac{R_i \omega_i - v_{wxi}}{R_i \omega_i}, \text{ during acceleration} \end{cases} \tag{10}$$

where, v_{wxi} expresses the velocity at tire center in the tire forward direction. According to the mentioned vehicle planar model, the slip angles of each wheel are

$$\begin{cases} \alpha_{fl} = -\delta_{fl} + \arctan\left(\frac{v_y + \dot{\psi}a}{v_x - \dot{\psi}d}\right) \approx -\delta_{fl} + \frac{v_y + \dot{\psi}a}{v_x - \dot{\psi}d} \\ \alpha_{fr} = -\delta_{fr} + \arctan\left(\frac{v_y + \dot{\psi}a}{v_x + \dot{\psi}d}\right) \approx -\delta_{fl} + \frac{v_y + \dot{\psi}a}{v_x + \dot{\psi}d} \\ \alpha_{fr} = -\delta_{rl} + \arctan\left(\frac{v_y - \dot{\psi}a}{v_x - \dot{\psi}d}\right) \approx -\delta_{fr} + \frac{v_y - \dot{\psi}a}{v_x - \dot{\psi}d} \\ \alpha_{rr} = -\delta_{rr} + \arctan\left(\frac{v_y - \dot{\psi}a}{v_x + \dot{\psi}d}\right) \approx -\delta_{rr} + \frac{v_y - \dot{\psi}a}{v_x + \dot{\psi}d} \end{cases} \tag{11}$$

Then, the lateral wheel slip is expressed as

$$\sigma_{yi} = \frac{v_{wxi}}{R_i \omega_i} \tan \alpha_i, \tag{12}$$

and we have the resultant slip

$$\sigma_i = \sqrt{\sigma_{xi}^2 + \sigma_{yi}^2} \tag{13}$$

2.3. Driver Model

As to calculate the steering wheel input δ_{sw}, a standard preview-based controller in CarSim is used as a driver model, which minimized the vehicle deviations from the desired path for a given preview time. Further, the desired yaw motion is generated through a reference model, which describes the ideal vehicle responses based on vehicle speed and δ_{sw}, in the form of

$$\frac{\omega_{z,des}(s)}{\delta_{sw}(s)} = \frac{k_r \overline{\omega}_{z,des}}{(T_1 s + 1)(T_2 s + 1)}, \tag{14}$$

$$\overline{\omega}_{z,des} = \frac{v_x}{(a+b) + \frac{(bC_r - aC_f)mv_x^2}{2C_r C_f (a+b)}} \tag{15}$$

where k_r denotes the gain of the reference model. In this work, we set $k_r = 0.05$. T_1 and T_2 are tunning parameters. C_f and C_r represent the lateral stiffness of a single front/rear tire. Desired lateral velocity $v_{y,des}$ is set to zero to avoid unnecessary tire lateral slip. Since the vehicle rollover can easily lead to fatal traffic accidents, it is necessary to prevent it through braking under rollover propensity quantitatively described by a Rollover Index (RI) that depends on the vehicle lateral acceleration and body roll angle [10]. When the index reaches the target warning value RI_{th}, brake control is adopted to prevent the vehicle from rolling over dangers. The RI is in the form of

$$\begin{cases} RI = C_1\left(\frac{|\varphi_t|\dot{\varphi}_{th} + \varphi_{th}|\dot{\varphi}_t|}{\varphi_{th}\dot{\varphi}_{th}}\right) + C_2\left(\frac{|a_y|}{a_{y,c}}\right) + (1 - C_1 - C_2)\frac{|\varphi_t|}{\sqrt{\varphi_t^2 + \dot{\varphi}_t^2}}, & \text{if } \varphi(\dot{\varphi} - k\varphi) \geq 0 \\ RI = 0, & \text{if } \varphi(\dot{\varphi} - k\varphi) < 0 \end{cases} \tag{16}$$

When the RI reaches the defined target value RI_{th}, we control the lateral acceleration by braking to reduce the RI down to RI_{th}, and the desired lateral acceleration is described as

$$\left|a_{y,des}\right| = \frac{a_{y,c}}{C_2}[RI_{des} - C_1\left(\frac{|\varphi_t|\dot{\varphi}_{th} + \varphi_{th}|\dot{\varphi}_t|}{\varphi_{th}\dot{\varphi}_{th}}\right) - (1 - C_1 - C_2)\frac{|\varphi_t|}{\sqrt{\varphi_t^2 + \dot{\varphi}_t^2}}] \tag{17}$$

Considering the relationship between vehicle lateral acceleration and driving speed

$$a_{y,des} = \dot{v}_y + v_{x,des}\omega_z, \tag{18}$$

$$a_y = \dot{v}_y + v_x\omega_z \tag{19}$$

It follows that

$$v_{x,des} = v_x + \frac{1}{\omega_z}\left(a_{y,des} - a_y\right) \tag{20}$$

Otherwise, we consider the desired longitudinal speed as a constant cruising value when the RI was controlled within a reasonable range. Then the desired vehicle speed becomes

$$v_{x,des} = v_{x,cons} \tag{21}$$

3. Design of the Control System

For a UCC design with all four wheels having independent torque, suspension, and steering-by-wire functions, the vehicle works as a redundantly actuated system. A hierarchical control structure shown in Figure 3 is presented to coordinate different control subsystems by the allocation of tire forces.

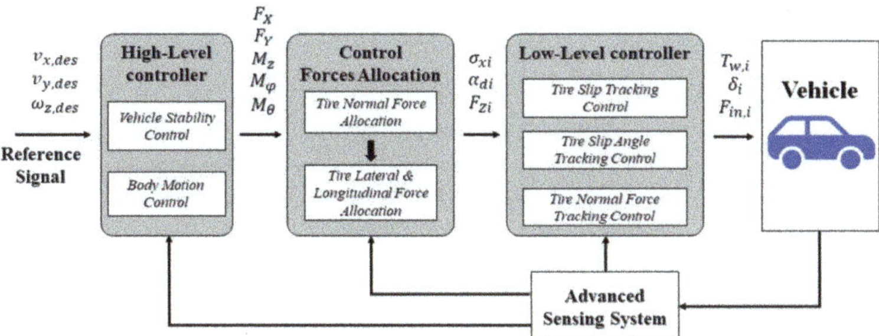

Figure 3. The architecture of unified vehicle dynamics control.

In this scheme, the high-level controller determines the resultant force/moment affected on the vehicle according to the control targets, including both the vehicle lateral stability and the spatial body motion (roll and pitch). Regarding the desired body motion dynamics, the target values of lateral and longitudinal transfer among the normal forces of every tire can be obtained. With the allocated normal forces through active suspension control, the CA algorithm distributes the resultant lateral and longitudinal forces to every wheel, which are finally realized through the closed-loop tracking control methods by considering the nonlinear tire model. In this work, we ignore the tire transient dynamics and assume that the desired tire longitudinal/lateral forces are well tracked through the variation of wheel steering angle δ_i and the driving/braking torque $T_{w,i}$. The steering angle of each wheel can be obtained through (11) with the desired value of wheel slip angle. According to the desired wheel longitudinal forces, $T_{w,i}$ are calculated as,

$$T_{w,i} = J_{w,i}\dot{\omega}_{w,i} + F_{x,i}r_{w,i},\tag{22}$$

where $J_{w,i}$ include the wheel-side inertia. $r_{w,i}$ and $\omega_{w,i}$ are the radius and the angular speed of the wheels, respectively. The values of δ_i and $T_{w,i}$ are treated as the inputs of the vehicle model in simulations. We also assume that all necessary quantitires in the control design can be practically measured or estimated.

3.1. High-Level Robust Controller Design

Parameter uncertainty is a common issue that requires robustness in control design. Compared with the actual vehicle model, common parameters with uncertainty include vehicle mass m, the inertia of moment I_{xx}, I_{yy}, I_{zz}, rolling center position e_r, pitching center position e_p, unmodeled dynamics such as suspension dynamics, and disturbance like wind gust and road roughness. Addressing the vehicle non-linearity and mentioned system uncertainties, a sliding mode controller is proposed for system robustness based on the simplified dynamics. If we define the system states as

$$x_i = [v_x, v_y, \dot{\psi}, \dot{\varphi}, \dot{\theta}]$$

The vehicle system can be given in state-space form as

$$\begin{cases} \dot{x}_1 = x_2x_3 - \frac{1}{m}F_r - \frac{m_s}{m}e_r x_3 x_4 + \frac{1}{m}u_1 + \Delta_1 \\ \dot{x}_2 = -x_1x_3 + \frac{m_s}{m}e_r \dot{x}_4 + \frac{1}{m}u_2 + \Delta_2 \\ \dot{x}_3 = \frac{1}{I_{zz}}u_3 + \Delta_3 \\ \dot{x}_4 = \frac{a_y[m_s(h_r+e_r)+m_u h_a]}{I_{xx}} + \frac{m_s g e_r \varphi}{I_{xx}} + \frac{1}{I_{xx}}u_4 + \Delta_4 \\ \dot{x}_5 = -\frac{a_x[m_s(h_p+e_p)+m_u h_a]}{I_{yy}} + \frac{m_s g e_p \theta}{I_{yy}} + \frac{1}{I_{yy}}u_5 + \Delta_5 \end{cases}\tag{23}$$

where Δ_i stand for unmodeled dynamics. The control inputs are defined as

$$(u_1, u_2, u_3, u_4, u_5) = \left(F_X, F_Y, M_z, M_\varphi, M_\theta\right)$$

These equations can be treated as five single-input-single-output systems, then the control inputs $u_1, \ldots, u_5$ become decoupled. Next, we select the velocity error as the sliding surface for state x_1, x_2, and x_3 for the path following purpose. Considering that the body roll and pitch motions should not be avoided completely for tire contact stability, we design dynamics sliding surfaces for the other two states.

$$\begin{cases} S_n = e_n = x_n - x_{n,des}, \ n = 1, 2, 3 \\ \qquad S_n = x_4 + C_\varphi \varphi, \ n = 4 \\ \qquad S_n = x_5 + C_\theta \theta, \ n = 5 \end{cases} \tag{24}$$

In this paper, we turn the coefficients as $C_\varphi = C_\theta = 1$. For each channel, the Lyapunov function candidate is in the form of

$$V_n = \frac{1}{2} S_n^2 \tag{25}$$

To achieve the attractive behavior of the sliding surface in a finite time period, it follows that

$$\dot{V}_n = S_n \dot{S}_n \le -\eta_n |S_n| \tag{26}$$

where the value of $\eta_n(>0)$ decides the speed of sliding surface convergence. The attractive equations can be given as

$$\dot{S}_n = -K_n Sign(S_n) \tag{27}$$

Taking channel one as example, we have

$$x_2 x_3 - \frac{F_r}{m} - \frac{m_s}{m} e_r \dot{x}_3 x_4 + \frac{1}{m} u_1 + \Delta_1 - \dot{x}_{1,des} = -K_1 Sign(S_1) \tag{28}$$

In practical use, some system parameters remain unknown to control design. Thus, the desired control efforts can only be expressed based on the nominal values, such as

$$u_1 = \overline{m}\left[-x_2 x_3 + \frac{F_r}{\overline{m}} + \frac{\overline{m}_s}{\overline{m}} \overline{e}_r \dot{x}_3 x_4 + \dot{x}_{1,des} - \overline{\Delta}_1 - K_1 Sign(S_1)\right] \tag{29}$$

Substituting this value into (26), it follows that

$$\dot{V}_1 = S_1 \left\{ \left(1 - \frac{\overline{m}}{m}\right) x_2 x_3 + \left(\frac{\overline{m}_s \overline{e}_r}{m} - \frac{m_s e_r}{m}\right) \dot{x}_3 x_4 + \left(\frac{\overline{m}}{m} - 1\right) \dot{x}_{1,des} + \left(\Delta_1 - \frac{\overline{m}}{m} \overline{\Delta}_1\right) - \frac{\overline{m}}{m} K_1 Sign(S_1) \right\} \tag{30}$$

Here, we select the nominal values as

$$\begin{cases} \overline{m} = \sqrt{m_{min} m_{max}}, \ \beta_m^{-1} \le \frac{\overline{m}}{m} \le \beta_m \\ \overline{m}_s = \sqrt{m_{s,min} m_{s,max}}, \ \beta_{ms}^{-1} \le \frac{\overline{m}_s}{m_s} \le \beta_{ms} \\ \overline{e}_r = \sqrt{e_{r,min} e_{r,max}}, \ \beta_{er}^{-1} \le \frac{\overline{e}_r}{e_r} \le \beta_{er} \end{cases} \tag{31}$$

where,

$$\beta_m = \sqrt{\frac{m_{max}}{m_{min}}}, \beta_{ms} = \sqrt{\frac{m_{s,max}}{m_{s,min}}}, \beta_{er} = \sqrt{\frac{e_{r,max}}{e_{r,min}}}$$

Then all factors in (30) have corresponding upper bounds

$$\begin{cases} \left|1 - \frac{\overline{m}}{m}\right| \le \max\left\{\left|1 - \beta_m^{-1}\right|, \left|1 - \beta_m\right|\right\} = \overline{\beta}_{11} \\ \left|\frac{\overline{m}_s \overline{e}_r}{m} - \frac{m_s e_r}{m}\right| \le \max\left\{\left|1 - \beta_{ms}^{-1} \beta_{er}^{-1}\right|, \left|1 - \beta_{ms} \beta_{er}\right|\right\} e_{r,max} = \overline{\beta}_{12} \\ \left|\Delta_1 - \frac{\overline{m}}{m} \overline{\Delta}_1\right| \le \max\left\{\left|\Delta_{1,max}\right|, \left|\Delta_{1,min}\right|\right\} + \beta_m \left|\overline{\Delta}_1\right| = \overline{\beta}_{13} \end{cases} \tag{32}$$

Due to the assumption that all the vehicle states and their derivates are physically upper bounded, Equation (30) becomes

$$\dot{V}_1 \le \left[\overline{\beta}_{11}|x_2x_3|_m + \overline{\beta}_{12}|\dot{x}_3x_4|_m + \overline{\beta}_{11}|\dot{x}_{1,des}|_m + \overline{\beta}_{13} - \beta_m^{-1}K_1\right]|S_1| \le -\eta_1|S_1| \tag{33}$$

In order to achieve the convergence inequality, it is then sufficient to have

$$K_1 \ge \beta_m\left(\overline{\beta}_{11}|x_2x_3|_m + \overline{\beta}_{12}|\dot{x}_3x_4|_m + \overline{\beta}_{11}|\dot{x}_{1,des}|_m + \overline{\beta}_{13} + \eta_1\right) \tag{34}$$

To avoid the chattering effects caused by the switching function, it is replaced by a saturation function, which is a continuous approximation with the thickness of ϕ_1, and then the control law of channel 1 becomes

$$u_1 = \overline{m}\left[-x_2x_3 + \frac{1}{2\overline{m}}CAx_1^2 + \frac{\overline{m}_s}{\overline{m}}\overline{e}_r\dot{x}_3x_4 + \dot{x}_{1,des} - \overline{\Delta}_1 - K_1Sat\left(\frac{S_1}{\phi_1}\right)\right] \tag{35}$$

Through the same process, the control laws for other sliding surfaces are

$$u_2 = \overline{m}\left[x_1x_3 + \frac{\overline{m}_s}{\overline{m}}\overline{e}_r\dot{x}_4 + \dot{x}_{2,des} - \overline{\Delta}_2 - K_2Sat\left(\frac{S_2}{\phi_2}\right)\right] \tag{36}$$

$$u_3 = \overline{I}_{zz}\left[\dot{\omega}_{z,des} - \overline{\Delta}_3 - K_3Sat\left(\frac{S_3}{\phi_3}\right)\right] \tag{37}$$

$$u_4 = -a_y\left[\overline{m}_s\left(\overline{h}_r + \overline{e}_r\right) + \overline{m}_u h_a\right] - \overline{m}_s\overline{e}_s g\varphi + \overline{I}_{xx}\left[-C_\varphi x_4 - \overline{\Delta}_4 - K_4Sat\left(\frac{S_4}{\phi_4}\right)\right] \tag{38}$$

$$u_5 = a_x\left[\overline{m}_s\left(\overline{h}_p + \overline{e}_p\right) + \overline{m}_u h_a\right] - \overline{m}_s\overline{e}_p g\theta + \overline{I}_{yy}\left[-C_\theta x_5 - \overline{\Delta}_5 - K_5Sat\left(\frac{S_5}{\phi_5}\right)\right] \tag{39}$$

For every channel, by choosing the K_n to be sufficiently large, the convergence inequalities can be guaranteed, and the sliding surfaces are designed to be attractive. In this paper, we choose $K_1 = 0.02$, $K_2 = 0.2$, $K_3 = 0.01$, $K_4 = 0.1$, and $K_5 = 0.1$. Through the appropriate CA method, the high-level control efforts u_1–u_5 are distributed into four wheels. However, we cannot always achieve these control efforts considering that the tire forces are generated through the contact behavior between tires and road and are physically bounded.

3.2. Control Allocation Algorithm

The hierarchical control algorithm with a redundant set of actuators frequently contains a high-level controller to generate virtual control efforts and a control allocation algorithm to coordinate the actuators to produce the desired control values together. If the control efforts require forces beyond the capabilities of the actuators due to saturation or physical limitations, the CA algorithm should be able to lower its performance and search for a control input vector that minimized the error. Additionally, redundant actuators can be utilized to provide fault tolerance for safety-critical conditions such as normal force losing. For real-time applications, the computational effort is also an essential property for choosing the CA algorithm. In this viewpoint, we applied a fixed-point CA method to solve the CA problem in constraint system control. Based on the nonlinear vehicle and tire models, the generalized force/moment can be expressed using a set of nonlinear functions of control variables

$$\mathbf{u_d} = F(\lambda, \mathbf{U}), \tag{40}$$

where, $\lambda = [F_{zi}, \mu_i, \delta_i]^T$ is a 12 × 1 vector configurating the CA problem and remains invariable during a CA computation step. μ_i is the friction coefficient between road and tires. $\mathbf{U} = [\alpha_i, \kappa_i]^T$ is a 8 × 1 control vector of slip angle and slip ratio of each tire. $\mathbf{u_d}$

denotes the desired generalized force, which can be rewritten using a first-order linearizing expression about an operating point,

$$F(\lambda, \mathbf{U}) \approx F(\lambda, \mathbf{U}_{k-1}) + \tfrac{\partial F}{\partial \mathbf{U}}(\lambda, \mathbf{U}_{k-1}) \cdot (\mathbf{U}_k - \mathbf{U}_{k-1}) \Rightarrow$$
$$F(\lambda, \mathbf{U}) - F(\lambda, \mathbf{U}_{k-1}) + \tfrac{\partial F}{\partial \mathbf{U}}(\lambda, \mathbf{U}_{k-1}) \cdot \mathbf{U}_{k-1} = \tfrac{\partial F}{\partial \mathbf{U}}(\lambda, \mathbf{U}_{k-1}) \cdot \mathbf{U}_k = \mathbf{B}_F \cdot \mathbf{U}_k \tag{41}$$

where, $\mathbf{B}_F$ is a Jacobian matrix defined by the configurating vector and the control vector of the last step, which describes the sensitivity of the control vector to desired value vector $\mathbf{u}'_d$, which is in the form of

$$\mathbf{u}'_d = \mathbf{u}_d - F(\lambda, \mathbf{U}_{k-1}) + \frac{\partial F}{\partial \mathbf{U}}(\lambda, \mathbf{U}_{k-1}) \cdot \mathbf{U}_{k-1} \tag{42}$$

Then, Equation (41) becomes

$$\mathbf{u}'_d \approx \mathbf{B}_F \cdot \mathbf{U}_k \tag{43}$$

The CA algorithm minimized the following objective function, including the CA errors and control effort. The optimization criteria can be given as

$$\min J = \frac{1}{2}(1 - \varepsilon)\Gamma^T \mathbf{W}_e \Gamma + \frac{1}{2}\varepsilon \mathbf{U}^T \mathbf{W}_U \mathbf{U} \tag{44}$$

where $\Gamma = (\mathbf{B}_F \mathbf{U} - \mathbf{u}'_d)$ is the control error vector, and the variation of control value $\mathbf{U} \in [\mathbf{U}_L, \mathbf{U}_U]$ is restrained by actuator rate limits and achievable value ranges. The two bounds are decided by

$$\begin{cases} \mathbf{U}_L = \max[\mathbf{U}_{min}, \mathbf{U}_{k-1} - \tau \cdot \mathbf{r}_{max}] \\ \mathbf{U}_U = \min[\mathbf{U}_{max}, \mathbf{U}_{k-1} + \tau \cdot \mathbf{r}_{max}] \end{cases} \tag{45}$$

where, τ represents the sampling time and $\mathbf{r}_{max}$ is the maximum rates of variable values. This constraint helps to ensure the process stability. ε is a number to balance the weight between control efforts error and actuation cost, which is chosen as 0.5. Suppose that we want to minimize $J:R^8 \rightarrow R$, a gradient descent algorithm has the iterative step

$$\mathbf{U}_{i+1} = \mathbf{U}_i - \eta \nabla J(\mathbf{U}_i) \tag{46}$$

$\nabla J(\mathbf{U})$ is the derivative of J at $\mathbf{U}$. Substituting the derivative of J into this equation. It follows that

$$\begin{cases} \mathbf{U}_{i+1} = sat\left[(1 - \varepsilon)\eta \mathbf{B}_F^T \mathbf{W}_e \mathbf{u}'_d - (\eta \mathbf{T} - \mathbf{I})\mathbf{U}_i\right] := I(\mathbf{U}_i) \\ \mathbf{T} = (1 - \varepsilon)\mathbf{B}_F^T \mathbf{W}_e \mathbf{B}_F + \varepsilon \mathbf{W}_U \end{cases} \tag{47}$$

where the step length parameter η is set to $\eta = 1/\|\mathbf{T}\|_F$, with $\|\cdot\|_T$ being the Frobenius norm of a matrix, which is more computationally efficient than induced norms. The saturation function sat can cut the elements of the control vector $\mathbf{U}_{(i+1)}$ at their limits. If J is convex and the operator I is convergent, then a point $\mathbf{U}_k$ is a fixed point of I if and only if $\nabla J(\mathbf{U}_k) = 0$, so if and only if $\mathbf{U}_k$ minimizes J, with the expression

$$\mathbf{U}_k = I(\mathbf{U}_k) \tag{48}$$

3.3. Tire Normal Forces Robust Tracking Control

Based on the mentioned tire models, the boundaries of friction ellipses are directly affected by the normal tire forces, which are the control vectors of the high-level strategy for inhibiting undesired vehicle motion. The tracking control based on the active suspension system is a challenging task because of the following reasons.

(1) In practical use, the wheel motions are under the effect of road roughness, especially considering the high unsprung mass introduced by the electric propulsion system,

such as the in-wheel motor. This kind of motion instability will cause the inaccuracy of tire load tracking.

(2) Quarter car model is the classic model which was widely utilized in suspension analysis and control synthesis. However, the spatial kinematics and dynamics considering the suspension geometry are relatively complicated, leading to the inaccuracy and uncertainty of model parameters. Though some analytic model of suspension geometry is presented and realized in simulation [33], the complex computation makes them less efficient in real-car implements.

(3) The active suspension control algorithm is always a trade-off between the vehicle ride comfort and tire-road adhesion stability, which is hard to be optimized simultaneously. Moreover, the tracking control requirement makes the problem more complex and increases the difficulty of controller design.

In this work, we used the traditional quarter car model and H-∞ robust control method to solve the modeling uncertainty and parameter variation problem. The model concept is illustrated through Figure 4, with M_b as the one-fourth sprung mass, M_w as the one-fourth unsprung mass, k_s as the equivalent suspension stiffness, b_s as the equivalent suspension damping, and k_t as the tire stiffness. The force from the vehicle body F_{in} is a known input. F_a is the active force returned by the active suspension controller.

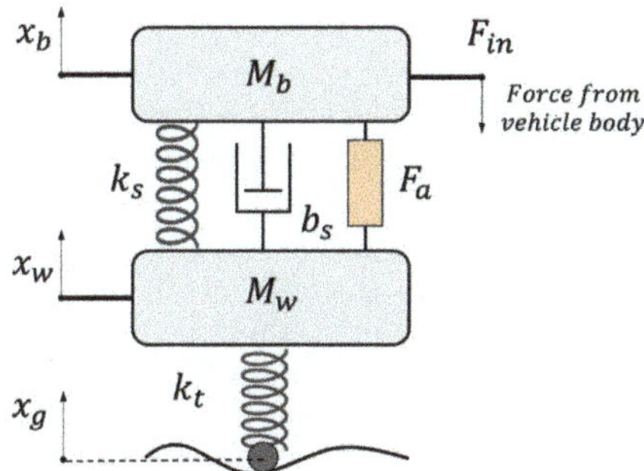

Figure 4. One-fourth active suspension model with body inertia force.

Based on the traditional quarter car model, an extra equivalent body inertia force F_{in} is imposed upon the sprung mass to reflect the coupling effect between the quarter model and the full-vehicle model. The value of F_{in} is supposed to be the desired normal tire force. Due to suspension geometry and ride comfort requirements, the equivalent suspension stiffness and damping are actually not constant values, which is treated as system uncertainties. The body inertia force and road profile are the outside disturbances to the system. However, we can still use their nominal values to establish the model. The states of the controlled system are defined as $[x_1, x_2, x_3, x_4] = [x_w, \dot{x}_w, x_b, \dot{x}_b]$, and generalized system inputs are $[u_1, u_2, u_3] = [F_{in}, x_g, F_a]$.

Then, the vertical dynamics of suspension can be given as

$$\begin{cases} \dot{x}_2 = -\left(\frac{k_s+k_t}{M_w}\right)x_1 - \frac{b_s}{M_w}x_2 + \frac{k_s}{M_w}x_3 + \frac{b_s}{M_w}x_4 + \frac{k_t}{M_w}u_2 - \frac{1}{M_w}u_3 \\ \dot{x}_4 = \frac{k_s}{M_b}x_1 + \frac{b_s}{M_b}x_2 - \frac{k_s}{M_b}x_3 - \frac{b_s}{M_b}x_4 - \frac{1}{M_b}u_1 + \frac{1}{M_b}u_3 \end{cases} \tag{49}$$

The system performance index is selected as

$$Y_{perf} = [\ddot{x}_b, e_n, F_a] \tag{50}$$

where the sprung mass acceleration is the index for ride comfort, and $e_n = F_{in} - k_t(x_g - x_w)$ expresses the normal force control error, defined as the tire contact stability index. F_a stands for the active suspension actuator force. The system output includes the performance index as well as the signals to be measured for feedback control, which may contain the body (unsprung mass) acceleration (BA) and tire dynamic load (DL). We use the linear fractional transformation method to define an extended state-space model to deal with the parameter uncertainties. The closed-loop system diagram is described in Figure 5 [34].

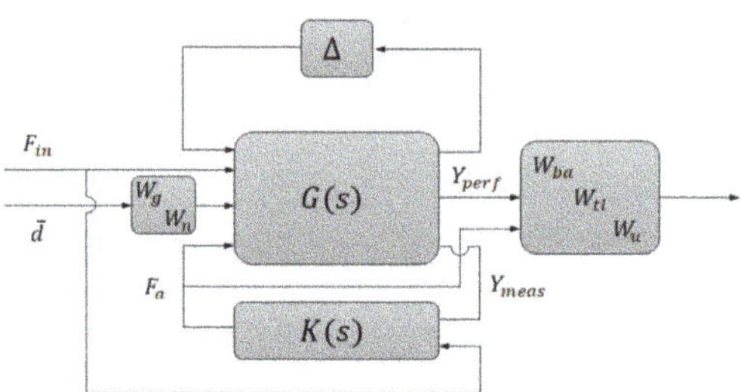

Figure 5. Robust tracking control of tire contact force.

The weighting functions of system performance index W_{ba}, W_{tl}, and control output and W_u, which define the target performance in the frequency domain, should be well tuned to acquire a stable and satisfactory controller. Here, we design and tune the weighting function according to the open-loop suspension performance, and the amplitude-frequency characteristics of BA and DL are compared with the inverse of the weighting function W_{ba} and W_{tl} in Figure 6. Furthermore, since the desired characteristic of BA and DL is hard to achieve simultaneously, we use different control structures to achieve the vehicle comfort target and tire stability target:

(1) Controller 1: For the state of tire contact force cannot reflect the dynamic behavior of sprung mass, only the sprung mass acceleration was selected as the feedback signal of the comfort-orientated control scheme.

(2) Controller 2: In the tire-stability-orientated control scheme, both the unsprung mass acceleration and tire contact force work as the feedback signals, with the former signal reflecting the desired tire normal force and the latter one providing the real normal force. The sensor reflects the DL could be tire-pressure based, for example.

In order to compare the performance of the closed-loop system of controller 1 and controller 2, the state-space system models are established in the MATLAB environment, where the corresponding H-∞ controllers are synthesized through coupled Riccati equations [35]. Further, a sinusoidal input was built to test the tracking performance of controller 2. Simulation results concerning body acceleration and tire dynamic load in the time domain are shown in Figure 7. Their root mean square values are listed in Table 1 for comparison. An A-class uneven road is simulated by filtered white noise according to ISO8608, with the vehicle speed chosen as 120 km/h.

Table 1. RMS values of body acceleration and tire dynamic load.

Suspension Performances	Passive Suspension (without Control)	Comfort-Orientated Control	Stability-Orientated Control
Body acceleration (m/s^2)	0.6732	0.2795 (↓58%)	0.9067 (↑35%)
Tire dynamic load (N)	242.7421	377.5729 (↑56%)	204.7990 (↓16%)

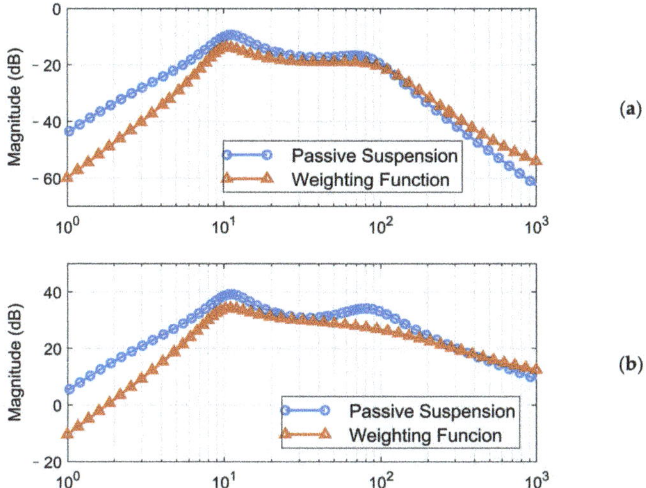

Figure 6. The inverse of system performance index weighting functions: (**a**) vertical acceleration; (**b**) tire dynamic load.

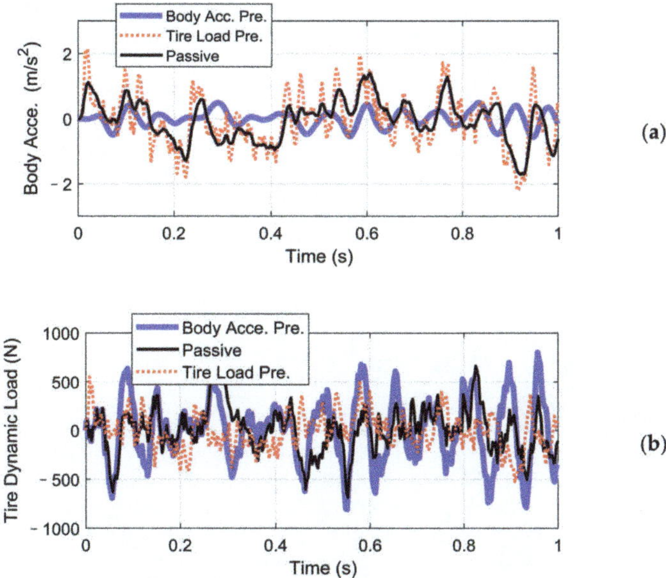

Figure 7. Performance of Controllers 1 and 2: (**a**) body acceleration; (**b**) tire dynamic load.

The simulation results illustrate that, under the comfort-oriented control, the body acceleration magnitude illustrated by the blue line was decreased, compared with the value without active control expressed by the black line. On the contrary, the average tire DL was increased. Reversely, the stability-oriented control inhibited the normal tire force tracking error but enlarged the body acceleration magnitude unavoidably. Results in Figure 8 indicated that controller 2 achieved good tracking performance under sinusoidal input signal with the suppression of road roughness disturbance.

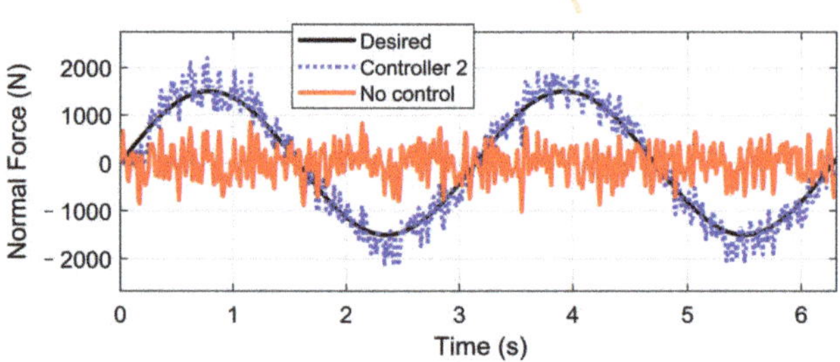

Figure 8. Normal force tracking of controller 2.

Although we only consider the tracking control of tire vertical forces, energy consumption of the active suspension system cannot be neglected. The tradeoff between vibration suppression and energy consumption should be considered in practical applications, which could be realized by adjusting the weighting funcion W_u.

4. Simulation Studies

In order to evaluate its overall performance, the UCC system with the application of the stability-oriented controller (controller 2) is implemented in a CarSim–Simulink co-simulation platform. The target vehicle is a B-class sports car. Its parameters are listed in Table 2. Compared with the traditional ESC approach, the dynamic performance and advantages of the proposed UCC method are interpreted through the following driving conditions.

Table 2. Vehicle parameters in the simulation.

Symbol	Description	Values and Units
m	Vehicle mass	1140 kg
C_D	Aerodynamic drag coefficient	0.34
a	Distance of front wheel axle from C.G.	1.165 m
b	Distance of rear wheel axle from C.G.	1.165 m
d	Half of the wheel base	0.7405 m
I_{zz}	Yaw inertia	996 kg m^2
m_s	Vehicle sprung mass	1020 kg
k_s	Suspension stiffness	33,972 N/m
b_s	Suspension damping	2000 N s/m
k_t	Tire stiffness	200,000 N/m

4.1. High-Speed Double Lane-Changing (DLC) on a Rough Road Surface

Double lane-changing is a standard test to evaluate the vehicle handling performance, which is usually done at a constant longitudinal speed. However, this maneuver becomes more challenging under a rough road surface at high speed because of the intensification of vertical vibration, and it is possible that the fluctuation of normal tire forces will influence the allocation accuracy of desired general longitudinal and lateral forces. In order to evaluate the dynamic performance of the proposed UCC system, a comparison simulation with the same controller was carried out based on the DLC test at 120 km/h speed under flat and rough road surfaces. The A-class uneven road profile is generated randomly. Since the effectiveness of active suspension control has been verified before, here we mainly focus on the lateral vehicle performances compared with the flat road situation. Simulations results are compared in Figures 9–13.

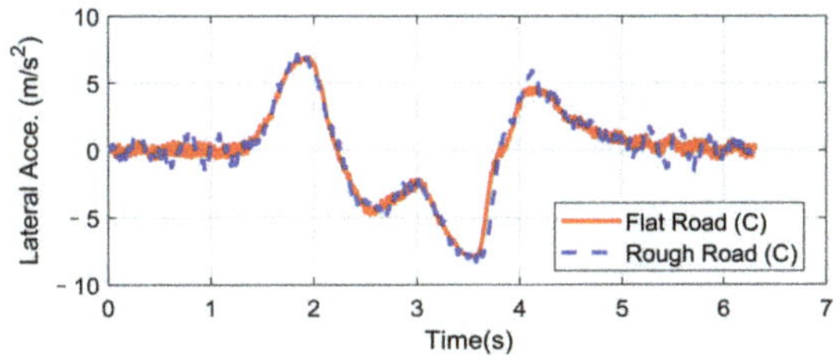

Figure 9. Lateral acceleration in high-speed DLC.

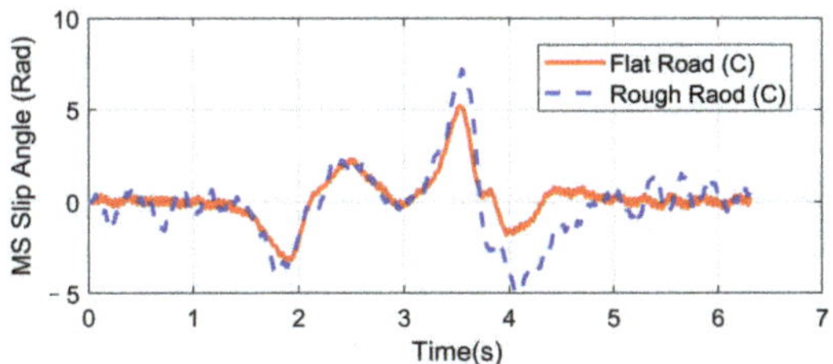

Figure 10. Vehicle slip angle in high-speed DLC.

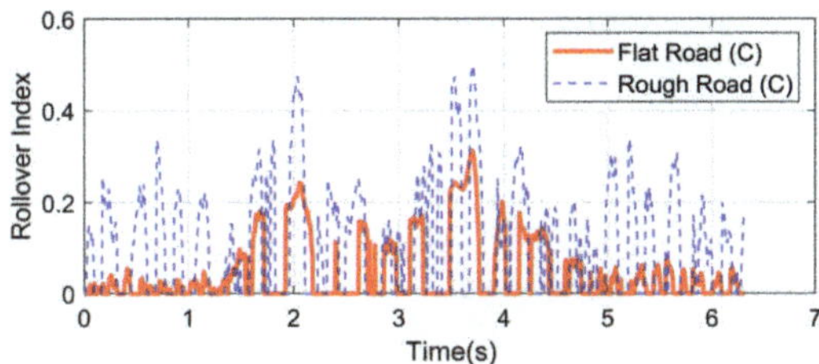

Figure 11. Rollover index in high-speed DLC.

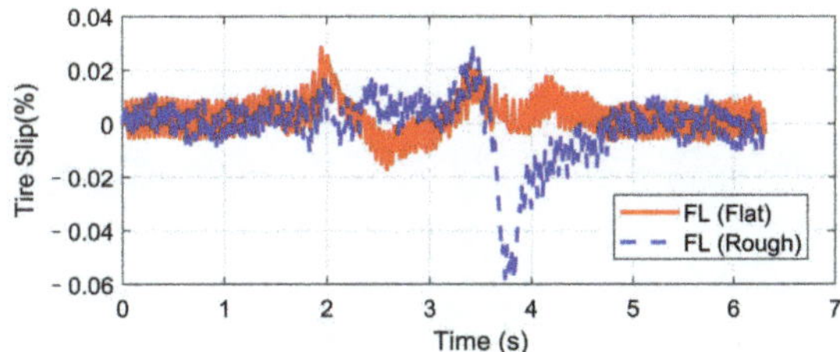

Figure 12. FL tire slip in high-speed DLC.

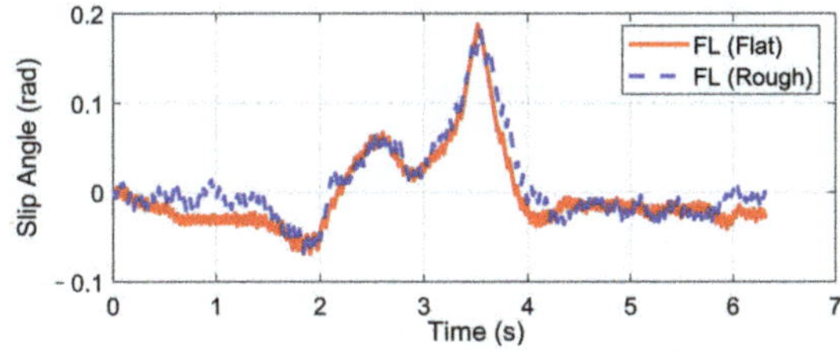

Figure 13. FL tire slip angle in high-speed DLC.

The simulation verified that the CA method could generate desired resultant forces for the vehicle to complete the DLC maneuver under road roughness. Besides, the results in Figures 9–11 demonstrate that the road disturbance generated slight undesired vehicle dynamics, including lateral and rolling motions. To investigate the effect of road roughness on the CA algorithm, the control values of the front left tire are compared in Figures 12 and 13.

We can see from the results that the UCC method has a certain degree of robustness due to the sliding mode controller. However, in the critical points of DLC process (around 2 s and 3.5 s), the vehicle lateral acceleration and yaw rate approached their maximum values, and the tire forces were close to the boundary. In order to achieve the required lateral force, slip angles of four tires should not have a significant variation, as shown in Figure 12. In these critical time points, the normal force fluctuation led to an extra yaw moment, which is supposed to be compensated by the longitudinal vehicle forces. As can be seen in Figure 13, the tire slip variation in such time points was more significant than at other times, and this effect was even more severe in rough road conditions. Since in the proposed UCC strategy, the desired forces are evenly allocated into four tires, the situations of the other three tires were similar and will not be discussed to avoid redundant expression.

4.2. High-Speed Double Lane-Changing (DLC) on a Flat Road Surface (Compared with ESC)

Based on the same DLC test, while the first scenario examined the UCC performance on the uneven road, this second one focused on the comparison with the traditional ESC method. The ESC system is an envelope stability controller based on direct yaw control, which can only control the longitudinal force distribution among four tires. The results in Figures 14–19 explain the superiority of the proposed UCC system on lateral vehicle dynamics compared with the ESC control strategy. The vehicle failed to pass this scenario without active control.

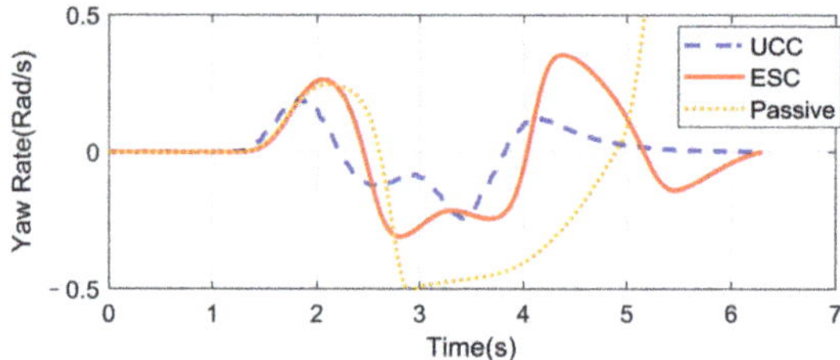

Figure 14. Yaw rate on flat road during high-speed DLC.

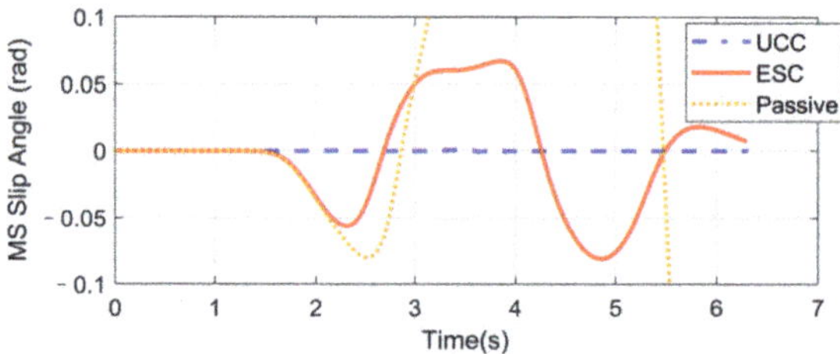

Figure 15. Vehicle slip angle on flat road during high-speed DLC.

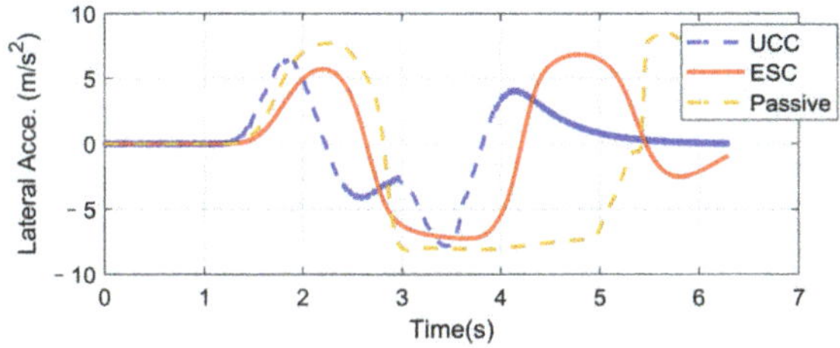

Figure 16. Lateral acceleration on flat road during high-speed DLC.

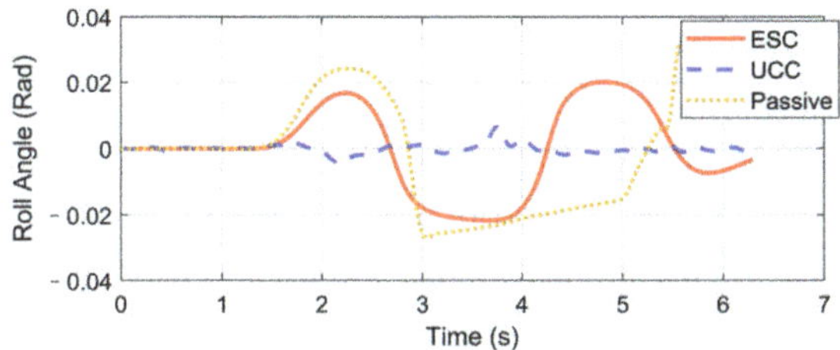

Figure 17. Body roll angle on flat road during high-speed DLC.

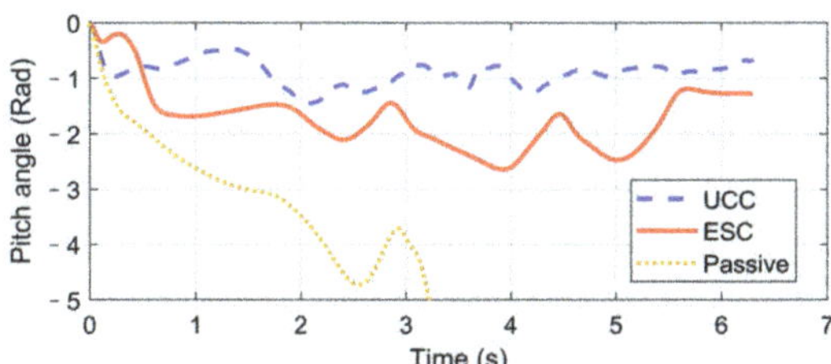

Figure 18. Body pitch angle on flat road during high-speed DLC.

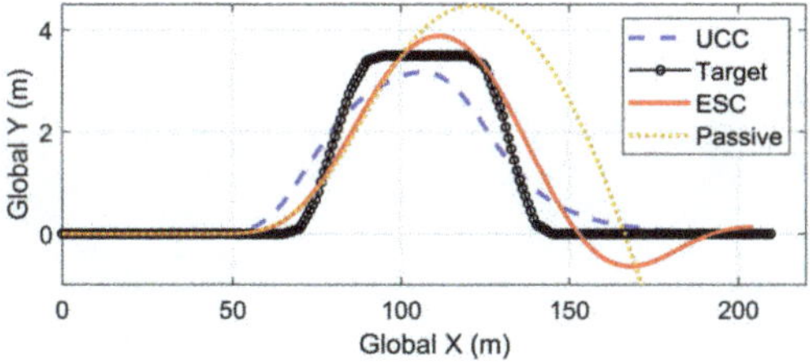

Figure 19. Vehicle path on flat road during high-speed DLC.

Figures 14–16 and Figure 19 show yaw rate, vehicle slip angle, lateral acceleration, and vehicle path, respectively, and illustrate that the UCC has better performance than the ESC with respect to the yaw stability. Therein the desired vehicle slip angle control was well done by the UCC control. This is due to the fact that the vehicle system with multiple actuators can make full use of the adhesion coefficient of each wheel to generate desired resultant force/moment. By comparing the vehicle rolling and pitch angles presented in Figures 17 and 18, one can clearly see that the proposed UCC maintained the body altitude effectively, enhancing the vehicle ride comfort due to active suspension control.

4.3. High-Speed Fishhook Maneuver

The Fishhook test is a dynamic test for predicting the dynamic rollover propensity. To begin the maneuver, the vehicle is driven in a straight line at the desired entrance speed, and then the driver releases the throttle and initiates the designed steering motion described in Figure 20. In this section, the steering maneuver of this test was adopted to verify the anti-rollover mechanism of the proposed UCC with the braking command triggered by the RI threshold set to 0.2. Figures 21–25 shows the results at entrance speed 120 km/h concerning relative vehicle states. If no control strategy is applied, the vehicle will lose control early at around 1.5 s, as shown in the figures.

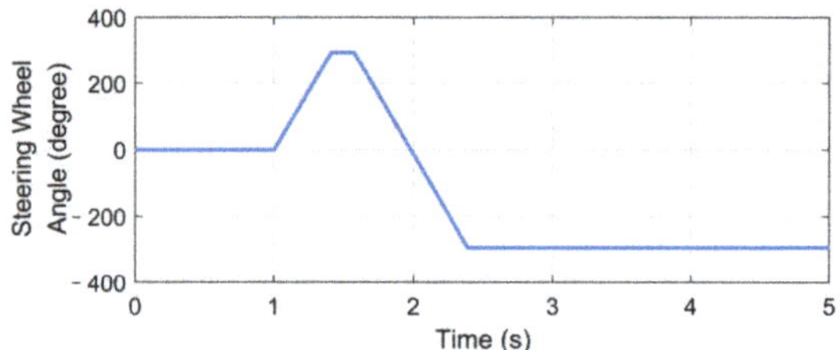

Figure 20. Steering angle in high-speed fishhook maneuver.

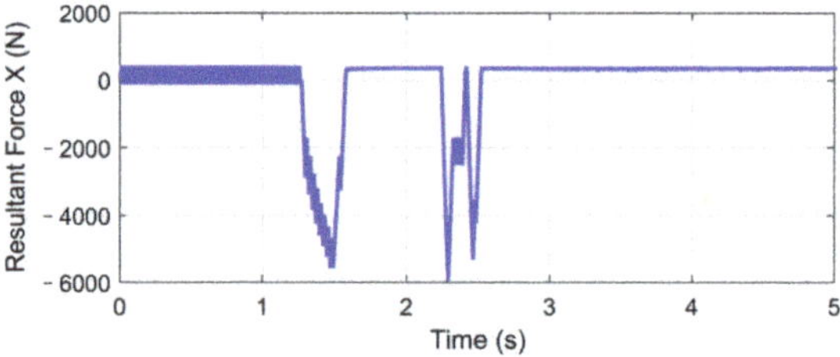

Figure 21. Longitudinal force in high-speed fishhook maneuver.

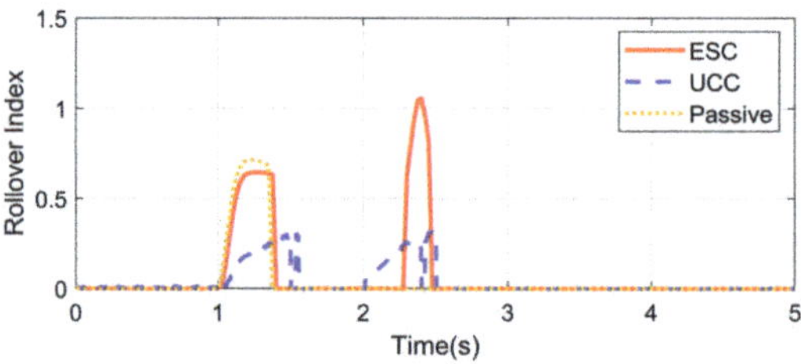

Figure 22. Rollover index in high-speed fishhook maneuver.

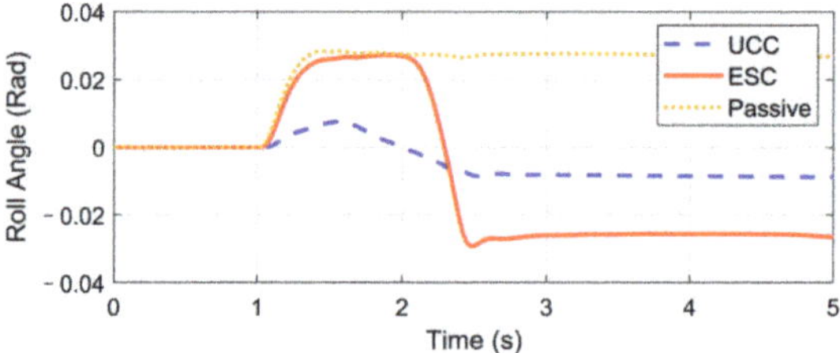

Figure 23. Body roll angle in high-speed fishhook maneuver.

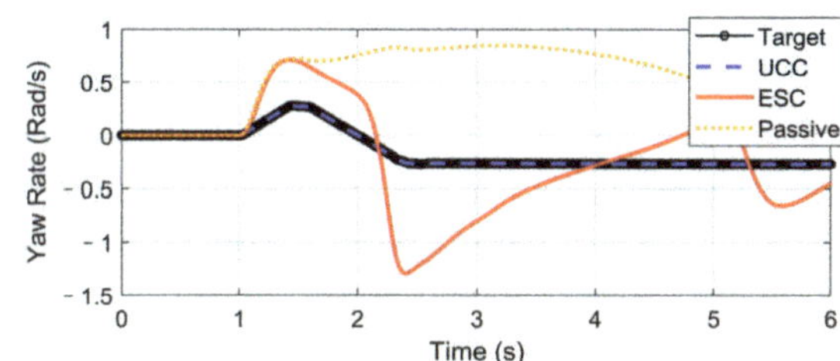

Figure 24. Yaw rate in high-speed fishhook maneuver.

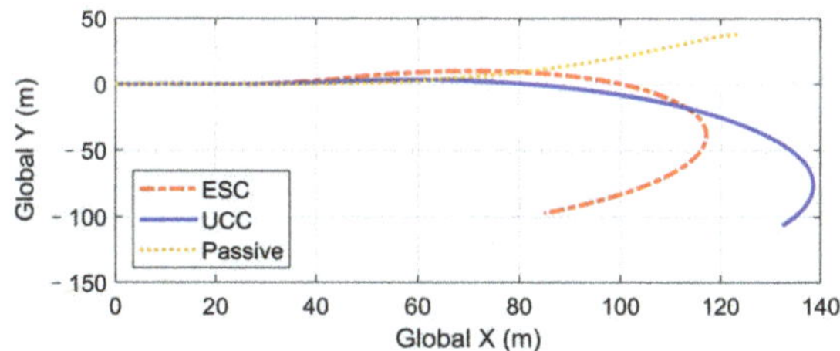

Figure 25. Vehicle global path in high-speed fishhook maneuver.

The result of vehicle RI clearly illustrates that, at time t = 1.5 and t = 2.3, when the steering angle reached its maximum value, the vehicle experienced the most apparent rollover propensity. Though the rollover did not happen actually due to the ESC control, the RI value still reached a dangerous level. The proposed UCC method can significantly reduce the RI value in the following two aspects.

(1) The body altitude control function of UCC reduced the vehicle roll angle, indicated in Figure 23.
(2) The braking force was triggered to inhibit the increase of lateral acceleration, which is shown in Figure 21.

We also observe from Figure 24 that the UCC method led to better dynamic performance and can achieve the desired yaw rate without losing stability. Additionally, the vehicle path curves shown in Figure 25 indicate that the undesired yaw moment generated by ESC control may promote the tendency of vehicle over-steering in high-speed curving situations, which can be avoided using the UCC with appropriate parameters tunning.

4.4. Tire Blow-Out in the Hard-Braking Process (Re-Configurable Control)

This scenario is designed for the condition where one tire blew out during hard braking. The initial vehicle speed is 120 km/h and a hard braking command (about −0.6 g) is given at 0 s, and then at the end of the first second, the front-left tire blew out and lost its lateral and longitudinal forces. The steering input is kept as zero, which keeps in line with the actual reaction. Based on different control strategies, three different conditions are considered, including UCC, ESC control, and passive vehicle. Results concerning vehicle and tire states are provided in Figures 26–30.

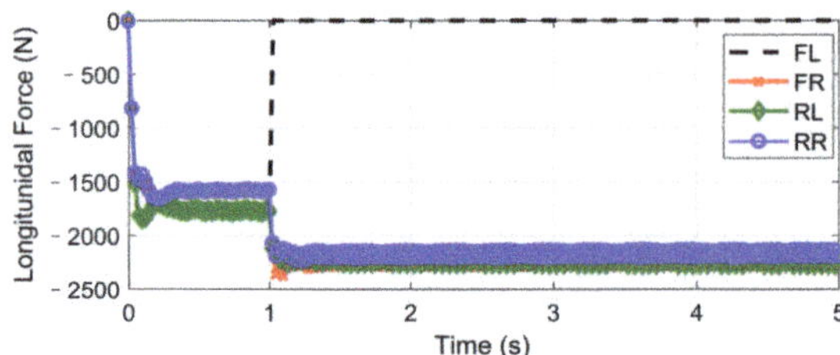

Figure 26. Longitudinal force in hard-braking.

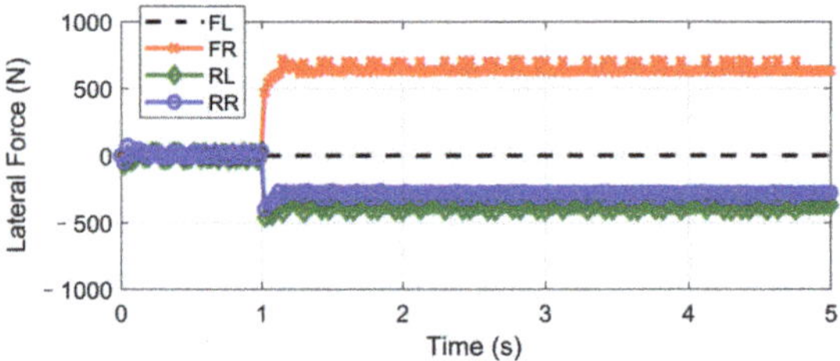

Figure 27. Lateral force in hard-braking.

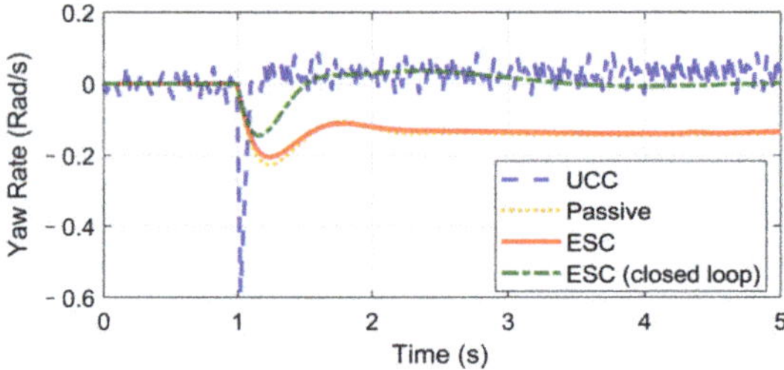

Figure 28. Yaw rate in hard-braking.

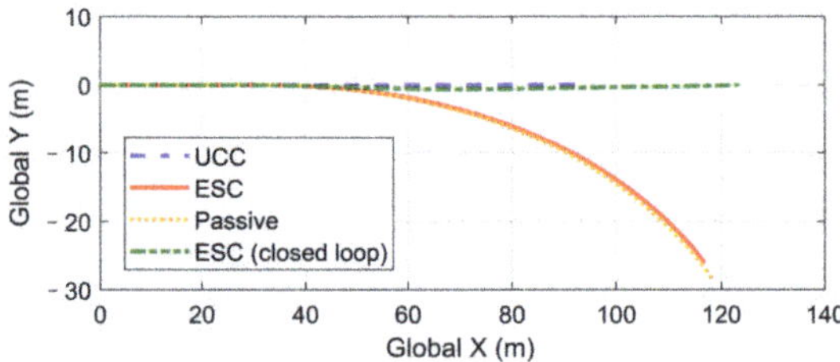

Figure 29. Vehicle path in hard-braking.

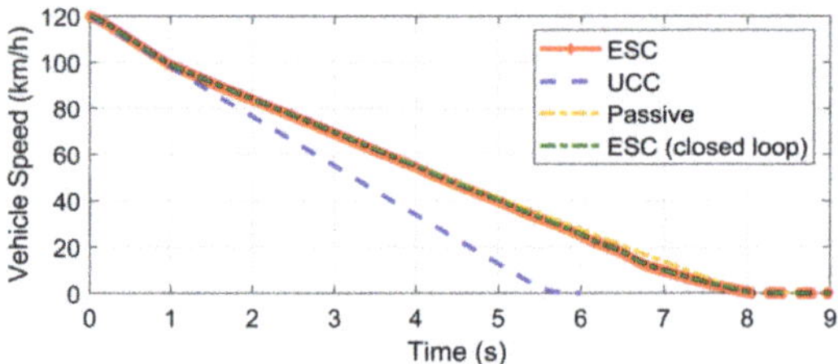

Figure 30. Vehicle speed in hard-braking.

Curves in Figures 26 and 27 described the re-allocation of tire longitudinal and lateral forces (in the body coordinate) after the front-left tire blew out through a reconfigurable UCC approach, keeping the desired resultant forces/moment unchanged. The vehicle yaw rate and path based on different control configurations are shown in Figures 28 and 29. These results illustrate the extra yaw moment due to the tire blowing-out and the lane departure behavior it caused for both the ESC control and passive situations. However, the prompt intervention of driver steering input (closed-loop ESC in figures) can reduce the danger to some extent accompanied by ESC, but it is highly demanding for non-professional drivers to respond appropriately under such emergencies.

On the contrary, the UCC just required the driver to keep straight ahead. Additionally, it is worth noticing in Figure 30 that the traditional ESC control cannot compensate for the lost tire brake forces. Thus, the brake distance was increased. However, the UCC reconfigurable method was able to hold the initial braking deceleration and inhibit the interference of extra yaw moment due to the re-allocation of tire forces. The proposed UCC is the only strategy that is able to keep the driver's intention for this emergency.

5. Conclusions

In this paper, a UCC strategy involving the vehicle yaw stability, body altitude, and tire contact stability was presented. A hierarchical control structure was adopted to realize the UCC, including high-level sliding mode control, fixed point CA, and a normal tire force robust tracking controller. Simulations of high-demanding driving situations concerning rough road surface, fast DLC, fishhook maneuver, and the tire blowing-out situation performed on a nonlinear full vehicle model have shown the effectiveness and advantages of the proposed control method. The following conclusions can be drawn:

(1) For a four-wheel independent drive–independent steering configuration, the proposed UCC method can effectively realize the tire planar force distribution and the desired vehicle motion, which proves to be a practical solution due to its simple feedback control rules and reconfigurable allocation.
(2) Considering the wheel vertical vibrations, we present a robust tire normal force tracking controller to address the tire contact stability issue. This work provides an innovative method for improving the vehicle driving stability and maintaining the desired body attitude simultaneously.

Author Contributions: Conceptualization, M.W. and W.W.; data curation, W.W.; formal analysis, W.W.; funding acquisition, X.C.; investigation, W.W.; methodology, X.C.; project administration, X.C.; resources, X.C.; software, W.W.; supervision, X.C. and W.W.; validation, M.W., X.C. and W.W.; writing—original draft, M.W.; writing—review and editing, W.W. All authors have read and agreed to the published version of the manuscript.

Funding: This project is partially supported by the National Key R&D Program of China (2018YFB0104800).

Informed Consent Statement: Informed consent was obtained from all subjects involved in the study.

Acknowledgments: The authors appreciate the reviewers for their helpful comments and suggestions in this study.

Conflicts of Interest: The authors declare no conflict of interest.

Abbreviations

UCC	Unified Chassis Control	ESC	Electric Stability Control
CG	Center of Gravity	RI	Rollover Index
BA	Body Acceleration	DL	Tire Dynamic Load
DLC	Doubel Lane Change	CA	Control Allocation

References

1. Murata, S. Innovation by in-wheel-motor drive unit. *Veh. Syst. Dyn.* **2012**, *50*, 807–830. [CrossRef]
2. Koehn, P.; Eckrich, M. Active steering-the BMW approach towards modern steering technology. *SAE Tech. Pap.* **2004**. [CrossRef]
3. Gysen, B.L.; Paulides, J.J.; Janssen, J.L.; Lomonova, E.A. Active electromagnetic suspension system for improved vehicle dynamics. *IEEE Trans. Veh. Technol.* **2010**, *59*, 1156–1163. [CrossRef]
4. Hoehn, B.R.; Stahl, K.; Gwinner, P.; Wiesbeck, F. Torque-vectoring driveline for electric vehicles. In *Proceedings of the International Federation of Automotive Engineering Societies FISITA 2012 World Automotive Congress*; Springer: Berlin/Heidelberg, Germany, 2013; pp. 585–593. [CrossRef]
5. Bouvin, J.L.; Hamrouni, E.; Moreau, X.; Benine-Neto, A.; Hernette, V.; Serrier, P.; Oustaloup, A. Hierarchical approach for global chassis control. In Proceedings of the 2018 European Control Conference (ECC), Limassol, Cyprus, 12–15 June 2018; pp. 2555–2560. [CrossRef]
6. Vivas-Lopez, C.A.; Tudon-Martinez, J.C.; Hernandez-Alcantara, D.; Morales-Menendez, R. Global chassis control system using suspension, steering, and braking subsystems. *Math. Probl. Eng.* **2015**, *2015*, 263424. [CrossRef]
7. Chokor, A.; Talj, R.; Doumiati, M.; Charara, A. A global chassis control system involving active suspensions, direct yaw control and active front steering. *IFAC-PapersOnLine* **2019**, *52*, 444–451. [CrossRef]
8. Cho, W.; Yoon, J.; Kim, J.; Hur, J.; Yi, K. An investigation into unified chassis control scheme for optimised vehicle stability and manoeuvrability. *Veh. Syst. Dyn.* **2008**, *46*, 87–105. [CrossRef]
9. Cho, W.; Choi, J.; Kim, C.; Choi, S.; Yi, K. Unified chassis control for the improvement of agility, maneuverability, and lateral stability. *IEEE Trans. Veh. Technol.* **2012**, *61*, 1008–1020. [CrossRef]
10. Yoon, J.; Cho, W.; Koo, B.; Yi, K. Unified chassis control for rollover prevention and lateral stability. *IEEE Trans. Veh. Technol.* **2009**, *58*, 596–609. [CrossRef]
11. Poussot-Vassal, C.; Sename, O.; Dugard, L.; Savaresi, S.M. Vehicle dynamic stability improvements through gain-scheduled steering and braking control. *Veh. Syst. Dyn.* **2011**, *49*, 1597–1621. [CrossRef]
12. He, J.; Crolla, D.A.; Levesley, M.C.; Manning, W.J. Coordination of active steering, driveline, and braking for integrated vehicle dynamics control. *Proc. Inst. Mech. Eng. Part D J. Automob. Eng.* **2006**, *220*, 1401–1420. [CrossRef]
13. Ono, E.; Hattori, Y.; Muragishi, Y.; Koibuchi, K. Vehicle dynamics integrated control for four-wheel-distributed steering and four-wheel-distributed traction/braking systems. *Veh. Syst. Dyn.* **2006**, *44*, 139–151. [CrossRef]
14. Zhao, J.; Wong, P.K.; Ma, X.; Xie, Z. Chassis integrated control for active suspension, active front steering and direct yaw moment systems using hierarchical strategy. *Veh. Syst. Dyn.* **2017**, *55*, 72–103. [CrossRef]
15. Mousavinejad, E.; Han, Q.L.; Yang, F.; Zhu, Y.; Vlacic, L. Integrated control of ground vehicles dynamics via advanced terminal sliding mode control. *Veh. Syst. Dyn.* **2017**, *55*, 268–294. [CrossRef]
16. Cheng, S.; Li, L.; Liu, C.Z.; Wu, X.; Fang, S.N.; Yong, J.W. Robust LMI-Based H-Infinite Controller Integrating AFS and DYC of Autonomous Vehicles with Parametric Uncertainties. *IEEE Trans. Syst. Man Cybern. Syst.* **2020**. [CrossRef]
17. Sun, J.; Cong, J.; Gu, L.; Dong, M. Fault-tolerant control for vehicle with vertical and lateral dynamics. *Proc. Inst. Mech. Eng. Part D J. Automob. Eng.* **2019**, *233*, 3165–3184. [CrossRef]
18. Wang, R.; Wang, J. Fault-tolerant control for electric ground vehicles with independently-actuated in-wheel motors. *J. Dyn. Syst. Meas. Control* **2012**, *134*. [CrossRef]
19. Ataei, M.; Khajepour, A.; Jeon, S. Model Predictive Control for integrated lateral stability, traction/braking control, and rollover prevention of electric vehicles. *Veh. Syst. Dyn.* **2020**, *58*, 49–73. [CrossRef]
20. Guo, J.; Luo, Y.; Li, K.; Dai, Y. Coordinated path-following and direct yaw-moment control of autonomous electric vehicles with sideslip angle estimation. *Mech. Syst. Signal Process.* **2018**, *105*, 183–199. [CrossRef]
21. Peng, H.; Wang, W.; An, Q.; Xiang, C.; Li, L. Path tracking and direct yaw moment coordinated control based on robust MPC with the finite time horizon for autonomous independent-drive vehicles. *IEEE Trans. Veh. Technol.* **2020**, *69*, 6053–6066. [CrossRef]
22. Pylypchuk, V.; Chen, S.K.; Moshchuk, N.; Litkouhi, B. Slip-Based Holistic Corner Control. In Proceedings of the ASME 2011 International Mechanical Engineering Congress and Exposition, Denver, CO, USA, 11–17 November 2011; pp. 337–342. [CrossRef]
23. Chen, S.K.; Ghoneim, Y.; Moshchuk, N.; Litkouhi, B.; Pylypchuk, V. Tire-Force Based Holistic Corner Control. In Proceedings of the ASME 2012 International Mechanical Engineering Congress and Exposition, Volume 11: Transportation Systems, Houston, TX, USA, 9–15 November 2012; pp. 133–140. [CrossRef]
24. Wang, J.; Longoria, R.G. Coordinated and reconfigurable vehicle dynamics control. *IEEE Trans. Control Syst. Technol.* **2009**, *17*, 723–732. [CrossRef]
25. Shim, T.; Margolis, D. Dynamic normal force control for vehicle stability enhancement. *Int. J. Veh. Auton. Syst.* **2005**, *3*, 1–14. [CrossRef]
26. Gáspár, P.; Szabó, Z.; Bokor, J. The design of an integrated control system in heavy vehicles based on an LPV method. In Proceedings of the 44th IEEE Conference on Decision and Control, Seville, Spain, 12–15 December 2005; pp. 6722–6727. [CrossRef]
27. Poussot-Vassal, C.; Sename, O.; Dugard, L.; Gaspar, P.; Szabo, Z.; Bokor, J. Attitude and handling improvements through gain-scheduled suspensions and brakes control. *Control Eng. Pract.* **2011**, *19*, 252–263. [CrossRef]
28. Hrovat, D. Influence of unsprung weight on vehicle ride quality. *J. Sound Vib.* **1988**, *124*, 497–516. [CrossRef]
29. Besselink, I.J.M.; Vos, R.; Nijmeijer, H. Influence of in-wheel motors on the ride comfort of electric vehicles. In Proceedings of the 10th International Symposium on Advanced Vehicle Control (AVEC10), Loughborough, UK, 22–26 August 2010; pp. 835–840.

30. Zhang, L.; Li, L.; Qi, B. Rollover prevention control for a four in-wheel motors drive electric vehicle on an uneven road. *Sci. China Technol. Sci.* **2018**, *61*, 934–948. [CrossRef]
31. Tan, D.; Wang, H.; Wang, Q. Study on the rollover characteristic of in-wheel-motor-driven electric vehicles considering road and electromagnetic excitation. *Shock Vib.* **2016**, *2016*, 1–13. [CrossRef]
32. Pacejka, H.B.; Bakker, E. The magic formula tyre model. *Veh. Syst. Dyn.* **1992**, *21*, 1–18. [CrossRef]
33. Hurel, J.; Mandow, A.; García-Cerezo, A. Kinematic and dynamic analysis of the McPherson suspension with a planar quarter-car model. *Veh. Syst. Dyn.* **2013**, *51*, 1422–1437. [CrossRef]
34. Wang, W.; Chen, X.; Wang, J. Unsprung Mass Effects on Electric Vehicle Dynamics based on Coordinated Control Scheme. In Proceedings of the 2019 American Control Conference (ACC), Philadelphia, PA, USA, 10–12 July 2019; pp. 971–976. [CrossRef]
35. Doyle, J.; Glover, K.; Khargonekar, P.; Francis, B. State-space solutions to standard H2 and H∞ control problems. In Proceedings of the 1988 American Control Conference, Atlanta, GA, USA, 15–17 June 1988; pp. 1691–1696. [CrossRef]

Article

Estimation of Vehicle Dynamic Parameters Based on the Two-Stage Estimation Method

Wenfei Li [1,2,3,*], **Huiyun Li** [1,2,3], **Kun Xu** [1,2,3], **Zhejun Huang** [1,2,3], **Ke Li** [1,2,3] and **Haiping Du** [4]

1. Shenzhen Institutes of Advanced Technology, Chinese Academy of Sciences, Shenzhen 518055, China; hy.li@siat.ac.cn (H.L.); kun.xu@siat.ac.cn (K.X.); zj.huang@siat.ac.cn (Z.H.); ke.li1@siat.ac.cn (K.L.)
2. CAS Key Laboratory of Human-Machine Intelligence-Synergy Systems, Shenzhen Institutes of Advanced Technology, Shenzhen 518055, China
3. Guangdong-Hong Kong-Macao Joint Laboratory of Human-Machine Intelligence-Synergy Systems, Shenzhen 518055, China
4. School of Electrical, Computer and Telecommunications Engineering, University of Wollongong, Wollongong 2522, Australia; hdu@uow.edu.au
*   Correspondence: wf.li1@siat.ac.cn; Tel.: +86-13811581559

Abstract: Vehicle dynamic parameters are of vital importance to establish feasible vehicle models which are used to provide active controls and automated driving control. However, most vehicle dynamics parameters are difficult to obtain directly. In this paper, a new method, which requires only conventional sensors, is proposed to estimate vehicle dynamic parameters. The influence of vehicle dynamic parameters on vehicle dynamics often involves coupling. To solve the problem of coupling, a two-stage estimation method, consisting of multiple-models and the Unscented Kalman Filter, is proposed in this paper. During the first stage, the longitudinal vehicle dynamics model is used. Through vehicle acceleration/deceleration, this model can be used to estimate the distance between the vehicle centroid and vehicle front, the height of vehicle centroid and tire longitudinal stiffness. The estimated parameter can be used in the second stage. During the second stage, a single-track with roll dynamics vehicle model is adopted. By making vehicle continuous steering, this vehicle model can be used to estimate tire cornering stiffness, the vehicle moment of inertia around the yaw axis and the moment of inertia around the longitudinal axis. The simulation results show that the proposed method is effective and vehicle dynamic parameters can be well estimated.

Keywords: vehicle dynamic parameters; Unscented Kalman Filter; multiple-model

Citation: Li, W.; Li, H.; Xu, K.; Huang, Z.; Li, K.; Du, H. Estimation of Vehicle Dynamic Parameters Based on the Two-Stage Estimation Method. *Sensors* **2021**, *21*, 3711. https://doi.org/10.3390/s21113711

Academic Editor: Felipe Jiménez

Received: 20 April 2021
Accepted: 22 May 2021
Published: 26 May 2021

Publisher's Note: MDPI stays neutral with regard to jurisdictional claims in published maps and institutional affiliations.

Copyright: © 2021 by the authors. Licensee MDPI, Basel, Switzerland. This article is an open access article distributed under the terms and conditions of the Creative Commons Attribution (CC BY) license (https://creativecommons.org/licenses/by/4.0/).

1. Introduction

Nowadays, modern road vehicles are using an increasing number of active systems to improve vehicle safety, passenger comfort, vehicle performance and energy efficiency. Advanced Driver Assistance Systems (ADAS), as well as Automated Driving (AD) technologies, are being increasingly implemented in vehicles, aiming for improved driving safety and passenger comfort [1,2]. In addition, the autonomous driving test rig is also an important method to test autonomous driving control algorithms (as shown in Figure 1, it is an autonomous driving test rig proposed by our research group) [3–5]. The implementation of these fields greatly depends on accurate vehicle dynamic parameters. Vehicle dynamic parameters are also important for vehicle modeling. Thus, vehicle dynamic parameters are important for vehicle design and testing. The vehicle dynamic parameters (VDPs), such as the vehicle mass, moment of inertia and position of the vehicle centroid, affect the closed-loop behavior of active safety systems and play an important role [6]. It is necessary to determine the VDPs to obtain real vehicle responses. Some of the VDPs can be easily measured such as the mass, the track width or the wheelbase. However, other parameters are unknown and difficult to be measured directly, such as the distance from vehicle centroid to the front axis. The moment of inertia around each axis can be measured by special equipment which is extremely costly. In contrast, the estimation method is a less intrusive and expensive way to obtain VDPs. The VDPs can be estimated by combining the estimation algorithm with some cheap sensors such as Inertial Measurement Unit (IMU), Global Positioning System (GPS), wheel speed sensors and steering angle sensor [7].

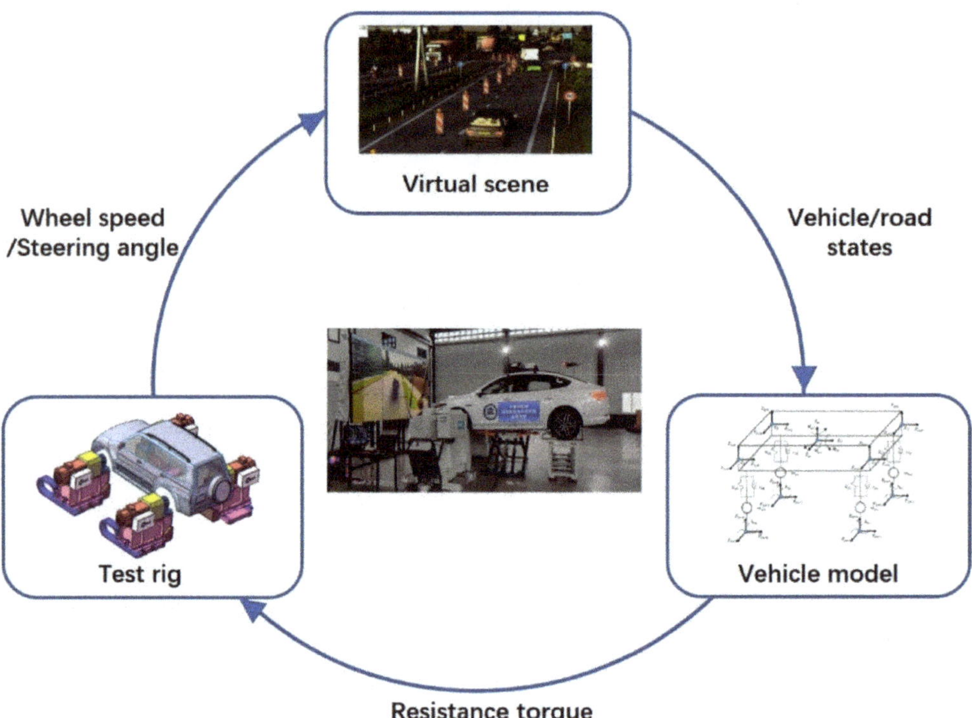

Figure 1. Autonomous driving test rig.

To obtain VDPs, many different methods have been proposed. In [8], a novel model-based parameter identification approach using optimized excitation trajectory is proposed to identify the VDPs. However, this method needs test rigs, which is a huge cost. In addition, a variety of algorithms for VDPs estimation have been presented in works of literature [9–39]. The influence of VDPs on vehicle dynamics often involves coupling. Most of the papers only study the estimation part of some parameters and the other parameters are treated as being easily measured or obtained. In actual applications, this strategy is not feasible. In real applications, all VDPs need to be obtained through simple sensors and estimation strategies. Since the VDPs are always coupled with the vehicle states, the state-parameter joint and dual estimation methods [9,10] have become increasingly prevalent and have been studied by many researchers. Some researchers use the Dual Kalman Filter (DKF) to identify the VDPs and the vehicle states simultaneously. Besides, VDPs estimation is usually classified based on the parameters of interest and the vehicle dynamics model used. In [11], common onboard sensors which are able to measure the lateral acceleration and yaw rate and a non-linear vehicle model are used. Augmented Extended Kalman Filtering is used to estimate motion states and tire cornering stiffness based on a non-linear vehicle model and sensor. Sideslip and roll angles of electric are estimated using lateral tire force sensors through RLS and the Kalman Filter based on the Single-track model in [12]. Sprung mass, yaw moment of inertia and longitudinal position of the center of gravity are identified through a dual unscented Kalman Filter in [13]. In [14], a four-wheel nonlinear vehicle model with roll dynamics and a correlation between the inertial parameters is used for a dual Unscented Kalman Filter to simultaneously identify the inertial parameters and the vehicle state. A local observability analysis on the nonlinear vehicle model is used to activate and deactivate different modes of the proposed algorithm. A Dual Extended Kalman Filter (DEKF) is used to estimate both vehicle states and vehicle parameters such as the vehicle mass, moment of inertia about the vertical axis and distance between the center of gravity and the front axle [15]. An extended Kalman Filter-based estimator adopting a dynamic vehicle model for determining the vehicle's longitudinal and lateral velocity as well as the yaw rate is proposed in [16]. In [17], a novel approach based on combined H_∞ and extended Kalman Filter (H_∞-EKF) is used to estimate the center of gravity position of electric vehicles. To implement this estimation algorithm, a simplified vehicle dynamics model is applied to the filter formulation. The H_∞ estimator is employed to filter states by means of minimizing the influence of unexpected noise, whose statistics are unknown. Simultaneously, the other EKF estimator uses the states derived by the former filter to identify the position of the vehicle centroid. A methodology based on multiple-models and a switching method for real-time estimation of the position of vehicle centroid is proposed in [13]. The method uses the well-known simple linear vehicle models for lateral and roll dynamics and assumes the availability of lateral acceleration, the yaw rate, velocity, and steering angle measurements. As mentioned in previous research, the existing estimation methods are either expensive or only portions of the VDPs can be estimated. However, vehicle dynamics modeling needs to completely determine the completed VDPs, while the cost of VDPs acquisition should be as small as possible. Thus, a method that can obtain completed VDPs at low cost urgently needs to be proposed.

In order to obtain completed VDPs at a low cost, we propose a two-stage estimation method consisting of multiple-models and Unscented Kalman Filter to estimate VDPs. In the first stage, the vehicle is set to accelerate/decelerate and the longitudinal vehicle model is used. During this stage, the height of the vehicle centroid, tire longitudinal stiffness and the longitudinal position of the vehicle centroid are estimated by the Unscented Kalman Filter. After these parameters are estimated, these estimated parameters can be used in the second stage. In the second stage, a Single-track with roll dynamics vehicle model is adopted and the vehicle is set to continuous steering. Through vehicle steering, this model can be used to estimate tire cornering stiffness, the vehicle moment of inertia around the yaw axis and the moment of inertia around the longitudinal axis. After the two-stage estimation, all VDPs are estimated. The rest of the paper is organized as follows: vehicle dynamics model are shown in Section 2. The method used to estimate VDPs is provided in Section 3. Section 4 shows and discusses the simulation results. Finally, Section 5 delivers the conclusions and points towards future work.

2. Vehicle Model

The vehicle model used in this paper is a multiple-model approach which is based on a longitudinal vehicle dynamics model (as shown in Figure 2a) and a single-track with roll dynamics vehicle model (as shown in Figure 2b,c), which comprises: the motion in the longitudinal direction x, the longitudinal velocity; the motion in the lateral direction y or lateral velocity; the yaw around the vertical axis z, described by the yaw rate and roll with regard to the longitudinal axis x; and the roll rate [13]. Figure 2 illustrates the vehicle model adopted in this paper. The whole motion of the vehicle is a direct result of the forces (the aerodynamic forces and rolling resistance are neglected in this paper) that are generated between the road and tires. As shown in Figure 2b, the four-wheel vehicle dynamics model can be simplified as a single-track model. Other states that depend directly on these states can be derived, such as longitudinal and lateral accelerations. The tire states, such as the wheel slip angle, slip ratio and rotational velocities are also important. Tire-road friction force can be obtained based on tire states and tire stiffness. The vehicle states are also largely dependent on VDPs. VDPs include vehicle mass, moments of inertia around each axis and the position of the vehicle centroid.

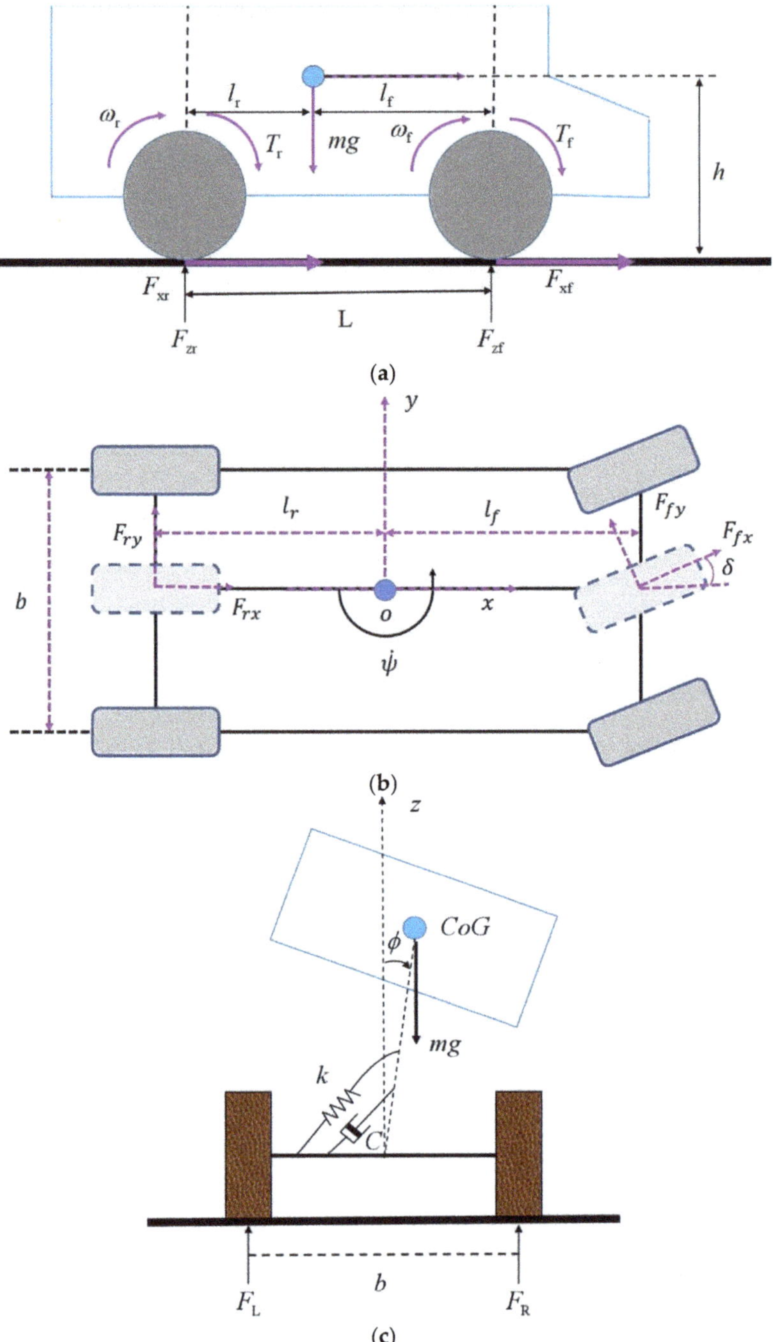

Figure 2. Vehicle model: (**a**) Longitudinal vehicle dynamics model; (**b**) Single-track vehicle model; (**c**) Vehicle roll dynamics.

The vehicle dynamic model can be described by differential equations. The vehicle model implemented here can be obtained from [17,18]. When the vehicle was accelerating or decelerating along the longitudinal direction, the longitudinal vehicle dynamics model was adopted. As shown in Figure 2a, the longitudinal vehicle dynamics model was built with the longitudinal motion, as well as the front and rear wheel rotations

$$m\dot{v}_x = F_{xf} + F_{xr} \tag{1}$$

$$J\dot{\omega}_f = T_f - rF_{xf} \tag{2a}$$

$$J\dot{\omega}_r = T_r - rF_{xr} \tag{2b}$$

where m is vehicle total mass, v_x represents the longitudinal vehicle velocity F_{xf} and F_{xr} represent the longitudinal forces of the front and rear tires. J is the wheel's moment of inertia. r is the equivalent radius of the front and rear tires. T_i, ω_i ($i = f, r$) represent the wheel torque and angular speed. The load distribution can be expressed by the vertical forces that act on each of the four wheels. These can be calculated as follows:

$$F_{zf} = mg\frac{l_r}{L} - ma_x\frac{h}{L} \tag{3a}$$

$$F_{zr} = mg\frac{l_f}{L} + ma_x\frac{h}{L} \tag{3b}$$

where F_{zf} and F_{zr} are vertical force of the front and rear wheels. a_x is the longitudinal accelerations, g is the gravitational constant, l_f is the distance between the vehicle centroid and vehicle front axis, l_r is the distance between the vehicle centroid and vehicle rear axis and h denotes the height of the vehicle centroid. L is the distance between the front axis and rear axis.

When the vehicle was being steered, the single-track with roll dynamics vehicle model was adopted. As shown in Figures 1c and 2b, the differential equations for the calculation of longitudinal and lateral acceleration are as follows:

$$\dot{v}_x = a_x + v_y\dot{\psi} \tag{4}$$

$$\dot{v}_y = a_y + v_x\dot{\psi} \tag{5}$$

$$a_x = \frac{1}{m}\left(F_{xf}cos\delta + F_{yf}sin\delta + F_{xr}\right) \tag{6}$$

$$a_y = \frac{1}{m}\left(F_{xf}sin\delta + F_{yf}cos\delta + F_{yr}\right) \tag{7}$$

Yaw and roll motion can be obtained from:

$$\ddot{\psi} = \frac{\Gamma}{I_z} \tag{8}$$

$$I_x\ddot{\phi} = mh\left(a_y + g\phi\right) - k_\phi\phi - c_\phi\dot{\phi} \tag{9}$$

where $\dot{\psi}$ is the yaw rate, $\dot{\phi}$ is the roll rate, I_z is the moment of inertia around the yaw axis, I_x is the moment of inertia around the longitudinal axis, k_ϕ is the roll stiffness, c_ϕ is the roll damping and a_y is lateral acceleration. Γ can be calculated as follows:

$$\Gamma = l_f(F_{xf}cos\delta + F_{yf}sin\delta) - l_rF_{yr} \tag{10}$$

where δ is the wheel steer angle while F_{yr} represents the lateral forces of the rear tires. There are many different approaches for achieving tire force, such as the so-called 'Magic Formula' by Pacejka [19], the tire model by Fiala [20] or the 'TMeasy' tyre model [21].

When the acceleration/deceleration strength of the vehicle is small and the steering angle is small, the tire force can be calculated as follows:

$$F_{xi} = C_\sigma s_i \tag{11a}$$

$$F_{yi} = C_\alpha \alpha_i \tag{11b}$$

where F_{xi}, F_{yi} $(i = f, r)$ represent the longitudinal and lateral tire forces, C_α denotes the tire cornering stiffness, C_σ denotes the tire longitudinal stiffness, s_i is the slip ratio and α_i is the slip angle. α_i can be presented as follows:

$$\alpha_f = \frac{v_y - l_f \dot{\psi}}{v_x} - \delta \tag{12a}$$

$$\alpha_r = \frac{v_y - l_r \dot{\psi}}{v_x} \tag{12b}$$

the slip ratio s_i $(i = f, r)$ can be presented as follows:

$$s_i = \frac{\omega_i r}{v_x} - 1 \tag{13}$$

When the acceleration/deceleration strength of the vehicle was small, the tire-road friction coefficient was proportional to the slip ratio rate [21]. Then the longitudinal tire force can also be presented as follows:

$$F_{xi} = F_{zi} C_K s_i \tag{14}$$

where C_K is the slip ratio rate. It is a constant value related to the road surface. When the road surface was different, C_K changed as well. From Equations (11a) and (14), it can be seen that tire longitudinal stiffness can be calculated based on the slip ratio rate and vertical force of the wheel. This means that the tire longitudinal stiffness can be obtained when the slip ratio rate is estimated.

3. Estimation Method

To adapt to non-linear problems in vehicle dynamics estimation, EKF is widely used for estimating different vehicle states. However, the accuracy of EKF-based estimation cannot be guaranteed due to linearization errors with Jacobian matrices when approximating non-linear systems [20–26]. More recently, additional attention has been paid to UKF estimation, which uses a set of sigma points to conduct non-linear transformation so that it can deal with strong non-linear estimation problems for vehicle dynamics systems [31]. The UKF, developed by Julier et al. [32] and refined by Wan and van der Merwe et al. [33] provides a new estimation approach. Unlike the EKF, the UKF approximates the probability density function of system states by implementing the Unscented Transformation (UT) instead of the system dynamics model. The UT captures the mean and covariance of the Gaussian random vector (GRV) to at least second-order accuracy through the use of a set of sample points. UKF is an effective method to estimate the states or the parameters of a discrete dynamic system. In this paper, we use UKF to estimate VDPs through a two-stage method. The frame diagram of the two-stage estimation method is shown in Figure 3. In this paper, we assume that the velocity of the vehicle can be measured by GPS, and the vehicle mass is known. The driving or braking torques of vehicle (T_i) can be obtained. Longitudinal acceleration a_x, lateral acceleration a_y and the yaw rate $\dot{\psi}$ can be measured by IMU. Rolling stiffness k_ϕ and roll damping c_ϕ are given by the manufacturer. The relevant parameters are listed in Table 1.

Table 1. Nomenclature.

Parameter	Description
m	Vehicle mass
g	Gravitational constant
I_x	The moment of inertia around the longitudinal axis
I_z	The moment of inertia around the yaw axis
b	Vehicle width
l_f	Distance between the vehicle centroid and vehicle front axis
l_r	Distance between the vehicle centroid and vehicle rear axis
r	Effective tire radius
h	Height of vehicle centroid
c_ϕ	Roll damping coefficient
k_ϕ	Roll stiffness
C_α	Tire cornering stiffness
C_K	Slip ratio rate
J	Wheel moment of inertia
v_x	Longitudinal vehicle velocity
F_{xf}	Longitudinal forces of the front tire
F_{xr}	Longitudinal forces of the rear tire
T_f	Front wheel torque
T_r	Rear wheel torque
F_{zf}	Vertical force of front wheel
F_{zr}	Vertical force of rear wheel
a_x	Longitudinal accelerations
$\dot{\psi}$	Yaw rate
$\dot{\phi}$	Roll rate
a_y	Lateral acceleration
δ	Wheel steer angle
F_{yr}	Lateral forces of the rear tires
C_σ	Tire longitudinal stiffness
S_i	Slip ratio
α_i	Slip angle

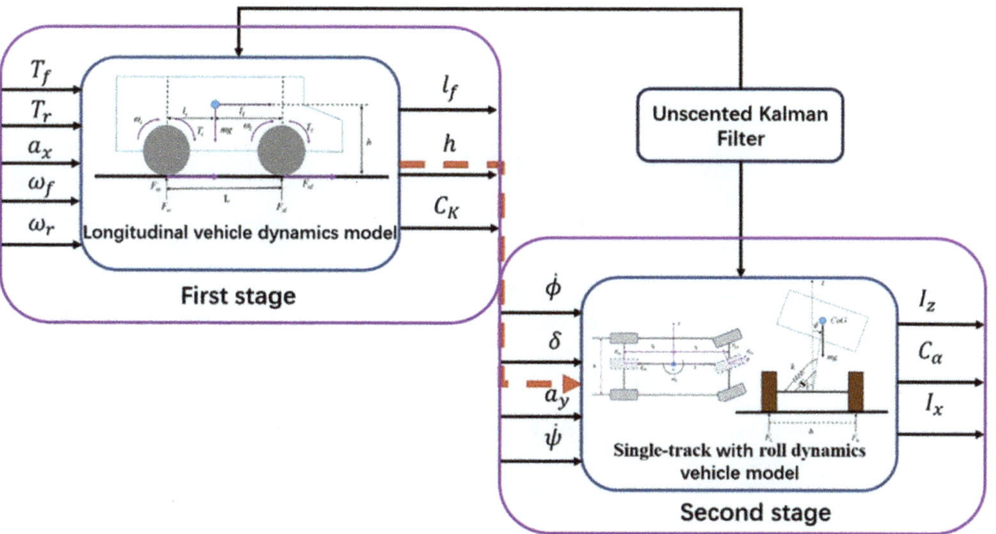

Figure 3. Two-stage estimation method.

As shown in Figure 3, the longitudinal vehicle dynamics model was adopted during the first stage. For the first stage, the longitudinal vehicle dynamics model could be described by Equations (1)–(3), (11a) and (14). To make sure the longitudinal vehicle dynamic model is able to reflect the real state of the vehicle, the absolute value of the front-wheel steering angle needed to be smaller than **0.62 deg** and the absolute value of yaw rate needed to be smaller than **1 deg/s** (as shown in Table 1). The inputs of the longitudinal vehicle dynamics model were wheel torque $T_i(i = f, r)$. The states of the longitudinal vehicle dynamics model included the angular speed $\omega_i(i = f, r)$ and longitudinal vehicle velocity v_x. The measurable outputs were the angular speed $\omega_i(i = f, r)$ and longitudinal vehicle velocity v_x. As shown in Figure 3, the distance between the vehicle centroid and vehicle front l_f, the height of vehicle centroid h and the tire longitudinal stiffness C_K are estimated parameters. To enable the VDPs to be estimated (persistent excitation requirement), specific command signals needed to be given to activate the corresponding parameters (As shown in Table 1).

When l_f, h and C_K were estimated during the first stage, these estimated VDPs could be used in the second stage (as shown in Figure 3). During the second stage, a single-track with roll dynamics vehicle model was used and is described by Equations (4)–(12). As shown in Figure 3, the inputs of the single-track with the roll dynamics vehicle model the wheel steer angle δ, lateral accelerations a_y, yaw rate $\dot{\psi}$ and the roll rate $\dot{\phi}$. The states of this model include longitudinal vehicle velocity v_x, lateral velocity v_y, longitudinal accelerations a_x, lateral acceleration a_y, yaw rate $\dot{\psi}$ and the roll rate $\dot{\phi}$. The estimated parameters are the moment of inertia around the yaw axis I_z, the moment of inertia around the longitudinal axis I_x and the tire cornering stiffness C_α. When the conditions of the second stage in Table 2 are met, VDPs (I_z, I_x, C_α) can be estimated.

Table 2. Two-stage estimation condition requirements.

First Stage: Linear Acceleration/Deceleration
• Longitudinal vehicle dynamic model
• | Front Wheel steering angel | <0.62 **deg**
• | Yaw Rate | <1deg/s
• Longitudinal acceleration/deceleration
Second stage: Continuous turn
• Single-track with roll dynamics vehicle model
• First stage estimation finished
• Longitudinal speed remains constant
• Continuous turn

As shown in Figure 3, UKF was used in both the first stage and second stage. UT is one of the most important parts of UKF. First, we introduce UT here. UT is shown in Table 2 [29].

When a system function is given as $y = f(x)$, x is the state and the dimension of x is L (as shown in Table 2). Given an L-dimensional GRV x with mean $\hat{x}$ and covariance P_x, the statistics of $y = f(x)$ were approximated by the selection of $2L + 1$ discrete sample points $\{\chi_i\}_{i=0}^{2L} = \{\hat{x} \ and \ \hat{x} \pm \sigma_j, j = 1, \ldots, L\}$ where σ_j is the ith column of the matrix $\sqrt{(L + \lambda)P_x}$. λ is a scaling parameter and depends on α, κ and L. The constant α determines the spread of sigma points about the mean $\hat{x}$. The constant κ is generally set to $3 - L$. The constant β was used to incorporate prior knowledge of the distribution. In this paper, $\alpha = 0.01$, $\beta = 2$.

As shown in Table 3, ω represents VDPs; x represents the states of dynamics; d represents the measured vector; u represents the input vector of the dynamic system. R_k^e is the measurement noise covariance. R_k^r is the processing noise covariance. The corresponding parameters are shown in Table 4 and Section 4. At different stages, these variables represented different parameters. The local observability was demonstrated by investigating the rank of the observability matrix [20]. If the observability matrix had the full column rank, it was said to be locally observable. Using the continuous state-space representation, the discretized state-space representation was written by the Euler's forward discretization in (15).

Table 3. UT.

UT Setup
$\lambda = \alpha^2(L+\kappa) - L$
$W_0^{(m)} = \frac{\lambda}{L+\lambda}$
$W_0^{(c)} = \frac{\lambda}{L+\lambda} + 1 - \alpha^2 + \beta$
$W_i^{(m)} = W_i^{(c)} = \frac{1}{2(L+\lambda)}, \quad i = 1,\ldots,2L$
$\gamma = \sqrt{L+\lambda}$

Table 4. UKF for VDPs estimation.

1: Initialize $\hat{\omega}_0^+$, $P_{\omega 0}^+$
$\hat{\omega}_0^+ = E[\omega(0)]$
$P_{\omega 0}^+ = E[(\omega(0) - \hat{\omega}_0^+)(\omega(0) - \hat{\omega}_0^+)]$
2: Prediction and sigma-point calculation:
$\hat{\omega}_k^- = \hat{\omega}_{k-1}^+$
$P_{\omega_k}^- = P_{\omega_{k-1}}^+ + R_{\omega_{k-1}}^r$
$W_{k \mid k-1} = [\, \hat{\omega}_k^- \quad \hat{\omega}_k^- + \gamma\sqrt{P_{\omega_k}^-} \quad \hat{\omega}_k^- - \gamma\sqrt{P_{\omega_k}^-} \,]$
$D_{k \mid k-1} = G(x_k, W_{k \mid k-1}, u_k)$
$\hat{d}_k^- = \sum_{i=0}^{2L} W_i^{(m)} D_{i,k \mid k-1}$
3: Update after the measurement of $d(k)$
$P_{d_k}^- = \sum_{i=0}^{2L} W_i^{(c)} (D_{i,k \mid k-1} - \hat{d}_k^-)(D_{i,k \mid k-1} - \hat{d}_k^-)^T + R_k^e$
$P_{\omega_k d_k}^- = \sum_{i=0}^{2L} W_i^{(c)} (W_{i,k \mid k-1} - \hat{\omega}_k^-)(D_{i,k \mid k-1} - \hat{d}_k^-)^T$
$K_k = P_{\omega_k d_k}^- (P_{d_k}^-)^{-1}$
$\hat{\omega}_k^+ = \hat{\omega}_k^- + K_k[d(k) - \hat{d}_k^-]$
$P_{\omega_k}^+ = P_{\omega_k}^- - K_k P_{d_k}^- K_k^T$

$$d(x_k) = x_{k-1} + T_s G(x_{k-1}, \omega_{k-1}, u_{k-1}) \tag{15}$$

where T_s is the sampling time. The observability matrix is the Jacobian of measurement vector d, with respect to the parameter vector ω. During the first stage, the Longitudinal vehicle dynamics model was used. According to Equations (1)–(3), (11a) and (14), the dynamic functions could be rewritten as:

$$m\dot{v}_x = \left(mg\frac{l_r}{L} - ma_x\frac{h}{L}\right)C_K s_f + \left(mg\frac{l_f}{L} + ma_x\frac{h}{L}\right)C_K s_r \tag{16}$$

$$J\dot{\omega}_f = T_f - r\left(mg\frac{l_r}{L} - ma_x\frac{h}{L}\right)C_K s_f \tag{17}$$

$$J\dot{\omega}_r = T_r - r\left(mg\frac{l_f}{L} + ma_x\frac{h}{L}\right)C_K s_r \tag{18}$$

and the measurement vector $\mathbf{d}$ was written as:

$$\mathbf{d} = \mathbf{G_1}(\mathbf{x}, \boldsymbol{\omega}_1, \mathbf{u}) = \begin{bmatrix} g_{11} \\ g_{12} \\ g_{13} \end{bmatrix} = \begin{bmatrix} \left(g\frac{l_r}{L} - a_x\frac{h}{L}\right)C_K s_f + \left(g\frac{l_f}{L} + a_x\frac{h}{L}\right)C_K s_r \\ \dfrac{T_f - r\left(mg\frac{l_r}{L} - ma_x\frac{h}{L}\right)C_K s_f}{J} \\ \dfrac{T_r - r\left(mg\frac{l_f}{L} + ma_x\frac{h}{L}\right)C_K s_r}{J} \end{bmatrix} \tag{19}$$

where ω_1 represents a constant vector of the vehicle inertial parameters. During the first stage, $\omega_1 = \begin{bmatrix} l_f & h & C_k \end{bmatrix}^T$. The observability matrix was defined as the Jacobian of G_1 with respect to the parameter vector ω_1. The Jacobian matrix, $C_1 = \nabla_{\omega_1} G_1$, was represented as:

$$C_1 = \nabla_{\omega_1} G_1 = \begin{bmatrix} \frac{\partial g_{11}}{\partial \omega_{11}} & \frac{\partial g_{11}}{\partial \omega_{12}} & \frac{\partial g_{11}}{\partial \omega_{13}} \\ \frac{\partial g_{12}}{\partial \omega_{11}} & \frac{\partial g_{12}}{\partial \omega_{12}} & \frac{\partial g_{12}}{\partial \omega_{13}} \\ \frac{\partial g_{13}}{\partial \omega_{11}} & \frac{\partial g_{13}}{\partial \omega_{12}} & \frac{\partial g_{13}}{\partial \omega_{13}} \end{bmatrix} \tag{20}$$

where

$$\frac{\partial g_{11}}{\partial \omega_{11}} = -\frac{g}{L}C_K s_f + \frac{g}{L}C_K s_r, \quad \frac{\partial g_{11}}{\partial \omega_{12}} = -\frac{a_x}{L}C_K s_f + \frac{a_x}{L}C_K s_r,$$

$$\frac{\partial g_{11}}{\partial \omega_{13}} = g s_f - g\frac{l_f}{L}s_f - a_x\frac{h}{L}s_f + g\frac{l_f}{L}s_r + a_x\frac{h}{L}s_r$$

$$\frac{\partial g_{12}}{\partial \omega_{11}} = \frac{rmg}{JL}C_K s_f, \quad \frac{\partial g_{12}}{\partial \omega_{12}} = \frac{rma_x C_K s_f}{JL}, \quad \frac{\partial g_{12}}{\partial \omega_{13}}$$

$$= -rmg\frac{1}{J}s_f + rmg\frac{l_f}{JL}s_f + ma_x r\frac{h}{JL}s_f$$

$$\frac{\partial g_{13}}{\partial \omega_{11}} = -\frac{rmg C_K s_r}{JL}, \quad \frac{\partial g_{13}}{\partial \omega_{12}} = -\frac{rma_x C_K s_r}{JL}, \quad \frac{\partial g_{13}}{\partial \omega_{13}}$$

$$= -r(mg\frac{l_r}{JL} + ma_x\frac{h}{JL})s_r$$

Then C_1 could be written as:

$$C_1 = \nabla_{\omega_1} G_1 =$$
$$\begin{bmatrix} -\frac{g}{L}C_K s_f + \frac{g}{L}C_K s_r & -\frac{a_x}{L}C_K s_f + \frac{a_x}{L}C_K s_r & g s_f - g\frac{l_f}{L}s_f - a_x\frac{h}{L}s_f + g\frac{l_f}{L}s_r + a_x\frac{h}{L}s_r \\ \frac{rmg C_K s_f}{JL} & \frac{rma_x C_K s_f}{JL} & -rmg\frac{1}{J}s_f + rmg\frac{l_f}{JL}s_f + ma_x r\frac{h}{JL}s_f \\ -\frac{rmg C_K s_r}{JL} & -\frac{rma_x C_K s_r}{JL} & -r(mg\frac{l_r}{JL} + ma_x\frac{h}{JL})s_r \end{bmatrix} \tag{21}$$

As shown in the above equation, the observability matrix was able to meet the requirement of the full column rank as long as the acceleration a_x was properly selected.

During the second stage, the measurement vector consisted of the longitudinal vehicle velocity, yaw rate and roll rate. According to Equations (4)–(14), the dynamic functions could be rewritten as:

$$\dot{v}_x = \frac{1}{m}\left(C_\sigma s_f \cos\delta + C_\alpha \alpha_f \sin\delta + C_\sigma s_r\right) + v_y \dot{\psi} \tag{22}$$

$$\ddot{\psi} = \frac{l_f\left(C_\sigma s_f \cos\delta + C_\alpha \alpha_f \sin\delta\right) - l_r C_\alpha \alpha_r}{I_z} \tag{23}$$

$$\ddot{\phi} = \frac{h\left(C_\sigma s_f \sin\delta + C_\alpha \alpha_f \cos\delta + C_\alpha \alpha_r\right) + mgh\phi - k_\phi \phi - c_\phi \dot{\phi}}{I_x} \tag{24}$$

then the measurement vector d was written as:

$$d=G_2(x,\omega_2)=\begin{bmatrix} g_{21} \\ g_{22} \\ g_{23} \end{bmatrix}$$

$$=\begin{bmatrix} \frac{1}{m}(C_\sigma s_f\cos\delta+C_\alpha\alpha_f\sin\delta+C_\sigma s_f)+v_y\dot\psi \\ \frac{l_f(C_\sigma s_f\cos\delta+C_\alpha\alpha_f\sin\delta)-l_rC_\alpha\alpha_r}{I_z} \\ \frac{h(C_\sigma s_f\sin\delta+C_\alpha\alpha_f\cos\delta+C_\alpha\alpha_r)+mgh\phi-k_\phi\phi-c_\phi\dot\phi}{I_x} \end{bmatrix} \tag{25}$$

where ω_2 represents a constant vector of the vehicle inertial parameters. During the second stage, $\omega_2=[\begin{array}{ccc} I_z & I_x & C_\alpha \end{array}]^T$. The observability matrix was defined as the Jacobian of G_2 with respect to the parameter vector ω_2. The Jacobian matrix, $C_2=\nabla_{\omega_2}G_2$, was represented as:

$$C_2=\nabla_{\omega_2}G_2=\begin{bmatrix} \frac{\partial g_{21}}{\partial\omega_{21}} & \frac{\partial g_{21}}{\partial\omega_{22}} & \frac{\partial g_{21}}{\partial\omega_{23}} \\ \frac{\partial g_{22}}{\partial\omega_{21}} & \frac{\partial g_{22}}{\partial\omega_{22}} & \frac{\partial g_{22}}{\partial\omega_{23}} \\ \frac{\partial g_{23}}{\partial\omega_{21}} & \frac{\partial g_{23}}{\partial\omega_{22}} & \frac{\partial g_{23}}{\partial\omega_{23}} \end{bmatrix} \tag{26}$$

where

$$\frac{\partial g_{21}}{\partial I_z}=0,\frac{\partial g_{21}}{\partial I_x}=0,\frac{\partial g_{21}}{\partial C_\alpha}=\frac{\alpha_f\sin\delta}{m}$$

$$\frac{\partial g_{22}}{\partial I_z}=-\frac{l_f(C_\sigma s_f\cos\delta+C_\alpha\alpha_f\sin\delta)-l_rC_\alpha\alpha_r}{I_z^2},\frac{\partial g_{22}}{\partial I_x}=0,\frac{\partial g_{22}}{\partial C_\alpha}=\frac{l_f\alpha_f\sin\delta-l_r\alpha_r}{I_z}.$$

$$\frac{\partial g_{23}}{\partial I_z}=0,\frac{\partial g_{23}}{\partial I_x}=-\frac{h(C_\sigma s_f\sin\delta+C_\alpha\alpha_f\cos\delta+C_\alpha\alpha_r)+mgh\phi-k_\phi\phi-c_\phi\dot\phi}{I_x^2},\frac{\partial g_{23}}{\partial C_\alpha}$$

$$=\frac{h\alpha_f\cos\delta+h\alpha_r}{I_x}$$

Then C_2 could be written as:
$$C_2=\nabla_{\omega_2}G_2$$

$$=\begin{bmatrix} 0 & 0 & \frac{\alpha_f\sin\delta}{m} \\ -\frac{l_f(C_\sigma s_f\cos\delta+C_\alpha\alpha_f\sin\delta)-l_rC_\alpha\alpha_r}{I_z^2} & 0 & \frac{l_f\alpha_f\sin\delta-l_r\alpha_r}{I_z} \\ 0 & -\frac{h(C_\sigma s_f\sin\delta+C_\alpha\alpha_f\cos\delta+C_\alpha\alpha_r)+mgh\phi-k_\phi\phi-c_\phi\dot\phi}{I_x^2} & \frac{h\alpha_f\cos\delta+h\alpha_r}{I_x} \end{bmatrix} \tag{27}$$

As shown in the above equation, the observability matrix was able to meet the requirement of full column rank as long as the steering angle δ was properly selected. Based on the above analysis, the VDPs can be estimated when the acceleration a_x and steering angle δ are designed according to Table 2.

In order to compare the estimation performance of the method proposed in this paper, we used the commonly used extended Kalman algorithm for comparison. The extended Kalman algorithm is shown in Table 5.

Table 5. EKF for VDPs estimation.

EKF Algorithm
1. Initialize $\hat{\omega}_0^+$, $P_{\omega_0}^+$ $\hat{\omega}_0^+ = E[\omega(0)]$ $P_{\omega_0}^+ = E\big[\big(\omega(0) - \hat{\omega}_0^+\big)\big(\omega(0) - \hat{\omega}_0^+\big)\big]$ 2. Prediction before the measurement of $d(k)$ $\hat{\omega}_k^- = \hat{\omega}_{k-1}^+$ $P_{\omega_k}^- = P_{\omega_{k-1}}^+ + R_{k-1}^r$ $\hat{d}_k^- = G\big(\hat{\omega}_{k-1}^+, s(k-1), u(k-1)\big)$ 3. Update after the measurement of $d(k)$ $K_k = P_{\omega_k}^- C_k^{-T}\big(C_k^- P_{\omega_k}^- C_k^{-T} + R_k^e\big)^{-1}$ $\hat{\omega}_k^+ = \hat{\omega}_k^- + K_k\big[d_k - \hat{d}_k^-\big]$ $P_{\omega_k}^+ = P_{\omega_k}^- - K_k C_k^- P_{\omega_k}^-$

The meanings of relevant parameters in the Table 5 are same as the meanings of relevant parameters in Table 4.

4. Simulation Results

The parameters of the vehicle model are shown in Table 6.

Table 6. Model parameters and definitions.

Parameter	Description	Value	Unit
m	Vehicle mass	1600	kg
g	Gravitational constant	9.8	m/s^2
I_x	The moment of inertia around the longitudinal axis	4175	$kg\ m^2$
I_z	The moment of inertia around the yaw axis	2000	$kg\ m^2$
b	Vehicle width	1.53	m
l_f	Distance between the vehicle centroid and vehicle front axis	1.4	m
l_r	Distance between the vehicle centroid and vehicle rear axis	1.1	m
r	Effective tire radius	0.3	m
h	Height of vehicle centroid	0.637	m
c_ϕ	Roll damping coefficient	5737	Nms/deg
k_ϕ	Roll stiffness	36,000	Nm/deg
C_α	Tire cornering stiffness	66,900	N/rad
C_K	Slip ratio rate	10	-
J	Wheel moment of inertia	0.6	$kg\ m^2$

As shown in Figure 3, the whole parameter estimation process was divided into two parts. The second stage of the estimation could only start after the first stage of the estimation was completed. Due to space limitations, the control of the vehicle is not discussed here. The vehicle control can refer to reference [36–39]. First, we estimated I_z, I_x, C_α. As shown in Table 1, the vehicle needed to be continuously accelerated and braked for VDPs estimation. For this paper, the vehicle speed command signal was set as shown in Figure 4. It is a sinusoidal signal with a period of 12.5 s and an amplitude of 10. It includes acceleration/deceleration and can meet the requirements of the first stage (as shown in Table 1).

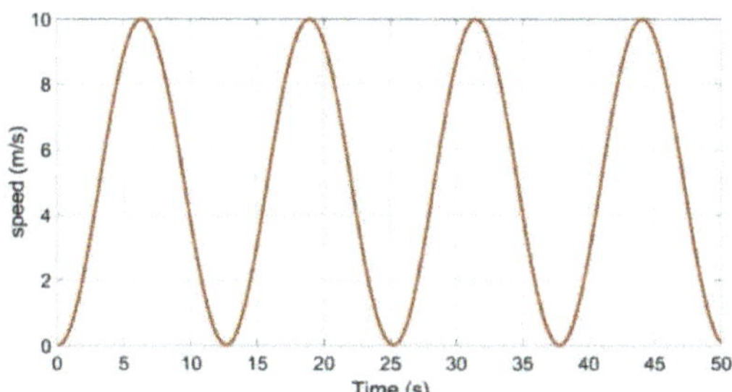

Figure 4. Vehicle longitudinal speed command.

VDPs can be estimated by making the car follow the command signal (as shown in Figure 4) to run for 4 cycles and 50 s. The simulation results are shown in Figure 5.

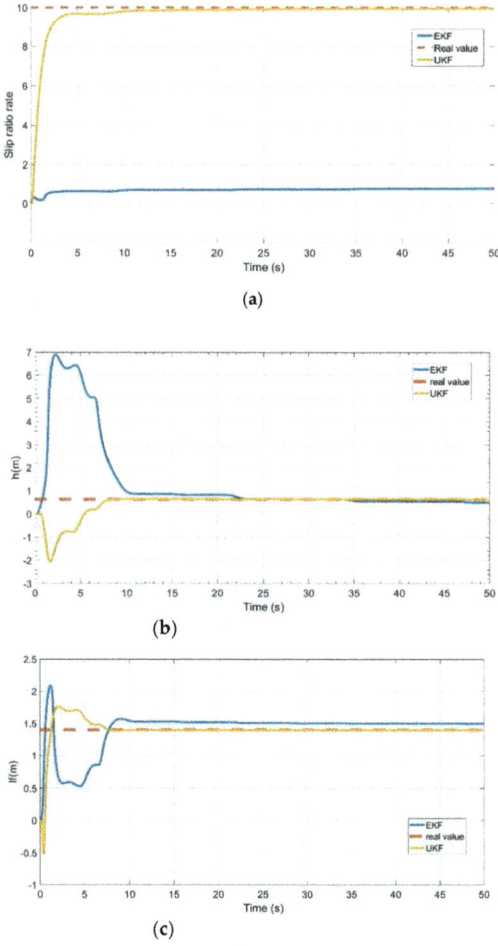

Figure 5. First stage VDPs estimation: (**a**) slip ratio rate estimation; (**b**) the height of vehicle centroid estimation; (**c**) estimation of the distance between the vehicle centroid and vehicle front axis.

As shown in Figure 5, the slip ratio rate C_σ, the height of vehicle centroid h and the distance between the vehicle centroid and vehicle front axis l_f were estimated and the estimated values approximated the real values in a short time (about 10 s) through the proposed method in this paper. Compared with the UKF used in this paper, the estimation error of EKF was larger (as shown in Figure 5). When C_σ, h and l_f were estimated, they were used in the second stage. To meet the requirement of the second stage (as shown in Table 2), the vehicle steering angle command signal was set as Figure 6.

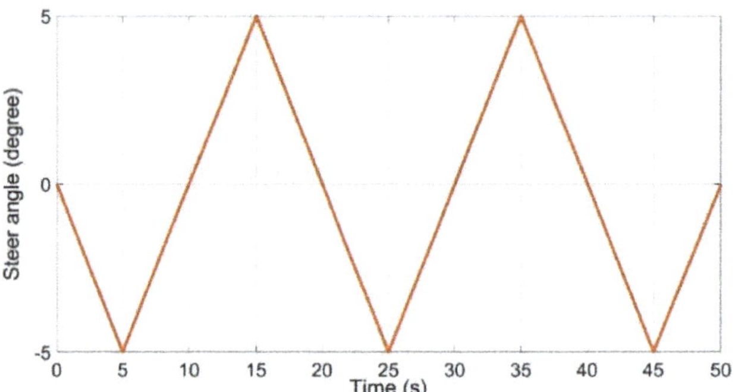

Figure 6. Vehicle steering angle command signal.

As shown in Figure 6, the vehicle steering angle command signal is a sawtooth wave with a period of 10 s and an amplitude of 5. Additionally, the vehicle operated at a speed of 10 m/s. The simulation results are shown in Figure 7.

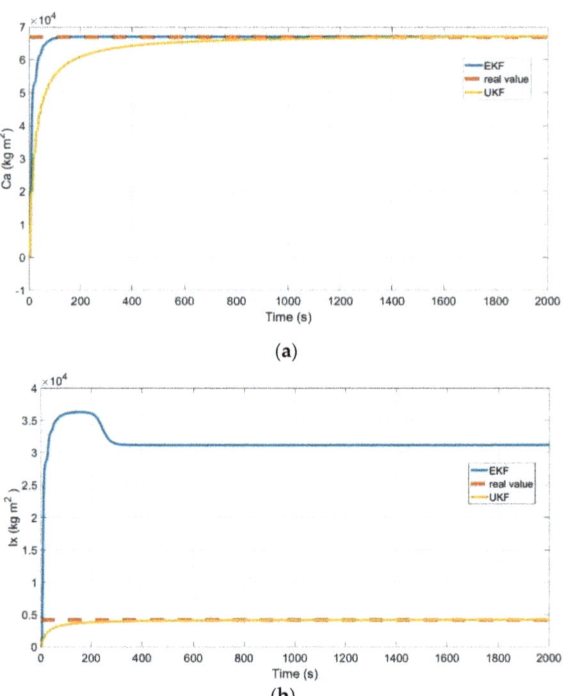

Figure 7. *Cont.*

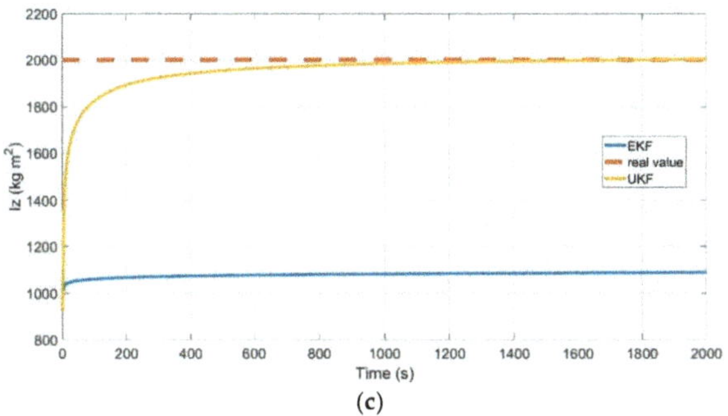

Figure 7. Second stage VDPs estimation. (**a**) the tire cornering stiffness estimation (**b**) the moment of inertia around the longitudinal axis estimation (**c**) the moment of inertia around the yaw axis estimation.

As shown in Figure 7, tire cornering stiffness C_α, the moment of inertia around the longitudinal axis I_x and the moment of inertia around the yaw axis I_y can be well estimated and the estimated error is small through the method proposed by us. However, the estimation error by EKF becomes larger compared with the first stage (as shown in Figure 7). The simulation results show that the proposed method is very capable of estimating the VDPs, and thus proves the effectiveness of the proposed method. This is mainly caused by two reasons. First, the estimated parameters with larger errors in the first stage are used in the second stage. Second, the parameter estimation in the second stage is a non-linear estimation (This can be seen in Equation (27)). The two simulation results prove that the method proposed in this paper can more accurately estimate the VDPs.

5. Discussion and Conclusions

In this paper, a new method is proposed to estimate VDPs. Different from other studies that only estimated portions of VDPs, the proposed two-stage estimation method which combines multiple-models and the Unscented Kalman Filter is able to estimate more VDPs. Because the states of a vehicle are affected by the tire stiffness, the tire stiffness is difficult to measure. The proposed estimation method is able to estimate VDPs and tire stiffness. The proposed two-stage estimation method also solves the problem that VDPs have a coupling effect on vehicle motion, which makes the VDPs difficult to estimate. For comparison, EKF is used. The simulation results prove that the proposed method not only can estimate VDPs but also that the estimation errors are small.

The proposed two-stage estimation method in this paper can obtain all the VDPs needed for vehicle dynamics modeling at one time. It is useful for vehicle modeling, control and autonomous driving control algorithm tests on a test rig. More and more artificial intelligence technologies are being widely used in autonomous driving. However, most intelligent control algorithms are trained using vehicle kinematics models. An intelligent control algorithm trained with the kinematics model cannot accurately reflect the state of a real vehicle on the road. In order to ensure the effectiveness of the intelligent control algorithm, the vehicle dynamics model needs to be used in the algorithm training process. However, VDPs provided by most vehicle and devices manufacturers are not complete. The method proposed in this paper can estimate most of the VDPs required for vehicle dynamics modeling. Then it can be used to develop intelligent control algorithms for autonomous vehicles.

Our current research work verifies the effectiveness of the method proposed in this paper from the simulation. It verifies the program in advance for the next step of real vehicle test verification. In addition, it is assumed that some vehicle states can be measured directly in this paper. However, they are difficult to obtain in real scenarios. In the future work, we will use a Dual Unscented Kalman Filter to estimate the unmeasurable states and VDPs simultaneously.

Author Contributions: Conceptualization, W.L.; methodology, W.L.; validation, Z.H. and K.L.; formal analysis, K.L.; investigation, W.L.; writing—original draft preparation, W.L.; writing—review and editing, Z.H.; supervision, H.L. and H.D.; project administration, H.L.; funding acquisition, H.L. and K.X. All authors have read and agreed to the published version of the manuscript.

Funding: This research was funded by the National Natural Science Foundation of China, Grant No. 62073311 and the Key Program of Natural Science Foundation of Shenzhen, Grant Nos. JCYJ20200109 115403807, JCYJ20200109115414354, Science and Technology Development Fun, Macao S.A.R. (FDCT), No.0015/2019/AKP.

Data Availability Statement: Not applicable.

Acknowledgments: This work was supported by grant from National Natural Science Foundation of China (Grant No. 62073311), the Key Program of Natural Science Foundation of Shenzhen (Grant Nos. JCYJ20200109115403807, JCYJ20200109115414354) and Science and Technology Development Fun, Macao S.A.R. (FDCT)(No.0015/2019/AKP). The authors would like to thank for support from National Natural Science Foundation of China, the Key Program of Natural Science Foundation of Shenzhen and Science and Technology Development Fun, Macao S.A.R. (FDCT).

Conflicts of Interest: The authors declare no conflict of interest.

References

1. Viehweger, M.; Vaseur, C.; Van Aalst, S.; Acosta, M.; Regolin, E.; Alatorre, A.; Desmet, W.; Naets, F.; Ivanov, V.; Ferrara, A.; et al. Vehicle state and tyre force estimation: Demonstrations and guidelines. *Veh. Syst. Dyn.* **2020**, *232*. [CrossRef]
2. Gong, X.; Suh, J.; Lin, C. A novel method for identifying inertial parameters of electric vehicles based on the dual H infinity filter. *Veh. Syst. Dyn.* **2019**, *58*, 28–48. [CrossRef]
3. Jian, W.; Bin, Z.J.X.; Jing, L.; Youde, L. Research on hardware-in-the-loop of automotive driving force control system based on dSPACE. *Automob. Technol.* **2009**, *9*, 14–18.
4. Heidrich, L.; Shyrokau, B.; Savitski, D.; Ivanov, V.; Augsburg, K.; Wang, D. Hardware-in-the-loop test rig for integrated vehicle control systems. *IFAC Proc. Vol.* **2013**, *46*, 683–688. [CrossRef]
5. Joshi, A. Powertrain and chassis hardware-in-the=loop (HIL) simulation of autonomous vehicle platform. *Intell. Connect. Veh. Symp.* **2017**. [CrossRef]
6. Jin, X.; Yin, G.; Chen, N. Advanced Estimation Techniques for Vehicle System Dynamic State: A Survey. *Sensors* **2019**, *19*, 4289. [CrossRef] [PubMed]
7. Guo, H.; Cao, D.; Chen, H.; Lv, C.; Wang, H.; Yang, S. Vehicle dynamic state estimation: State of the art schemes and per-spectives. *IEEE/CAA J. Autom. Sin.* **2018**, *5*, 418–431. [CrossRef]
8. Yao, D.; Ulbricht, P.; Tonutti, S.; Büttner, K.; Günther, P. A novel approach for experimental identification of vehicle dynamic parameters. *Proc. Inst. Mech. Eng. Part D J. Automob. Eng.* **2020**, *234*, 2634–2648. [CrossRef]
9. Yu, Z.; Wang, J. Simultaneous Estimation of Vehicle's Center of Gravity and Inertial Parameters Based on Acker-mann's Steering Geometry. *J. Dyn. Syst. Meas. Control* **2017**, *139*, 031006. [CrossRef]
10. Hong, S.; Erdogan, G.; Hedrick, K.; Borrelli, F. Tyre–road friction coefficient estimation based on tyre sensors and lateral tyre deflection: Modelling, simulations and experiments. *Veh. Syst. Dyn.* **2013**, *51*, 627–647. [CrossRef]
11. Hong, S.; Lee, C.; Borrelli, F.; Hedrick, J.K. A Novel Approach for Vehicle Inertial Parameter Identification Using a Dual Kalman Filter. *IEEE Trans. Intell. Transp. Syst.* **2014**, *16*, 151–161. [CrossRef]
12. Park, G.; Choi, S.B. An Integrated Observer for Real-Time Estimation of Vehicle Center of Gravity Height. *IEEE Trans. Intell. Transp. Syst.* **2020**, 1–12. [CrossRef]
13. Solmaz, S.; Akar, M.; Shorten, R.; Kalkkuhl, J. Real-time multiple-model estimation of centre of gravity position in automotive vehicles. *Veh. Syst. Dyn.* **2008**, *46*, 763–788. [CrossRef]
14. Vargas-Melendez, L.; Boada, B.L.; Boada, M.J.L.; Gauchia, A.; Diaz, V. Sensor Fusion Based on an Integrated Neural Network and Probability Density Function (PDF) Dual Kalman Filter for On-Line Estimation of Vehicle Parameters and States. *Sensors* **2017**, *17*, 987. [CrossRef] [PubMed]
15. Torres-Moreno, J.L.; Blanco-Claraco, J.L.; Giménez-Fernández, A.; Sanjurjo, E.; Naya, M.Á. Online Kinematic and Dynamic-State Estimation for Constrained Multibody Systems Based on IMUs. *Sensors* **2016**, *16*, 333. [CrossRef] [PubMed]
16. Rozyn, M.; Zhang, N. A method for estimation of vehicle inertial parameters. *Veh. Syst. Dyn.* **2010**, *48*, 547–565. [CrossRef]
17. Zhang, N.; Dong, G.M.; Du, H.P. Investigation into untripped rollover of light vehicles in the modified fishhook and the sine maneuvers. Part I: Vehicle modelling, roll and yaw instability. *Veh. Syst. Dyn.* **2008**, *46*, 271–293. [CrossRef]
18. Ning, D.; Sun, S.; Li, H.; Du, H.; Li, W. Active control of an innovative seat suspension system with acceleration measurement based friction estimation. *J. Sound Vib.* **2016**, *384*, 28–44. [CrossRef]

19. Du, H.; Zhang, N.; Ji, J.C.; Gao, W. Robust Fuzzy Control of an Active Magnetic Bearing Subject to Voltage Saturation. *IEEE Trans. Control Syst. Technol.* **2009**, *18*, 164–169. [CrossRef]
20. Li, B.; Du, H.; Li, W. A Potential Field Approach-Based Trajectory Control for Autonomous Electric Vehicles with In-Wheel Motors. *IEEE Trans. Intell. Transp. Syst.* **2017**, *18*, 2044–2055. [CrossRef]
21. Du, H.; Zhang, N. Constrained H∞ control of active suspension for a half-car model with a time delay in control. *Proc. Inst. Mech. Eng. Part D J. Automob. Eng.* **2008**, *222*, 665–684. [CrossRef]
22. Huang, X.; Wang, J. Real-Time Estimation of Center of Gravity Position for Lightweight Vehicles Using Combined AKF–EKF Method. *IEEE Trans. Veh. Technol.* **2014**, *63*, 4221–4231. [CrossRef]
23. Doumiati, M.; Victorino, A.; Lechner, D.; Baffet, G.; Charara, A. Observers for vehicle tyre/road forces estimation: Experimental validation. *Veh. Syst. Dyn.* **2010**, *48*, 1345–1378. [CrossRef]
24. Kistler. Measuring Systems and Sensors. 2019. Available online: https://www.kistler.com/en/ (accessed on 4 January 2019).
25. Wenzel, T.A.; Burnham, K.J.; Blundell, M.V.; Williams, R.A. Dual extended kalman filter for vehicle state and param-eter estimation. *Veh. Syst. Dyn.* **2006**, *44*, 153–171. [CrossRef]
26. Julier, S.J.; Uhlmann, J.K.; Durrant-Whyte, H.F. A new approach for filtering nonlinear systems. In Proceedings of the 1995 American Control Conference—ACC'95, Seattle, WA, USA, 21–23 June 1995; pp. 1628–1632.
27. Wan, E.A.; van der Merwe, R.; Nelson, A.T. Dual estimation and the unscented transformation. *Adv. Neural Inf. Process. Syst.* **2000**, *12*, 666–672.
28. Yoon, J.; Peng, H. Robust vehicle sideslip angle estimation through a disturbance rejection filter that integrates a magnetom-eter with GPS. *IEEE Trans. Intell. Transp. Syst.* **2014**, *15*, 191–204. [CrossRef]
29. Bavdekar, V.A.; Prakash, J.; Shah, S.L.; Gopaluni, R.B. Constrained dual ensemble Kalman filter for state and parameter estimation. In Proceedings of the 2013 American Control Conference, Washington, DC, USA, 17–19 June 2013; pp. 3093–3098.
30. Gadola, M.; Chindamo, D.; Romano, M.; Padula, F. Development and validation of a Kalman filter-based model for vehicle slip angle estimation. *Veh. Syst. Dyn.* **2013**, *52*, 68–84. [CrossRef]
31. Guo, H.; Chen, H.; Xu, F.; Wang, F.; Lu, G. Implementation of EKF for Vehicle Velocities Estimation on FPGA. *IEEE Trans. Ind. Electron.* **2012**, *60*, 3823–3835. [CrossRef]
32. Chen, J.; Song, J.; Li, L.; Jia, G.; Ran, X.; Yang, C. UKF-based adaptive variable structure observer for vehicle sideslip with dynamic correction. *IET Control Theory Appl.* **2016**, *10*, 1641–1652. [CrossRef]
33. Ren, H.; Chen, S.; Shim, T.; Wu, Z. Effective assessment of tyre–road friction coefficient using a hybrid estimator. *Veh. Syst. Dyn.* **2014**, *52*, 1047–1065. [CrossRef]
34. Wan, E.A.; Nelson, A.T. Dual Extended Kalman Filter Methods. In *Kalman Filtering and Neural Networks*; Haykin, S., Ed.; Wiley: New York, NY, USA, 2003; Chapter 5; pp. 123–173.
35. Wan, E.A.; Nelson, A.T. Neural dual extended Kalman filtering: Applications in speech enhancement and monaural blind signal separation. In Proceedings of the Neural Networks for Signal Processing Workshop, Amelia Island, FL, USA, 24–26 September 1997.
36. Wenzel, T.A.; Burnham, K.J.; Blundell, M.V.; Williams, R.A. Approach to vehicle state and parameter estimation using extended Kalman filtering. In Proceedings of the 7th International Symposium on Advanced Vehicle Control AVEC04, HAN University, Arnhem, NL, USA, 23–27 August 2004; pp. 725–730.
37. Wang, R.; Zhang, H.; Wang, J. Linear parameter-varying controller design for four-wheel independently actuated electric ground vehicles with active steering systems. *IEEE Trans. Control Syst. Technol.* **2013**, *22*, 1281–1296.
38. He, J.; Crolla, D.A.; Levesley, M.C.; Manning, W.J. Coordination of active steering, driveline, and braking for integrated vehicle dynamics control. *Proc. Inst. Mech. Eng. Part D J. Automob. Eng.* **2006**, *220*, 1401–1420. [CrossRef]
39. Manning, W.; Crolla, D.A. A review of yaw rate and sideslip controllers for passenger vehicles. *Trans. Inst. Meas. Control* **2007**, *29*, 117–135. [CrossRef]

sensors

MDPI

Article

A Novel V2V Cooperative Collision Warning System Using UWB/DR for Intelligent Vehicles

Mingyang Wang, Xinbo Chen, Baobao Jin, Pengyuan Lv, Wei Wang * and Yong Shen *

Institute of Intelligent Vehicles, School of Automotive Studies, Tongji University, No. 4800 Cao'an Highway, Jiading District, Shanghai 201804, China; wangmingyang@sina.cn (M.W.); austin_1@163.com (X.C.); jin_baobao2019@163.com (B.J.); lv_Pengyuan@tongji.edu.cn (P.L.)
* Correspondence: lazguronwang@gmail.com (W.W.); shenyong@tongji.edu.cn (Y.S.)

Abstract: The collision warning system (CWS) plays an essential role in vehicle active safety. However, traditional distance-measuring solutions, e.g., millimeter-wave radars, ultrasonic radars, and lidars, fail to reflect vehicles' relative attitude and motion trends. In this paper, we proposed a vehicle-to-vehicle (V2V) cooperative collision warning system (CCWS) consisting of an ultra-wideband (UWB) relative positioning/directing module and a dead reckoning (DR) module with wheel-speed sensors. Each vehicle has four UWB modules on the body corners and two wheel-speed sensors on the rear wheels in the presented configuration. An over-constrained localization method is proposed to calculate the relative position and orientation with the UWB data more accurately. Vehicle velocities and yaw rates are measured by wheel-speed sensors. An extended Kalman filter (EKF) is applied based on the relative kinematic model to combine the UWB and DR data. Finally, the time to collision (TTC) is estimated based on the predicted vehicle collision position. Furthermore, through UWB signals, vehicles can simultaneously communicate with each other and share information, e.g., velocity, yaw rate, which brings the potential for enhanced real-time performance. Simulation and experimental results show that the proposed method significantly improves the positioning, directing, and velocity estimating accuracy, and the proposed system can efficiently provide collision warning.

Keywords: collision warning system; ultra-wideband; dead reckoning; time to collision

Citation: Wang, M.; Chen, X.; Jin, B.; Lv, P.; Wang, W.; Shen, Y. A Novel V2V Cooperative Collision Warning System Using UWB/DR for Intelligent Vehicles. *Sensors* **2021**, *21*, 3485. https://doi.org/10.3390/s21103485

Academic Editor: Chao Huang

Received: 11 April 2021
Accepted: 14 May 2021
Published: 17 May 2021

Publisher's Note: MDPI stays neutral with regard to jurisdictional claims in published maps and institutional affiliations.

Copyright: © 2021 by the authors. Licensee MDPI, Basel, Switzerland. This article is an open access article distributed under the terms and conditions of the Creative Commons Attribution (CC BY) license (https://creativecommons.org/licenses/by/4.0/).

1. Introduction

The global status report on road safety 2018, launched by the WHO in December 2018, highlighted that the number of annual road traffic deaths had reached 1.35 million [1]. Two-vehicle and multi-vehicle collisions were the most severe types of accidents. Studies showed that more than 80% of road traffic accidents resulted from drivers' belated responses, and more than 65% resulted in rear-end collisions [2]. Researches indicate that more than 80% of accidents could have been averted if drivers had focused and driven correctly in three seconds before the accident [3].

In recent years, more and more researchers have focused on advanced driving assistance systems (ADAS) to raise consumers' awareness of safety devices and to reduce the risk of accidents caused by careless driving. As an essential component of the collision warning system, the forward collision warning system (FCWS), can measure the distance with the leading vehicle by itself and warn drivers when the distance between vehicles is less than the safe distance. At present, FCWS using active sensors, such as laser [4,5], radar [6], vision sensor [7–9], and infrared [10], has been widely studied. Sanberg et al. [11] presented a stereo vision-based CWS suited for real-time execution in a car. Hernandez et al. designed an object warning collision system for high-conflict vehicle-pedestrian zones using a laser [12]. Coelingh et al. [13,14] proposed a collision avoidance and automatic braking system using a car mounted with radar and camera. Srinivasa et al. [15] proposed an improved CWS combining data from a forward-looking camera and a radar. Although these sensors have high accuracy, they cannot work robustly in bad weather such as rain, snow, and fog and effectively identify dangerous vehicles in visual blind areas. Many advanced algorithms have been proposed to overcome the defects of the sensors [16,17]. However, these algorithms are always limited to particular scenarios, e.g., lane changing [18] and turning [19].

CCWS is an effective solution to this issue, which combines traditional CWS with vehicle-to-infrastructure (V2I) communication and V2V communication [20]. In CCWS, the sensor defects of a single vehicle are supplemented by acquiring information from other vehicles or infrastructures. A V2V-based system shares information among the on-board units (OBU) of vehicles. In V2I systems, accidents and hazardous events are detected by roadside units (RSU) and sent to the OBUs of vehicles [21]. Since vehicles can communicate directly through V2V without dependence on infrastructures, it is more suitable for CWS than V2I. Yang et al. [22] proposed a novel FCWS, which used license plate recognition and vehicle-to-vehicle (V2V) communication to warn the drivers of both vehicles. Xiang et al. [23] proposed an FCWS based on dedicated short-range communication (DSRC) and the global positioning system (GPS). Yang et al. [24] proposed an FCWS combining differential global positioning system (DGPS) and DSRC. Patra et al. [25] proposed a novel FCWS, in which GPS provides the relative positioning information, and vehicles communicate through a vehicular network integrated with smartphones. In general, CCWS can overcome the limitations of the in-vehicle sensor-based CWS by sharing information such as vehicle speed, location, and angle to surrounding vehicles. However, the current V2V based CWSs implement relative positioning and communication separately using different technologies, e.g., predicting collision warning based on radars but communicating through WIFI, which may affect the real-time performance.

To address this issue, the UWB-based CCWS seamlessly combines CWS and V2V without delay. UWB is a communication technology that uses nanosecond narrow pulse signal to transmit data and to measure distances, which has become an effective transmission technology in location-aware sensor networks [26]. Inherently, the UWB-based ranging technology has the advantages of high time resolution and can achieve centimeter-level ranging accuracy [27]. UWB is more adaptable to different environments than traditional sensors used in CWS [28]. There has also been some research on UWB-based CCWS. Sun et al. [29] proposed a UWB/INS (Inertial Navigation System)-based automatic guided vehicle (AGV) collision avoidance system. Liu et al. [30] designed a vehicle collision-avoidance system based on UWB wireless sensor networks. Marianna et al. used UWB to obtain distance information and calculated the collision time to provide collision warnings for workers [31]. Kianfar et al. presented a CWS for the underground mine, which predicted collisions using distances between workers and the mining vehicle measured by UWB [32]. In summary, the existing UWB-based CCWS mainly has two technical routes, which are based on absolute positioning and relative positioning, respectively. The former is hard to popularize due to the small coverage area and high cost of base stations. For the latter, most of the existing research only considers the relative distance between targets rather than the position and ignores the information such as relative velocity and orientation.

To deal with the above problems, a CCWS based on UWB and DR is proposed in this paper. In the proposed system, relative positioning and communication are implemented by UWB simultaneously, which contributes to better real-time performance. Four UWB modules are installed on each vehicle, which makes it possible to calculate not only two-dimension (2D) relative positions but also relative orientations. An over-constrained method is proposed to improve the positioning/directing accuracy. Then, the accuracy and stability of the system are further improved, and the TTC can be estimated with the integration of DR.

This paper is organized as follows: In Section 2, the three subsystems of CCWs are introduced. Section 3 carries on a simulation to evaluate the performance of the system. In Section 4, we conduct experiments and analyze the results. Finally, we summarize the conclusions in Section 5.

2. Algorithm and Modeling

The CWS consists of three parts, the UWB-based relative positioning and directing system, the DR system based on wheel-speed sensors, and the TTC estimation system. In the following sections, the UWB-based relative positioning/directing system is shortened to the UWB system. In this section, the UWB and DR subsystems are established. Then, an EKF-based fusion algorithm is proposed to integrate UWB with DR, which significantly improves the accuracy of relative position, orientation, and velocity. Finally, the TTC estimation method in several different collision scenarios is put forward.

2.1. The Relative Positioning and Directing System

According to the vehicle axis system regulated by ISO 8855: 2011 [33], as shown in Figure 1, the origin is located at the automotive rear axle center. The X-axis points to the forward of the vehicle, and the Y-axis points to the left. In this paper, all proposed systems are established based on this axis system.

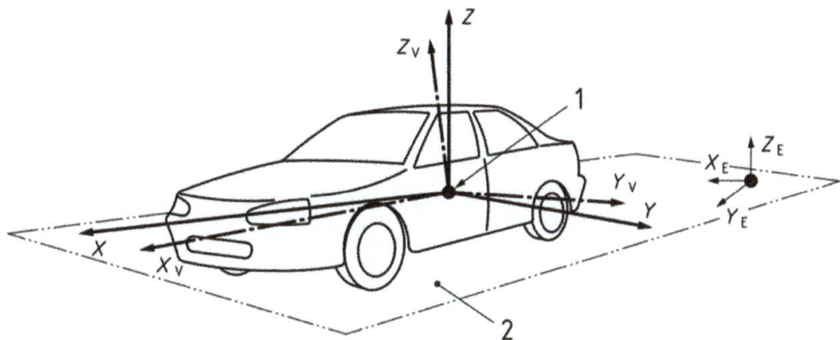

Figure 1. Vehicle axis system.

Figure 2 shows the UWB system model. XOY represents the coordinate system of vehicle 1. X′O′Y′ represents the coordinate system of vehicle 2. Points 1, 2, 3, and 4 represent the UWB modules on vehicle 1, and points M, N, P, and Q represent the UWB modules on vehicle 2. The coordinate of each UWB module in its own vehicle axis system is known when installed. As Figure 2, $X_K = [x_K, y_K]^T$ is defined as the position of module K in the axis system of vehicle 1 and $X'_K = [x'_K, y'_K]^T$ is defined as the position of module K in the axis system of vehicle 2, where $K = (1, 2, 3, 4, M, N, P, Q, C, O, O')$.

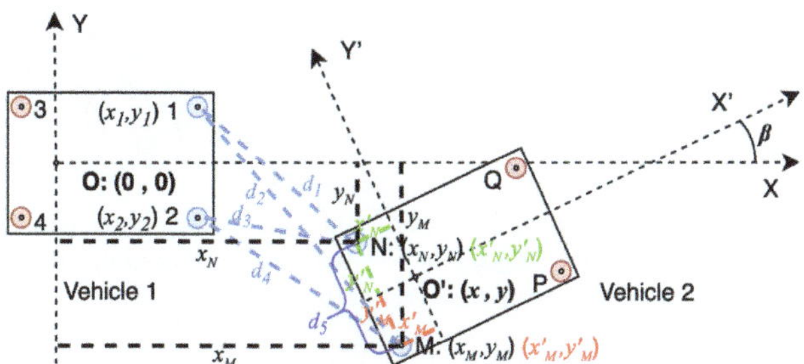

Figure 2. The UWB based relative positioning system model.

With the distances measured by UWB and the coordinates of UWB modules, the relative position and orientation $[x, y, \beta]^T$ can be calculated. $[x, y]^T$ is the position of vehicle 2 in the axis system of vehicle 1. β is the relative orientation, which means the intersection angle of the two vehicles' driving directions.

As the ranging precision of UWB is very sensitive to NLOS, not all UWB modules are necessary at the same time. Therefore, only four modules, two on each vehicle, in LOS are picked at the same time. The other modules are used to help distinguish multiple solutions. On account of the high time resolution and low multipath effect of UWB signals, it is not complex to distinguish NLOS and LOS signals.

Figure 2 shows a typical driving scenario. Vehicle 2 is changing lanes to the front of vehicle 1. Apparently, rear-end collision risk exists if vehicle 1 drives faster than vehicle 2 and does not brake. Since the CWS is especially necessary in this condition, we take it as an example to interpret our algorithm. In this case, points 1, 2, M, and N are in LOS. Define d_1, d_2, d_3, and d_4 as the real distances shown in Figure 2, and $\hat{d}_1$, $\hat{d}_2$, $\hat{d}_3$, and $\hat{d}_4$ as the corresponding measurements ranged by UWB. Other known parameters include $X_1 = [x_1, y_1]^T$, $X_2 = [x_2, y_2]^T$, $X_M' = [x_M', y_M']^T$, $X_N' = [x_N', y_N']^T$. Then, we have

$$\begin{cases} d_1 = \sqrt{(x_M - x_1)^2 + (y_M - y_1)^2} \\ d_2 = \sqrt{(x_M - x_2)^2 + (y_M - y_2)^2} \\ d_3 = \sqrt{(x_N - x_1)^2 + (y_N - y_1)^2} \\ d_4 = \sqrt{(x_N - x_1)^2 + (y_N - y_1)^2} \end{cases} . \tag{1}$$

As d_1, d_2, d_3, and d_4 are unknown, $\hat{d}_1$, $\hat{d}_2$, $\hat{d}_3$, and $\hat{d}_4$ are substituted into Equation (1) for the estimated positions of M and N, $\hat{X}_M = [\hat{x}_M, \hat{y}_M]^T$ and $\hat{X}_N = [\hat{x}_N, \hat{y}_N]^T$. Then, the estimated distance between M and N can be calculated by Equation (2).

$$\hat{d}_5 = \sqrt{(\hat{x}_M - \hat{x}_N)^2 + (\hat{y}_M - \hat{y}_N)^2} \tag{2}$$

However, when UWB modules are installed, the real distance between M and N is a determined constant, which can be calculated by Equation (3).

$$d_5 = \sqrt{(x_M' - x_N')^2 + (y_M' - y_N')^2} \tag{3}$$

When ranging error exists, $\hat{d}_5 \neq d_5$. In order to get the least square (LS) solutions that could better meet all the distances, we rewrite Equation (1) as Equation (4).

$$\begin{cases} d_1 = \sqrt{(x_M - x_1)^2 + (y_M - y_1)^2} \\ d_2 = \sqrt{(x_M - x_2)^2 + (y_M - y_2)^2} \\ d_3 = \sqrt{(x_N - x_1)^2 + (y_N - y_1)^2} \\ d_4 = \sqrt{(x_N - x_1)^2 + (y_N - y_1)^2} \\ d_5 = \sqrt{(x_M - x_N)^2 + (y_M - y_N)^2} \end{cases} \tag{4}$$

Significantly, it is an overdetermined nonlinear equation set with five equations and four unknowns. When ranging error exists, the equation set does not have exact solutions. We define function g as shown in Equation (5).

$$g(x_M, y_M, x_N, y_N) = \left(\hat{d}_1 - \sqrt{(x_M - x_1)^2 + (y_M - y_1)^2} \right)^2$$
$$+ \left(\hat{d}_2 - \sqrt{(x_M - x_2)^2 + (y_M - y_2)^2} \right)^2$$
$$+ \left(\hat{d}_3 - \sqrt{(x_N - x_1)^2 + (y_N - y_1)^2} \right)^2 \tag{5}$$
$$+ \left(\hat{d}_4 - \sqrt{(x_N - x_2)^2 + (y_N - y_2)^2} \right)^2$$
$$+ \left(\hat{d}_5 - \sqrt{(x_M - x_N)^2 + (y_M - y_N)^2} \right)^2$$

Then, the positioning algorithm is converted to an optimization problem with the optimized objective function g. According to the first-order necessary condition of optimization problems, the partial derivative of the function g should be zero, which is

$$\frac{\partial g}{\partial x_M} = \frac{\partial g}{\partial y_M} = \frac{\partial g}{\partial x_N} = \frac{\partial g}{\partial y_N} = 0. \tag{6}$$

Several sets of local optimal solutions may be derived from Equation (6). Define $[x_M^*, y_M^*, x_N^*, y_N^*]$ as the global LS solution that minimizes the objective function g. Then, we have

$$[x_M^*, y_M^*, x_N^*, y_N^*]^T = \text{argmin}[g(x_M, y_M, x_N, y_N)]. \tag{7}$$

The solutions of Equation (7) are much more accurate than those of Equation (1). It will be proved later by simulation in Section 3. When no real solutions can be solved from Equation (7), we can go back to Equation (1) for solutions instead.

In the example scenario, we can get two sets of solutions that are symmetric about the line determined by point 1 and point 2, as shown in Figure 3. Dealing with this, ranging information between other UWB modules can be drawn. For example, in Figure 3, distances $\overline{M4}$ and $\overline{Q2}$ can be used to distinguish the two sets of solutions.

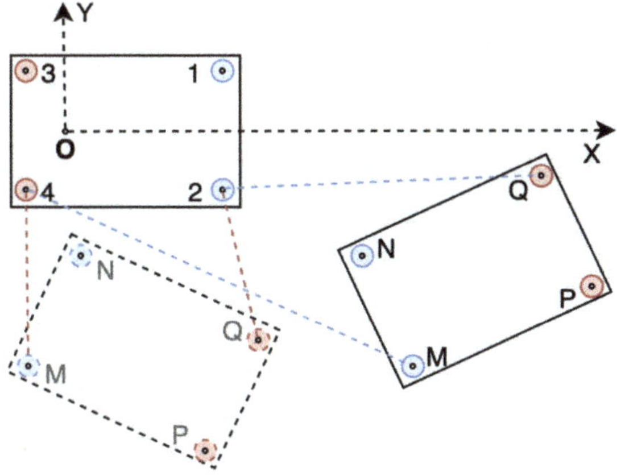

Figure 3. Two sets of solutions.

After $[x_M^*, y_M^*, x_N^*, y_N^*]$ is solved, the relative orientation β and position $[x, y]^T$ can be derived as

$$\beta = \text{atan2}(y_M - y_N, x_M - x_N) - \frac{\pi}{2}$$
$$\begin{bmatrix} x \\ y \end{bmatrix} = \begin{bmatrix} \bar{x}^* \\ \bar{y}^* \end{bmatrix} - \begin{bmatrix} \cos(\beta) & -\sin(\beta) \\ \sin(\beta) & \cos(\beta) \end{bmatrix} \begin{bmatrix} \bar{x}' \\ \bar{y}' \end{bmatrix} \tag{8}$$

where $\begin{bmatrix} \bar{x}^* \\ \bar{y}^* \end{bmatrix} = \frac{1}{2}\begin{bmatrix} x_M^* + x_N^* \\ y_M^* + y_N^* \end{bmatrix}$, $\begin{bmatrix} \bar{x}' \\ \bar{y}' \end{bmatrix} = \frac{1}{2}\begin{bmatrix} x_M' + x_N' \\ y_M' + y_N' \end{bmatrix}$, $\mathrm{atan2}(y, x) = 2\arctan\left(\frac{y}{\sqrt{x^2+y^2}+x}\right)$.

2.2. The DR System Based on Wheel Speed Sensors

The proposed system consists of four wheel-speed sensors, which are installed on the rear wheels of two vehicles. According to the Ackerman steering model shown in Figure 4, the instantaneous center of a vehicle is located on the line of the rear axle. The velocity v, yaw rate ω, and tuning radius r can be derived as shown in Equation (9).

$$\begin{cases} v = \frac{v_r + v_l}{2} \\ \omega = \frac{v_r - v_l}{L} \\ r = \frac{v}{\omega} \end{cases} \tag{9}$$

where v_r denotes the speed of the right wheel, v_l represents the speed of the left wheel, and L indicates the rear wheelbase.

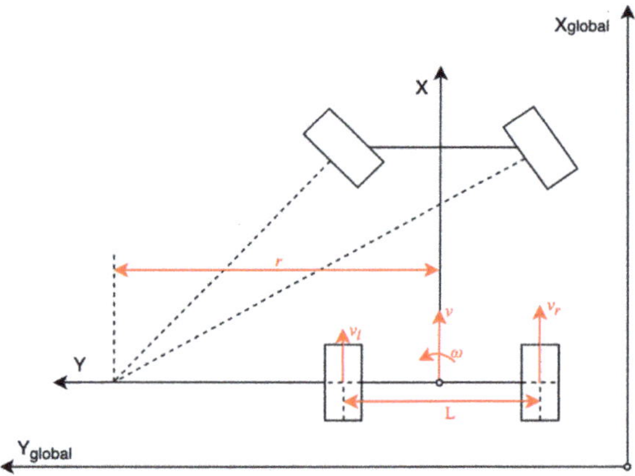

Figure 4. Ackerman steering model.

Then, the position $[x_{t+\Delta t}, y_{t+\Delta t}]^T$ and yaw angle $yaw_{t+\Delta t}$ of the vehicle in the global axis system at time $t + \Delta t$ can be reckoned by $[x_t, y_t]^T$, v_t, and yam_t at time t as shown in Equation (10).

$$\begin{bmatrix} x_{t+\Delta t} \\ y_{t+\Delta t} \\ yaw_{t+\Delta t} \end{bmatrix} = \begin{bmatrix} x_t + v_t\Delta t\cos(yaw_t) \\ y_t + v_t\Delta t\sin(yaw_t) \\ yaw_t + \omega\Delta t \end{bmatrix} \tag{10}$$

2.3. The EKF Based UWB/DR Fusion Model

We define X_k as the state vector at time k. It contains the relative position/orientation $P_k = [x_k, y_k, \beta_k]^T$, as well as yaw rates and velocities of the two vehicles $S_k = [\omega_{1_k}, \omega_{2_k}, v_{1_k}, v_{2_k}]^T$, which can be expressed as Equation (11).

$$X_k = [x_k, y_k, \beta_k, \omega_{1_k}, \omega_{2_k}, v_{1_k}, v_{2_k}]^T \tag{11}$$

We define Δt as the time period from time $k - 1$ to time k. X_k can be predicted by X_{k-1} based on the relative kinematics model shown in Figure 5. The state equation can be expressed on the basis of Equation (10) as Equation (12).

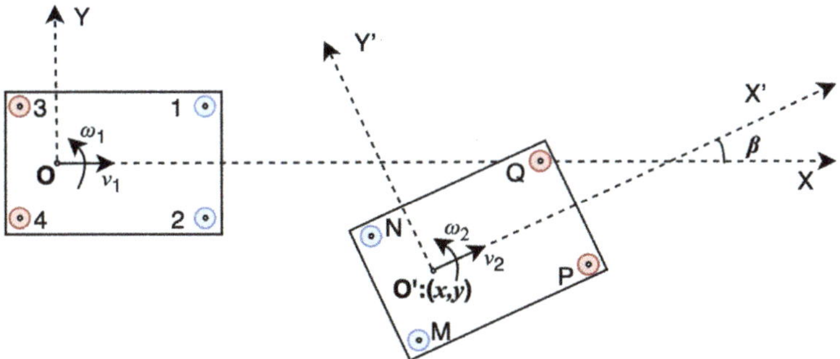

Figure 5. The relative kinematic model.

$$
X_k = \begin{bmatrix} x_k \\ y_k \\ \beta_k \\ \omega_{1_k} \\ \omega_{2_k} \\ v_{1_k} \\ v_{2_k} \end{bmatrix} = f(X_{k-1}, W) = \begin{bmatrix} C\cos(\theta) + D\sin(\theta) \\ -C\sin(\theta) + D\cos(\theta) \\ \beta_{k-1} - \theta + \omega_{2_{k-1}}\Delta t + \frac{1}{2}W_{\omega_2}\Delta t^2 \\ \omega_{1_{k-1}} + W_{\omega_1}\Delta t \\ \omega_{2_{k-1}} + W_{\omega_2}\Delta t \\ v_{1_{k-1}} + W_{v_1}\Delta t \\ v_{2_{k-1}} + W_{v_2}\Delta t \end{bmatrix}, \tag{12}
$$

where

$C = x_{k-1} - v_{1_{k-1}}t + v_{2_{k-1}}\cos(\beta_{k-1})\Delta t - W_{v_1}\omega\Delta t^2/2 + W_{v_2}\cos(\beta_{k-1})\Delta t^2/2,$

$D = y_{k-1} + v_{2_{k-1}}\sin(\beta_{k-1})\Delta t + W_{v_2}\sin(\beta_{k-1})\Delta t^2/2,$

$\theta = \omega_{1_{k-1}}\Delta t + W_{\omega_1}\Delta t^2/2.$

Then, the transition matrix of the state vector A can be derived as Equation (13).

$$
A = \frac{\partial f}{\partial X} = \begin{bmatrix} \cos(\theta) & \sin(\theta) & A_{1,3} & A_{1,4} & 0 & -\cos(\theta)\Delta t & A_{1,7} \\ -\sin(\theta) & \cos(\theta) & A_{2,3} & A_{2,4} & 0 & \sin(\theta)\Delta t & A_{2,7} \\ 0 & 0 & 1 & -\Delta t & \Delta t & 0 & 0 \\ 0 & 0 & 0 & 1 & 0 & 0 & 0 \\ 0 & 0 & 0 & 0 & 1 & 0 & 0 \\ 0 & 0 & 0 & 0 & 0 & 1 & 0 \\ 0 & 0 & 0 & 0 & 0 & 0 & 1 \end{bmatrix}, \tag{13}
$$

where

$A_{1,3} = -\cos(\theta)\sin(\beta_{k-1})v_{2_{k-1}}\Delta t + \sin(\theta)\cos(\beta_{k-1})v_{2_{k-1}}\Delta t,$

$A_{1,4} = -C\sin(\theta)\Delta t + D\cos(\theta)\Delta t,$

$A_{1,7} = \cos(\theta)\cos(\beta_{k-1})\Delta t + \sin(\theta)\sin(\beta_{k-1})\Delta t,$

$A_{2,3} = \sin(\theta)\sin(\beta_{k-1})v_{2_{k-1}}\Delta t + \cos(\theta)\cos(\beta_{k-1})v_{2_{k-1}}\Delta t,$

$A_{2,4} = -C\cos(\theta)\Delta t - D\sin(\theta)\Delta t,$

$A_{2,7} = -\sin(\theta)\cos(\beta_{k-1})\Delta t + \cos(\theta)\sin(\beta_{k-1})\Delta t.$

Similarly, the transition matrix of process noise is:

$$
G = \frac{\partial f}{\partial W} = \begin{bmatrix} G_{1,1} & 0 & G_{1,3} & G_{1,4} \\ G_{2,1} & 0 & G_{2,3} & G_{2,4} \\ -\Delta t^2/2 & \Delta t^2/2 & 0 & 0 \\ \Delta t & 0 & 0 & 0 \\ 0 & \Delta t & 0 & 0 \\ 0 & 0 & \Delta t & 0 \\ 0 & 0 & 0 & \Delta t \end{bmatrix}, \tag{14}
$$

where
$$G_{1,1} = [-C\sin(\theta) + D\cos(\theta)]\Delta t^2/2,$$
$$G_{1,3} = -\cos(\theta)\Delta t^2/2,$$
$$G_{1,4} = [\cos(\theta)\cos(\beta_{k-1}) + \sin(\theta)\sin(\beta_{k-1})]\Delta t^2/2,$$
$$G_{2,1} = [-C\cos(\theta) - D\sin(\theta)]\Delta t^2/2,$$
$$G_{2,3} = \sin(\theta)\Delta t^2/2,$$
$$G_{2,4} = [-\sin(\theta)\cos(\beta_{k-1}) + \cos(\theta)\sin(\beta_{k-1})]\,\Delta t^2/2.$$

The error covariance matrix Q of process noise consists of error covariances of speeds and yaw rates, that is:

$$Q = \mathrm{cov}(W) = \begin{bmatrix} \sigma_{\omega_1}^2 & 0 & 0 & 0 \\ 0 & \sigma_{\omega_2}^2 & 0 & 0 \\ 0 & 0 & \sigma_{v_1}^2 & 0 \\ 0 & 0 & 0 & \sigma_{v_2}^2 \end{bmatrix}. \tag{15}$$

Thus, the predicting process of the model is:

$$\hat{X}_k^- = f(\hat{X}_{k-1})$$
$$P_k^- = AP_{k-1}A + GQG. \tag{16}$$

We define Z_k as the observation vector, containing the relative position and orientation of vehicle 2 measured by the UWB system, four wheel-speeds measured by the DR system, and the observation noise V_k. Then, the observation equation can be expressed as Equation (17).

$$Z_k = \begin{bmatrix} x_{UWB,k} \\ y_{UWB,k} \\ \beta_{UWB,k} \\ v_{r1,k} \\ v_{l1,k} \\ v_{r2,k} \\ v_{l2,k} \end{bmatrix} = HX_k + V_k \tag{17}$$

Referring to Equation (9), the velocities and yaw rates of the two vehicles can be expressed by the velocities measured by wheel-speed sensors as Equation (18).

$$\begin{bmatrix} v_{1r} \\ v_{1l} \\ v_{2r} \\ v_{2l} \end{bmatrix} = \begin{bmatrix} L_1/2 & 0 & 1 & 0 \\ -L_1/2 & 0 & 1 & 0 \\ 0 & L_2/2 & 0 & 1 \\ 0 & -L_2/2 & 0 & 1 \end{bmatrix} \begin{bmatrix} \omega_1 \\ \omega_2 \\ v_1 \\ v_2 \end{bmatrix} \tag{18}$$

Then, the Jacobian matrix H is obtained as Equation (19).

$$H = \begin{bmatrix} 1 & 0 & 0 & 0 & 0 & 0 & 0 \\ 0 & 1 & 0 & 0 & 0 & 0 & 0 \\ 0 & 0 & 1 & 0 & 0 & 0 & 0 \\ 0 & 0 & 0 & L_1/2 & 0 & 1 & 0 \\ 0 & 0 & 0 & -L_1/2 & 0 & 1 & 0 \\ 0 & 0 & 0 & 0 & L_2/2 & 0 & 1 \\ 0 & 0 & 0 & 0 & -L_2/2 & 0 & 1 \end{bmatrix} \tag{19}$$

The estimating process is:

$$K_k = P_k^- H^T (HP_k^- H^T + R)^{-1}$$
$$X^- = \hat{X}_k^- + K_k(Z_k - H\hat{X}_k^-)$$
$$P_k = P_k^- - K_k HP_k^-. \tag{20}$$

In Equation (20), R represents the error covariance matrix of Z_k. It can be divided into the error covariance matrix of the UWB system R_{UWB} and the error covariance matrix of the DR system R_{DR}. That is:

$$R = \text{cov}(V_k) = \begin{bmatrix} R_{UWB} & 0 \\ 0 & R_{DR} \end{bmatrix},$$ (21)

where $R_{UWB} = \begin{bmatrix} \sigma_x^2 & 0 & 0 \\ 0 & \sigma_y^2 & 0 \\ 0 & 0 & \sigma_\beta^2 \end{bmatrix}$, $R_{DR} = \begin{bmatrix} \sigma_{v_{r1}}^2 & 0 & 0 & 0 \\ 0 & \sigma_{v_{l1}}^2 & 0 & 0 \\ 0 & 0 & \sigma_{v_{r2}}^2 & 0 \\ 0 & 0 & 0 & \sigma_{v_{l2}}^2 \end{bmatrix}.$

R_{DR} is decided by measurement errors of the wheel-speed sensors directly, whereas R_{UWB} is decided by positioning and directing errors, which is indirectly decided by the ranging error of UWB modules. Define $D = [d_1, d_2, d_3, d_4]$. On the basis of Equation (5), we can derive the relationship between the deviation D and the deviation of UWB modules' position X_M and X_N as Equation (22). d_5 is ignored because it is not a measurement but a constant, which means $dd_5 = 0$.

$$dD = \begin{bmatrix} dd_1 \\ dd_2 \\ dd_3 \\ dd_4 \end{bmatrix} = \frac{\partial D}{\partial(x_M, y_M, x_N, y_N)} \begin{bmatrix} dx_M \\ dy_M \\ dx_N \\ dy_N \end{bmatrix} = F_D \begin{bmatrix} dx_M \\ dy_M \\ dx_N \\ dy_N \end{bmatrix},$$ (22)

F_D can be derived as Equation (23).

$$F_D = \begin{bmatrix} \frac{x_M - x_1}{d_1} & \frac{y_M - y_1}{d_1} & 0 & 0 \\ \frac{x_M - x_2}{d_2} & \frac{y_M - y_2}{d_2} & 0 & 0 \\ 0 & 0 & \frac{x_N - x_1}{d_3} & \frac{y_N - y_1}{d_3} \\ 0 & 0 & \frac{x_N - x_2}{d_4} & \frac{y_N - y_2}{d_4} \end{bmatrix}.$$ (23)

where $d_1 = \hat{d}_1, d_2 = \hat{d}_2, d_3 = \hat{d}_3, d_4 = \hat{d}_4, x_M = x_M^*, y_M = y_M^*, x_N = x_N^*, y_N = y_N^*$.

From Equation (8), we can get the relationship between the deviation of the vehicle position and orientation $X_{UWB} = [x, y, \beta]$ and the deviation of the UWB modules' position X_M and X_N as Equation (24).

$$dX_{UWB} = \begin{bmatrix} x \\ y \\ \beta \end{bmatrix} = \frac{\partial X_{UWB}}{\partial(x_M, y_M, x_N, y_N)} \begin{bmatrix} dx_M \\ dy_M \\ dx_N \\ dy_N \end{bmatrix} = F_{X_{UWB}} \begin{bmatrix} dx_M \\ dy_M \\ dx_N \\ dy_N \end{bmatrix}.$$ (24)

$F_{X_{UWB}}$ can be derived as Equation (25).

$$F_{X_{UWB}} = \begin{bmatrix} F_{1,1} & F_{1,2} & F_{1,3} & F_{1,4} \\ F_{2,1} & F_{2,2} & F_{2,3} & F_{2,4} \\ -\frac{(y_M - y_N)}{d_5^2} & \frac{(x_M - x_N)}{d_5^2} & \frac{(y_M - y_N)}{d_5^2} & -\frac{(x_M - x_N)}{d_5^2} \end{bmatrix},$$ (25)

where

$F_{1,1} = 1/2 + [x_M' \cos(\beta) - y_M' \sin(\beta)](y_M - y_N)/d_5^2,$
$F_{1,2} = -[x_M' \cos(\beta) - y_M' \sin(\beta)](x_M - x_N)/d_5^2,$
$F_{1,3} = 1/2 - [x_M' \cos(\beta) - y_M' \sin(\beta)](y_M - y_N)/d_5^2,$
$F_{1,4} = [x_M' \cos(\beta) - y_M' \sin(\beta)](x_M - x_N)/d_5^2,$
$F_{2,1} = [y_M' \cos(\beta) + x_M' \sin(\beta)](y_M - y_N)/d_5^2,$
$F_{2,2} = 1/2 - [y_M' \cos(\beta) + x_M' \sin(\beta)](x_M - x_N)/d_5^2,$

$$F_{2,3} = -\left[y'_M \cos(\beta) + x'_M \sin(\beta)\right](y_M - y_N)/d_5^2,$$
$$F_{2,4} = 1/2 + \left[y'_M \cos(\beta) + x'_M \sin(\beta)\right](x_M - x_N)/d_5^2,$$
$$x'_{MN} = (x'_M + x'_N)/2, \quad y'_{MN} = (y'_M + y'_N)/2,$$
$$x_M = x^*_M, \quad y_M = y^*_M, \quad x_N = x^*_N, \quad y_N = y^*_N.$$

Then, R_{UWB} can be expressed as Equation (26).

$$R_{UWB} = F_{X_{UWB}} \left(F_D^T F_D\right)^{-1} F_D^T R_D F_D \left(F_D^T F_D\right)^{-1} F_{X_{UWB}}^T, \tag{26}$$

where $R_D = \mathrm{diag}\left(\sigma_{d_1}^2, \sigma_{d_2}^2, \sigma_{d_3}^2, \sigma_{d_4}^2, \sigma_{d_5}^2\right)$ is determined directly by UWB ranging error covariance.

2.4. The Collision Warning Model

CWS mainly works in two ways, headway measurement warning (HMW) and TTC-based warning [34]. Both of them need to measure the distance to the front vehicle but estimate the collision time with different speeds as Equation (27).

$$Headway\ Collision\ Time = \frac{Headway}{v_{Rear\,Vehicle}}$$
$$TTC = \frac{Headway}{v_{Rear\,Vehicle} - v_{Front\,Vehicle}} \tag{27}$$

The TTC-based system takes relative velocity into account, so it provides a more accurate collision warning. In this paper, the proposed system allows vehicles to share information through UWB, such as velocities. The TTC method is apparently the better choice.

Two vehicles driving on the road have the probability of collisions in various types, such as head-to-head collision, rear-end collision, and side collision. Different kinds of collisions may happen at different times. That means all cases need to be taken into account in order to obtain the exact TTC. Before establishing the collision warning model, we simplified the shape of a vehicle as a rectangle. With this assumption, all kinds of collisions can be described as point-to-edge collisions. Edge-to-edges collisions and point-to-point collisions are also covered by point-to-edge collisions, as shown in Figure 6.

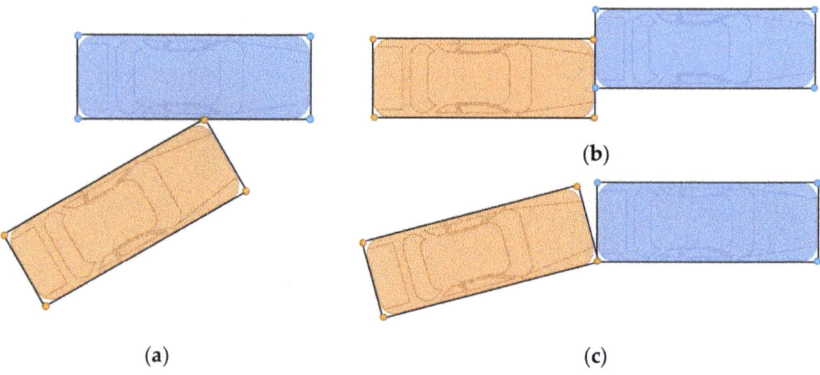

Figure 6. Collision types. (**a**) Point-to-edge collision; (**b**) Edge-to-edge collision; (**c**) Point-to-point collision.

After unifying different collision types, TTC can be calculated in the same way. We take the collision type shown in Figure 7 as an example. In this case, the front left corner of vehicle 2 collides on the right edge of vehicle 1. As we defined in Section 2.1, the coordinate of a point in the axis system of vehicle 1 is expressed as $X_k = [x_k, y_k]^T$, and $X_k' = [x_k', y_k']^T$ in the axis system of vehicle 2. R_i ($i = 1,2,3,4$) represents the four corners of vehicle 1. F_i ($i = 1,2,3,4$) represents the four corners of vehicle 2. Therefore, the coordinate of R_i

is $X_{R_i} = \left[x_{R_i}, y_{R_i}\right]^T$, which is known by measuring the size of the vehicle 1. Similarly, $X'_{F_i} = \left[x'_{F_i}, y'_{F_i}\right]^T$ is also known by measuring the size of vehicle 2. The relative position of $X = [x, y]^T$ and the relative orientation β are estimated by the UWB/DR system. Then, the coordinates of vehicle 2's corners in the axis system of vehicle 1 can be derived as Equation (28).

$$\left[X_{F_1}, X_{F_2}, X_{F_3}, X_{F_4}\right] = R\left[X'_{F_1}, X'_{F_2}, X'_{F_3}, X'_{F_4}\right] + X[1,1,1,1] \tag{28}$$

where $R = \begin{bmatrix} \cos(\beta) & -\sin(\beta) \\ \sin(\beta) & \cos(\beta) \end{bmatrix}$.

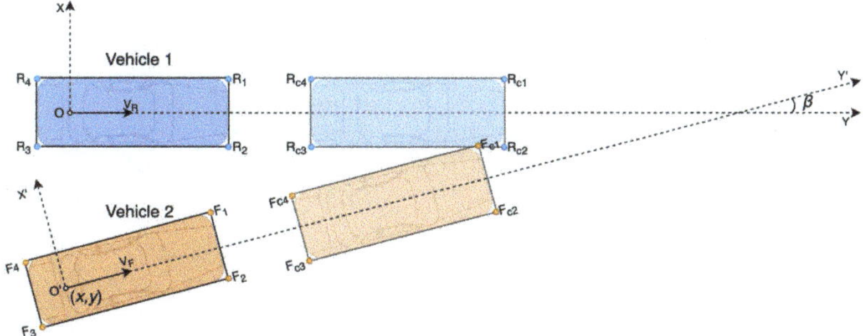

Figure 7. The collision warning model.

We define all the points at the collision time as R_{C_i} and F_{C_i}, and their coordinates as $X_{R_{C_i}} = \left[x_{R_{C_i}}, y_{R_{C_i}}\right]^T$, $X_{F_{C_i}} = \left[x_{F_{C_i}}, y_{F_{C_i}}\right]^T$. The velocity vectors of the two vehicles are known for the UWB/DR system, which are $V_R = [v_R \cos(\beta_R), v_R \cos(\beta_R)]^T$ $(\beta_R = 0)$ and $V_F = [v_F \cos(\beta_F), v_F \sin(\beta_F)]^T$ $(\beta_F = \beta)$. Assume that point F_i collides on the edge between R_j and R_k at time $t_{F_i,R_{jk}}$. Then $X_{F_{C_i}}$, $X_{R_{C_j}}$, and $X_{R_{C_k}}$ can be expressed as Equation (29).

$$\begin{aligned} X_{F_{C_i}} &= X_{F_i} + V_F t_{F_i,R_{jk}} \\ X_{R_{C_j}} &= X_{R_j} + V_R t_{F_i,R_{jk}} \\ X_{R_{C_k}} &= X_{R_k} + V_R t_{F_i,R_{jk}} \end{aligned} \tag{29}$$

Point F_i collides on the edge between R_j and R_k means F_{c_i} is on the segment $\overline{R_{C_j}R_{C_k}}$, which can be expressed as Equation (30).

$$\vec{F_{C_i}R_{C_j}} \cdot \vec{F_{C_i}R_{C_k}} = -\|\vec{F_{C_i}R_{C_j}}\|\|\vec{F_{C_i}R_{C_k}}\| \tag{30}$$

Solution t of Equation (30) is the collision time under the condition that corners of vehicle 2 collide on edges of vehicle 1, including 16 different conditions altogether. In the other 16 cases in which the corners of vehicle 1 collide on the edges of vehicle 2, the collision times can be calculated similarly. Thirty-two collision times can be calculated in total. Ignoring negative values, the minimum of the rest value is TCC. That is:

$$\begin{aligned} TTC &= \min\left(t_{R_i,F_{jk}}, t_{F_i,R_{jk}}\right), \ (i = 1,2,3,4; jk = 12,23,34,41), \\ &t_{R_i,F_{jk}} \geq 0, t_{F_i,R_{jk}} \geq 0. \end{aligned} \tag{31}$$

When $TTC \to \infty$ or $TTC < 0$, there is no risk of collision.

3. Simulation

In this section, simulation is conducted to evaluate our algorithm. Firstly, the accuracy of the UWB positioning and directing system is validated by comparing the algorithm with and without the constraint of d_5. Secondly, the accuracy of the UWB/DR fusion model based on EKF is compared to the accuracy of UWB and DR separately. Finally, plenty of driving scenarios are generated to evaluate the success rate of the CWS.

3.1. Simulation of the Overconstrained UWB Positioning and Directing System

In Section 2.1, a relative positioning/directing algorithm with the constraint of d_5 is proposed. Its performance is simulated in this section. Firstly, a driving scenario is established in the driving scenario designer of MATLAB as shown in Figure 8. The blue cube represents vehicle 1, and the red cube represents vehicle 2. The lines in blue and red denote their driving track. Kinematic parameters of vehicle and positions of UWB modules and wheel sensors in their own vehicle axis system are defined in the model. The UWB ranging error is set to $\sigma_d = 0.05$ m, and the wheel speed error is set to $\sigma_v = 0.2$ m/s referring to the sensors we will use in experiments. Calculating results of our algorithm are compared to the real values exported by the model.

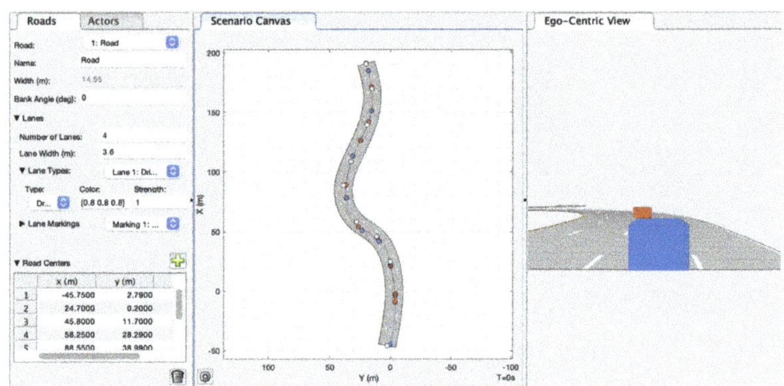

Figure 8. The virtual scenario in the driving scenario designer.

Solutions of the algorithm with and without the constraint of d_5 are compared in Figure 9 and Table 1. The improvement of accuracy with the derivation of d_5 is very intuitive, especially for x and β. In Table 1, the root mean square error (RMSE) is recommended to compare their accuracy quantitatively.

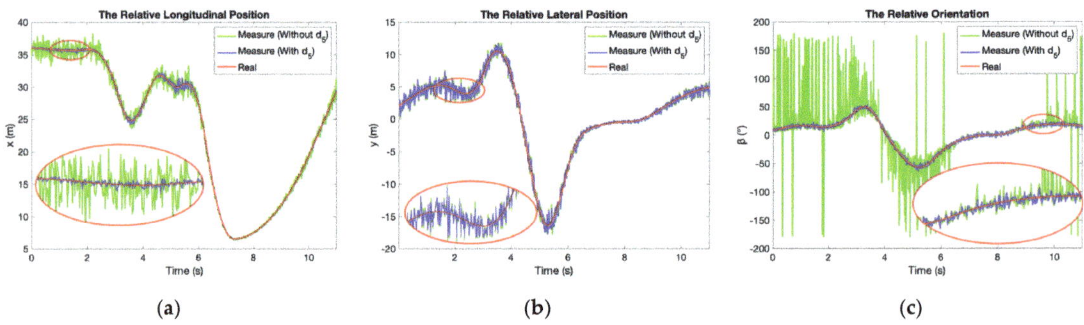

Figure 9. Comparison of relative positioning and directing algorithm with and without the constraint of d_5: (**a**) The relative longitudinal position x; (**b**) The relative lateral position y; (**c**) The relative orientation β.

Table 1. RMSE of the algorithm with and without d_5.

Algorithm	$RMSE_x$ (m)	$RMSE_y$ (m)	$RMSE_\beta$ (°)
With d_5	0.70	0.73	46.29
Without d_5	0.21	0.58	2.42

3.2. Simulation of the UWB/DR Fusion Algorithm

We also take the scenario in Section 3.1 as an example to validate the performance of the UWB/DR fusion algorithm. The comparison results are shown in Figure 10 and Table 2. The proposed UWB/DR fusion method based on EKF significantly improves the accuracy and stability of positioning and directing.

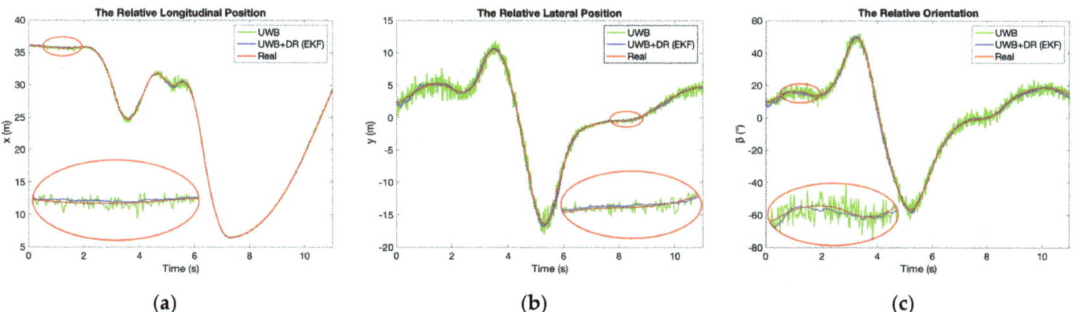

Figure 10. Comparison of positioning and directing performance using UWB and fusion of UWB/DR: (**a**) The relative longitudinal position x; (**b**) The relative lateral position y; (**c**) The relative orientation β.

Table 2. RMSE of position and orientation estimated by UWB and UWB/DR.

Algorithm	$RMSE_x$ (m)	$RMSE_y$ (m)	$RMSE_\beta$ (°)
UWB	0.21	0.58	2.42
UWB + DR (EKF)	0.06	0.17	0.83

Figures 11 and 12 and Table 3 compare the accuracy of yaw rates and velocities estimated by UWB/DR to DR. They are improved significantly as well, which contributes to the better prediction accuracy of TTC in the next section.

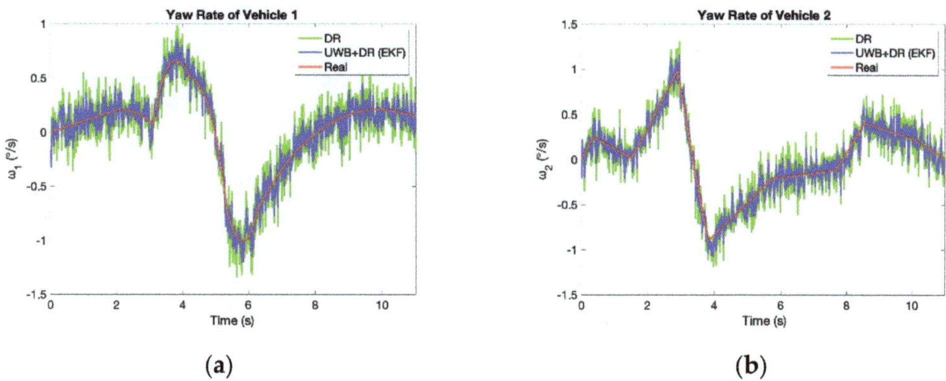

Figure 11. Comparison of yaw rates measured by DR and estimated by UWB/DR: (**a**) Yaw rate of vehicle 1; (**b**) Yaw rate of vehicle 2.

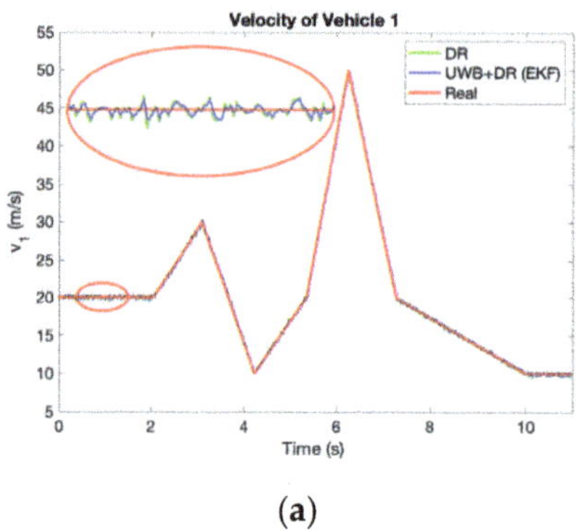
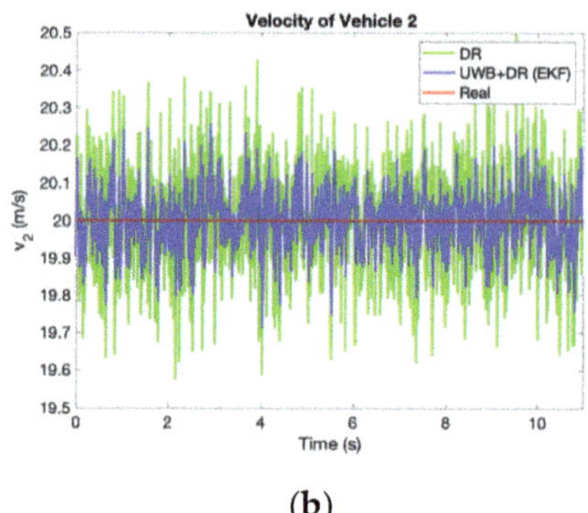

(**a**)   (**b**)

Figure 12. Comparison of velocities measured by DR and estimated by UWB/DR: (**a**) Velocity of vehicle 1; (**b**) Velocity of vehicle 2.

Table 3. RMSE of velocities and yaw rates estimated by DR and UWB/DR.

Algorithm	$RMSE_{\omega_1}$ (°/s)	$RMSE_{\omega_2}$ (°/s)	$RMSE_{v_1}$ (m/s)	$RMSE_{v_2}$ (m/s)
DR	8.92	8.71	0.14	0.15
UWB + DR (EKF)	5.07	4.60	0.12	0.08

3.3. Simulation of CWS based on TTC Estimation

In this section, we generate plenty of driving scenarios with different velocities, relative positions, and relative orientations, as shown in Figure 13. The ranges of parameters are set as outlined in Table 4.

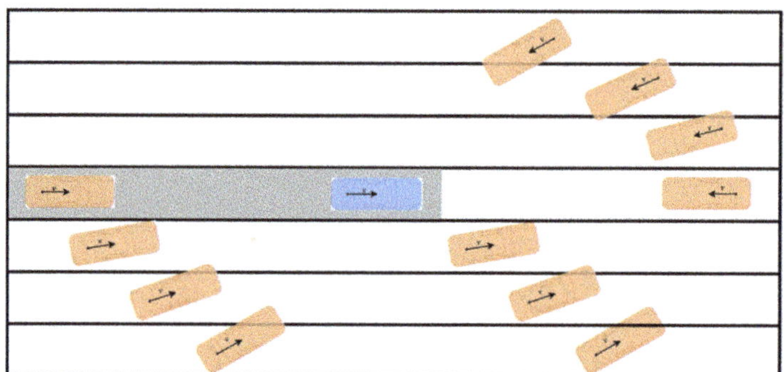

Figure 13. TTC simulating scenarios.

Table 4. Ranges of parameters in TTC simulation.

Parameters	Range
v_1 & v_2 (km/h)	0~75
x (m)	−200~200
y (m)	−15~15
β (°)	0~360

TTC_{real} is certain when a scenario is established, and TTC_{est} estimated by CWS is calculated every 10 ms. The collision warning threshold is set to 3.0 s. It means that when $TTC_{est} \leq 3.0$ s, the CWS will send an alert. $TTC_{err} = TTC_{est} - TTC_{real}$ denotes the TTC error at the warning time as Figure 14.

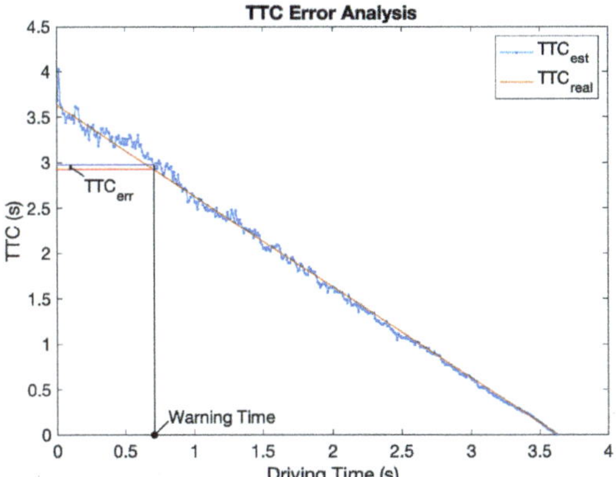

Figure 14. TTC estimation error.

In order to guarantee driving safety, we set 2.7 s as the latest warning time. If the system does not work when the vehicle is colliding within 2.7 s, the collision warning evaluation is failed. In addition, in order not to disturb the driver too much, if the system sends alerts when vehicles have no risk of collision within 4 s, we regard the warning as false. Then, TTC_{err} can be divided into three conditions corresponding to three evaluations of collision warning as Equation (32).

$$TTC_{err} \begin{cases} > 0.3 & \text{Failed} \\ \in [-1, 0.3] & \text{Correct} \\ < -1 & \text{False} \end{cases} \tag{32}$$

- "Failed" denotes warning too late or not warning;
- "Correct" denotes warning in the proper time period;
- "False" denotes warning too early or warning by mistake.

The scenario marked with gray background is the typical rear-end collision scenario, which is the most critical function of a collision warning system. One hundred and ninety-six rear-end collision scenarios are generated, and Table 5 shows the results. In all the 196 simulation scenarios, two of them behave false, which means that the collision warning is triggered too early. All of the others perform correctly. It shows the reliability of the proposed CWS in the most common rear-end collision scenarios.

Table 5. Collision warning evaluation in rear-end scenarios.

Evaluation	Quantity
Failed	0
Correct	194
False	2

Then, we emulate other scenarios in which two vehicles drive in any lanes from any positions to any directions defined in Figure 13 and Table 4. Results are shown in Table 6. The scenarios with initial TTC_{real} less than 3 s will not be considered. In the remaining 10,823 scenarios, 10,593 of them perform correctly. The collision warning success rate is 97.9%.

Table 6. Collision warning evaluation in random scenarios.

Evaluation	Quantity
Failed	0
Correct	10,596
False	227

4. Experiments

In this section, experiments are divided into two parts: straight driving experiments and curved driving experiments. The straight driving experiments are conducted referring to JT/T883-2014, which describes the standard experiments for FCWS, published by the Ministry of Transport of the People's Republic of China (MOT). As JT/T883-2014 only regulates straight driving experiments, to further validate the performance of our system, curved driving experiments are conducted in addition. Since the CWS is implemented based on the UWB/DR relative positioning/directing system, the positioning/directing accuracy can reflect the performance of the CWS. Therefore, in the curved experiments, we drive through complex routes and compare the positioning/directing accuracy to the parameters of a commercial millimeter-wave radar (MMWR) used for collision warning.

4.1. Experimental Equipment and Environment

Figure 15 shows the equipment used in the experiments. Two vehicles are required in the experiments for relative positioning and directing. UWB modules are installed on the corners of the vehicles. Four wheel-speed sensors designed by our team are installed on the centers of the wheels. The wheel-speed measurements are transmitted to a receiver inside the vehicle wirelessly, which receives the velocity information from the four wheels and then sends it to the controller area network (CAN) bus. In the proposed system, only the speeds of the rear wheels are used. UWB modules are also developed by our team based on DW1000. Two vehicles share data through UWB. All data are transferred to the CAN bus and recorded by the computer using a USB-CAN adapter. A computing terminal receives sensor data from the CAN bus and calculates the relative position, direction, velocity, and TTC. Results from the computing terminal are compared to the measurement of a high-precision integrated positioning system, which combines dual-antenna real-time kinematic (RTK)-GPS and INS. The long-range radio (LoRa) antenna is used to receive differential signals from the RTK-GPS base station, which is installed in the testing ground. A total station is used to measure the relative coordinates of the UWB modules to the main RTK-GPS antenna. It should be noted that the main GPS antenna is not right above the center of the rear wheels. The deviation needs to be derived from measurements of the total station and compensated in the algorithm.

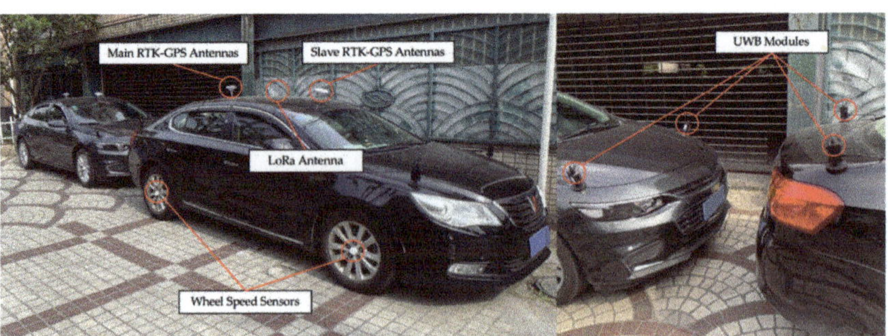

Figure 15. Experimental Equipment.

Figure 16 shows the testing ground in which we conduct experiments. The driving routes of the two types of experiments are also marked in Figure 16.

Figure 16. The testing ground and vehicle driving routes.

4.2. Straight Driving Experiments

According to JT/T883-2014 [35], experiments for FCWS consist of three tests. Each test needs repeating seven times. Only if five of them were passed, and no two consecutive failed tests exist could the test be evaluated as passed. In the standard experiments, the headway distances, velocities, and accelerations of vehicles need controlling around specific values, so we design software as shown in Figure 17, with necessary parameters displayed, which helps drivers better control vehicles and records necessary data. The TTC derived from the data of the RTK-GPS/INS is recognized as real TTC.

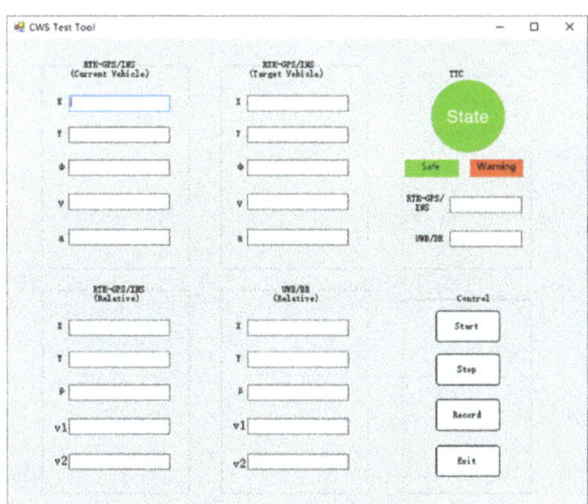

Figure 17. Vehicle state display software.

4.2.1. Test 1

Test 1 is designed as shown in Figure 18. The rear vehicle drives at the speed of 72 km/h toward the parked front vehicle from an inertial headway distance of 150 m. If the collision warning system is triggered before the real TTC is 2.7 s, the test is passed. Otherwise, the test is failed.

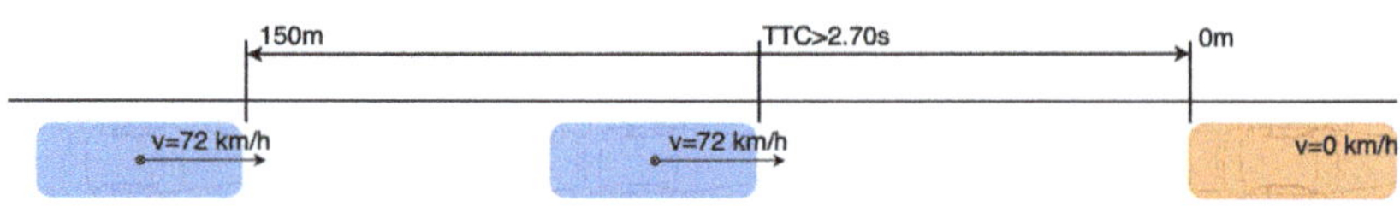

Figure 18. Test 1 in the straight driving experiments.

4.2.2. Test 2

Test 2 is designed as shown in Figure 19. The rear vehicle drives at the speed of 72 km/h toward the front vehicle, which drives at the speed of 32 km/h, from an initial headway distance of 150 m. If the collision warning system is triggered before the real TTC is 2.1 s, the test is passed. Otherwise, the test is failed.

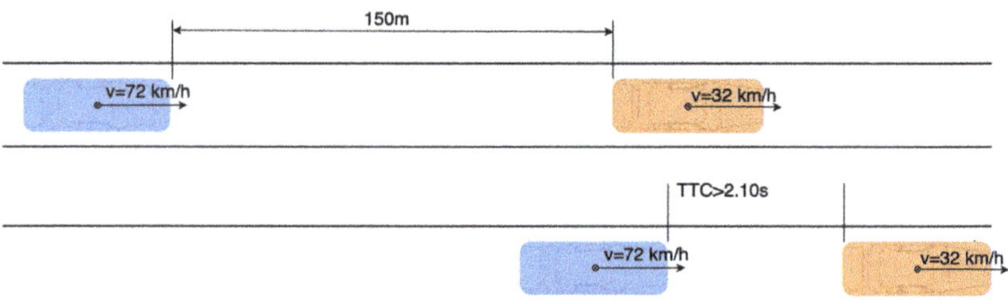

Figure 19. Test 2 in the straight driving experiments.

4.2.3. Test 3

Test 3 is designed as Figure 20. The rear vehicle drives at the speed of 72 km/h toward the front vehicle, which drives at the speed of 32 km/h and decelerates with the acceleration of -0.3 g. If the collision warning system is triggered before the real TTC is 2.4 s, the test is passed. Otherwise, the test is failed.

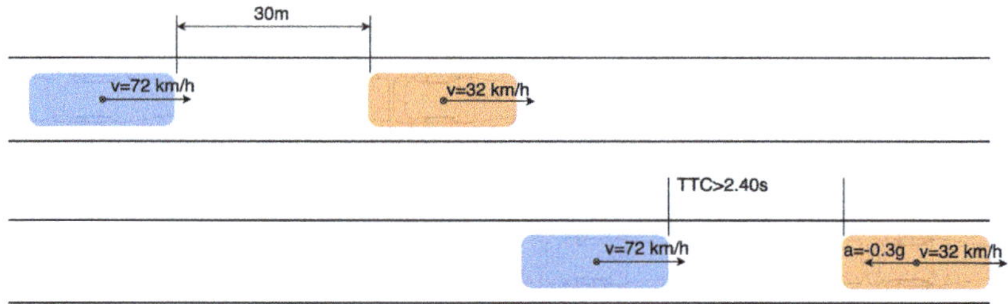

Figure 20. Test 3 in the straight driving experiments.

4.2.4. Results Analysis of the Straight Driving Experiments

During each test, two TTC values are calculated: (1) TTC_{real}, which is derived from the RTK-GPS/INS information; (2) TTC_{CWS}, which is estimated using the UWB/DR measurements. Since the terminating conditions in the three experiments are different, to satisfy all the three tests and reserve some margin, we set the warning TTC_{CWS} to 3.0 s. The software in Figure 17 will send a warning when either TTC_{real} or TTC_{CWS} reaches its marginal value. If TTC_{real} reaches the regulated marginal value when TTC_{CWS} is still greater than 3.0 s, the test is terminated and evaluated as failed. In the standard, only the minimum threshold of the collision warning time is regulated, whereas the maximum threshold is not. In other words, the standard only cares about "how safe" the warning is, with no consideration of "how accurate" it is. However, as we explained in Section 3, too early warnings are annoying and offensive, so we set 4.0 s as the upper limit. If the CWS is triggered when

$TTC_{real} > 4.0$ s, we also regard the test as failed. According to JT/T883-2014, each test needs repeating seven times. Tables 7–9 show the results of the three tests, respectively.

Table 7. Results of Test 1.

	1	2	3	4	5	6	7
TTC(CWS)	2.9987	2.9907	2.9759	2.9814	2.9729	2.9722	2.9799
TTC(Real)	3.0047	3.0069	2.9925	2.9963	2.9902	3.0136	3.0219
Evaluation	Pass	Pass	Pass	Pass	Pass	Pass	Pass

Table 8. Results of Test 2.

	1	2	3	4	5	6	7
TTC(CWS)	2.9863	2.9810	2.9987	2.9804	2.9954	2.9899	2.9673
TTC(Real)	3.0423	3.1166	3.0245	3.0354	3.1283	3.1269	3.0226
Evaluation	Pass	Pass	Pass	Pass	Pass	Pass	Pass

Table 9. Results of Test 3.

	1	2	3	4	5	6	7
TTC(CWS)	2.9782	2.9905	2.9947	2.9789	2.9942	2.8623	2.8958
TTC(Real)	2.7560	2.8110	2.7877	2.6851	2.6511	2.5831	2.8975
Evaluation	Pass	Pass	Pass	Pass	Pass	Pass	Pass

According to Tables 7–9, all the tests were passed, which proves that the proposed system can satisfy the requirement of MOT and has the ability to provide collision warning for vehicles in time.

4.3. Curved Driving Experiments

JT/T883-2014 only regulates the straight driving experiments but does not request or give advice to curved driving experiments. However, to further validate the superiority of our system, we conduct curved driving experiments and compare its accuracy to a commercial MMWR, Aptiv (Electronically Scanning RADAR) ESR 2.5, which is used in CWS. The MMWR measures the relative distance, relative azimuth, and relative velocity. Table 10 shows the accuracy of Aptiv ESR 2.5 according to its datasheet. ρ, θ, and v represent the relative distance, azimuth angle, and velocity, respectively. The MMWR has two working modes, middle-distance mode and long-distance mode, and the accuracies are different.

Table 10. The accuracy of the MMWR.

Mode	Coverage (m)	$RMSE_\rho$ (m)	$RMSE_\theta$ (°)	$RMSE_v$ (m/s)
Middle Distance	50	0.25	1	0.12
Long Distance	100	0.5	0.5	0.12

In order to facilitate comparison, the curved experiments are also divided into a middle-distance experiment under vehicle distances within 50 m and a long-distance experiment under vehicle distances within 100 m. The estimated values of the relative position $[x, y]$ are converted to the polar coordinate $[\rho, \theta]$, and the velocities of the two vehicles $[v_1, v_2]$ are converted to the relative velocity v, as shown in Figure 21. In addition, the relative orientation β cannot be measured by MMWR directly.

4.3.1. Middle-Distance Experiments

The proposed CWS and MMWR are all dynamic systems, so the vehicle distance is not kept to a constant value but changes in the experiment. During the middle-distance experiment, the vehicle distance changes between 10 and 50 m. In our system, the vehicle distance represents the distance between the real axle centers of the two vehicles, so it cannot be zero.

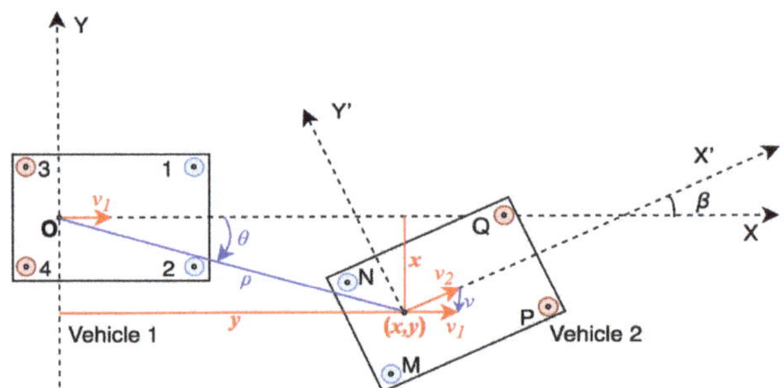

Figure 21. The transformation from Cartesian coordinates to polar coordinates.

4.3.2. Long-Distance Experiments

During the long-distance experiment, the vehicle distances change between 10 and 100 m.

4.3.3. Results Analysis of the Curved Experiments

According to Figures 22 and 23, the accuracy of the proposed system improves significantly after fusion, which reaches the same conclusion as the simulation results shown in Figure 12. The relative position is described as Cartesian coordinate $[x, y]$ in the simulation but as polar coordinate $[\rho, \theta]$ in the experiments. Both x and y improve after fusion as shown in Table 2, whereas only θ without ρ improves after fusion according to Tables 11 and 12. That is because the accuracy improvement of θ can contribute to better accuracy of both x and y, as Figure 21. Therefore, the experimental results are consistent with the simulation. The comparison result of the proposed system and the MMWR is shown in Table 13, which combines Tables 11 and 12 with Table 10.

Table 11. The accuracy of the MMWR.

Mode	$RMSE_\rho$ (m)	$RMSE_\theta$ (°)	$RMSE_v$ (m/s)	$RMSE_\beta$ (°)
No Fusion	0.14	0.76	0.22	1.84
Fusion	0.14	0.31	0.11	0.39

Table 12. The accuracy of the MMWR.

Mode	$RMSE_\rho$ (m)	$RMSE_\theta$ (°)	$RMSE_v$ (m/s)	$RMSE_\beta$ (°)
No Fusion	0.18	0.77	0.22	1.86
Fusion	0.17	0.31	0.12	0.40

Table 13. Accuracy comparison of the proposed system and the MMWR.

Mode	System	$RMSE_\rho$ (m)	$RMSE_\theta$ (°)	$RMSE_v$ (m/s)	$RMSE_\beta$ (°)
Middle Distance	MMWR	0.25	1	0.12	None
	Proposed System (No Fusion)	0.14	0.76	0.22	1.84
	Proposed System (Fusion)	0.14	0.31	0.11	0.39
Long Distance	MMWR	0.5	0.5	0.12	None
	Proposed System (No Fusion)	0.18	0.77	0.24	1.86
	Proposed System (Fusion)	0.17	0.31	0.12	0.40

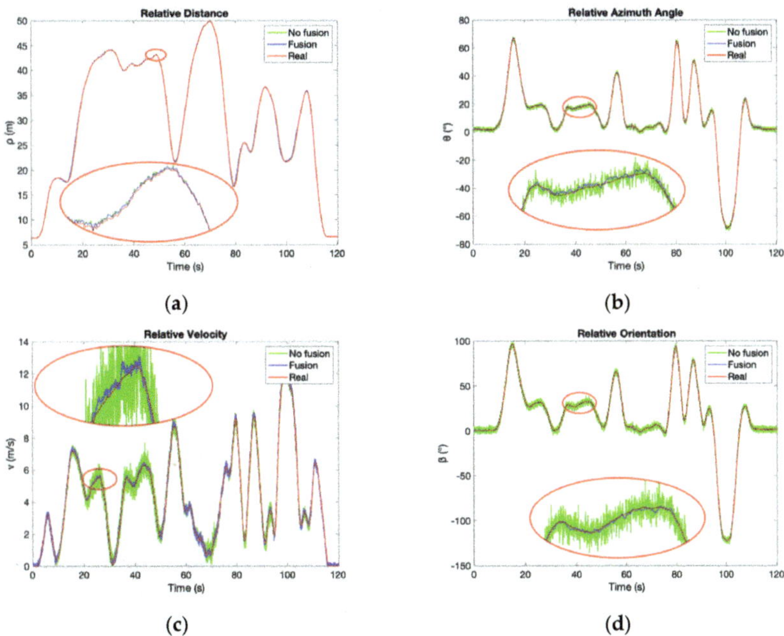

Figure 22. The results of the middle-distance experiments. (**a**) Relative distance; (**b**) Relative azimuth angle; (**c**) Relative velocity; (**d**) Relative orientation.

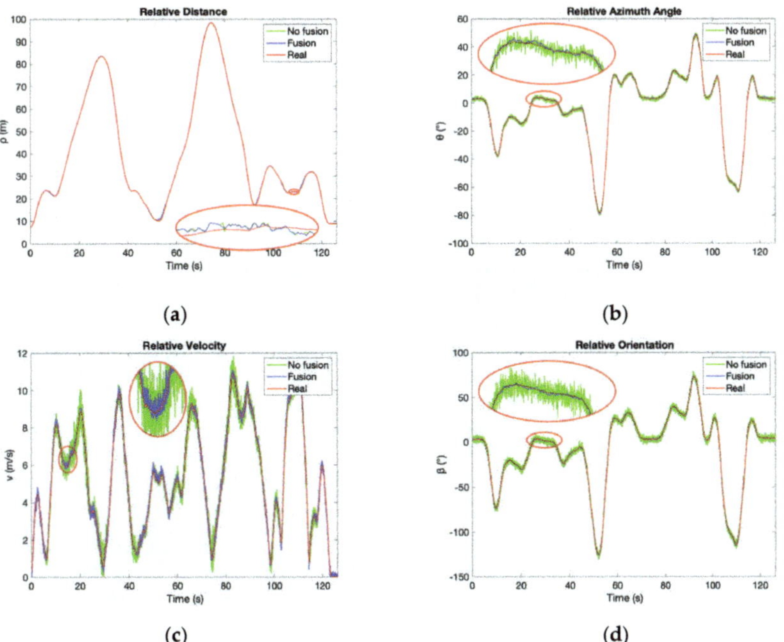

Figure 23. The results of the long-distance experiments. (**a**) Relative distance; (**b**) Relative azimuth angle; (**c**) Relative velocity; (**d**) Relative orientation.

The distance accuracy of the proposed system is always much better than the MMWR, no matter with or without fusion. The azimuth accuracy without fusion is about 0.76° in both experiments, which is better than the middle-distance MMWR but is inferior to the long-distance MMWR. However, velocity accuracy without fusion is worse than the MMWR in both modes. As for the fusion system, the accuracy of relative distance and azimuth performs significantly better than the MMWR, and the relative velocity accuracy also improves to a similar level as the MMWR in both middle and long-distance modes. Table 14 shows the accuracy enhanced rates of the proposed system to the MMWR.

Table 14. Enhanced rate of the proposed system to the MMWR.

Mode	$RMSE_\rho$ (m)	$RMSE_\theta$ (°)	$RMSE_v$ (m/s)
Middle Distance	44%	69%	8%
Long Distance	66%	38%	0%

In addition, the proposed system can provide the relative orientation, which is not available directly in the MMWR system.

5. Conclusions

In this paper, we proposed a CWS combining UWB and DR. An improved relative positioning/directing algorithm based on UWB is presented, and a DR model based on the speeds of the rear wheels is established. Then, a fusion algorithm using EKF is proposed to improve the accuracy of relative position, orientation, and velocity. Afterwards, the advantage of the proposed system is preliminarily verified by simulation. Finally, experiments are conducted to further validate the performance of our system, and the experiment results are compared to a commercial MMWR used in CWS. The main conclusions are summarized as follows:

- The proposed relative positioning/directing algorithm with an additional distance constraint significantly improves the relative positioning/directing accuracy, especially the directing accuracy, as shown in Figure 9 and Table 1.
- The fusion method significantly improves the relative positioning/directing accuracy and slightly improves the velocity accuracy according to the simulation and experiment results.
- The proposed CWS passes the regulated tests in JT/T883-2014 published by MOT, which proves the feasibility of the proposed system.
- In middle-distance mode up to 50 m, compared to the MMWR, the proposed system improves the relative positioning/directing accuracy by 44%, 69%, and 8%, respectively, in the relative distance, azimuth angle, and velocity. As for in long-distance mode, the enhanced rate is 66% and 38%, respectively, for the relative distance and azimuth angle. The relative velocity accuracy of the proposed system is similar to the MMWR.
- In both middle and long-distance modes, the proposed system can provide relative orientations with errors no more than 0.4° RMSE, which is not available directly in MMWR systems, but it is very beneficial to the CWS.

The inadequacy of the proposed system is the velocity accuracy. Although it performs at the same level as MMWR in terms of velocity accuracy, it can be further improved. To facilitate comparison of the proposed system and the MMWR, the velocity data of the experiments are shown as relative velocity. We also analyze the accuracy of absolute velocities of two vehicles. In the middle-distance experiment, $RMSE_{v_1} = 0.16$ m/s and $RMSE_{v_2} = 0.13$ m/s, and in the long-distance experiment, $RMSE_{v_1} = 0.16$ m/s and $RMSE_{v_2} = 0.16$ m/s. Both of them are inferior to the simulation results. It is because the DR system is established based on a theoretical Ackerman steering model, which ignores the stiffness of suspensions and tires. In the actual situation, vehicle dynamic parameters such as side-slip angles will also affect the precision of the algorithm. Therefore, our research direction in the future is the system with a more accurate vehicle dynamic model and with more sensors integrated such as IMU and GPS.

Author Contributions: Conceptualization, M.W. and Y.S.; Funding acquisition, X.C. and Y.S.; Investigation, M.W.; Methodology, M.W. and P.L.; Software, M.W., B.J. and P.L.; Validation, M.W. and B.J.; Writing—original draft, M.W.; Writing—review and editing, W.W. and Y.S. All authors have read and agreed to the published version of the manuscript.

Funding: This research was funded by National Key R&D Program of China (Grant No. 2018YFB0104802) and Industry University Research Project of Shanghai Automotive Industry Science and Technology Development Foundation (Grant No. 1705).

Informed Consent Statement: Informed consent was obtained from all subjects involved in the study.

Acknowledgments: The authors are grateful to the subjects in the experiment and appreciate the reviewers for their helpful comments and suggestions in this study.

Conflicts of Interest: The authors declare no conflict of interest.

References

1. World Health Organization (WHO). Global Status Report on Road Safety 2018. Available online: https://www.who.int/violence_injury_prevention/road_safety_status/2018/en/ (accessed on 7 October 2020).
2. Kusano, K.D.; Gabler, H.C. Safety Benefits of Forward Collision Warning, Brake Assist, and Autonomous Braking Systems in Rear-End Collisions. *IEEE Trans. Intell. Transp. Syst.* **2012**, *13*, 1546–1555. [CrossRef]
3. Owens, J.M.; Dingus, T.A.; Guo, F.; Fang, Y.; Perez, M.; McClafferty, J.; Tefft, B. Prevalence of Drowsy Driving Crashes: Estimates from a Large-Scale Naturalistic Driving Study. AAA Foundation for Traffic Safety: Wahsington, WA., USA, 2018.
4. Ewald, A.; Willhoeft, V. Laser Scanners for Obstacle Detection in Automotive Applications. In Proceedings of the IEEE Intelligent Vehicles Symposium 2000 (Cat. No.00TH8511), Dearborn, MI, USA, 5 October 2000; pp. 682–687.
5. Chen, S.-K.; Parikh, J.S. Developing a Forward Collision Warning System Simulation. In Proceedings of the IEEE Intelligent Vehicles Symposium 2000 (Cat. No.00TH8511), Dearborn, MI, USA, 5 October 2000; pp. 338–343.
6. Kim, J.; Han, D.S.; Senouci, B. Radar and Vision Sensor Fusion for Object Detection in Autonomous Vehicle Surroundings. In Proceedings of the 2018 Tenth International Conference on Ubiquitous and Future Networks (ICUFN), Prague, Czech Republic, 3–6 July 2018; pp. 76–78.
7. Peng, W.; Zhiqiang, L. The Study of Intelligent Vehicle Anti-Collision Forewarning Technology by Multi-Information Detection. In Proceedings of the 2013 Third International Conference on Intelligent System Design and Engineering Applications, Hong Kong, China, 16–18 January 2013; pp. 1557–1561.
8. Srinivasa, N. Vision-Based Vehicle Detection and Tracking Method for Forward Collision Warning in Automobiles. In Proceedings of the Intelligent Vehicle Symposium, 2002. IEEE, Versailles, France, 17–21 June 2003; Volume 2, pp. 626–631.
9. Liu, J.-F.; Su, Y.-F.; Ko, M.-K.; Yu, P.-N. Development of a Vision-Based Driver Assistance System with Lane Departure Warning and Forward Collision Warning Functions. In Proceedings of the 2008 Digital Image Computing: Techniques and Applications, Canberra, Australia, 1–3 December 2008; pp. 480–485.
10. Shieh, W.-Y.; Hsu, C.-C.J.; Chen, H.-C.; Wang, T.-H.; Chen, C.-C. Construction of Infrared Signal-Direction Discriminator for Intervehicle Communication. *IEEE Trans. Veh. Technol.* **2015**, *64*, 2436–2447. [CrossRef]
11. Sanberg, W.P.; Dubbelman, G. From Stixels to Asteroids: Towards a Collision Warning System Using Stereo Vision. *Electron. Imaging* **2019**, *2019*, 34-1–34-7. [CrossRef]
12. Hernandez, D.C.; Filonenko, A.; Hariyono, J.; Shahbaz, A. Kang-Hyun Jo Laser Based Collision Warning System for High Conflict Vehicle-Pedestrian Zones. In Proceedings of the 2016 IEEE 25th International Symposium on Industrial Electronics (ISIE), Santa Clara, CA, USA, 8–10 June 2016; pp. 935–939.
13. Coelingh, E.; Jakobsson, L.; Lind, H.; Lindman, M. Collision Warning with Auto Brake—A Real-Life Safety Perspective, Innovations for Safety: Opportunities and Challenges. In Proceedings of the 20th International Technical Conference on the Enhanced Safety of Vehicles (ESV), Lyon, France, 18–21 June 2017.
14. Coelingh, E.; Eidehall, A.; Bengtsson, M. Collision Warning with Full Auto Brake and Pedestrian Detection—A Practical Example of Automatic Emergency Braking. In Proceedings of the 13th International IEEE Conference on Intelligent Transportation Systems, Funchal, Portugal, 19–22 September 2010; pp. 155–160.
15. Srinivasa, N.; Chen, Y.; Daniell, C. A Fusion System for Real-Time Forward Collision Warning in Automobiles. In Proceedings of the 2003 IEEE International Conference on Intelligent Transportation Systems, Shanghai, China, 12–15 October 2003; pp. 457–462.
16. Huang, C.; Lv, C.; Hang, P.; Xing, Y. Toward Safe and Personalized Autonomous Driving: Decision-Making and Motion Control with DPF and CDT Techniques. *IEEE ASME Trans. Mechatron.* **2021**, *26*, 611–620. [CrossRef]
17. Huang, C.; Lv, C.; Hang, P.; Hu, Z.; Xing, Y. Human-Machine Adaptive Shared Control for Safe Automated Driving under Automation Degradation. *arXiv* **2021**, arXiv:2103.04563.

18. Huang, C.; Huang, H.; Hang, P.; Gao, H.; Wu, J.; Huang, Z.; Lv, C. Personalized Trajectory Planning and Control of Lane-Change Maneuvers for Autonomous Driving. *IEEE Trans. Veh. Technol.* **2021**, 1. [CrossRef]

19. Hang, P.; Lv, C. Human-Like Decision Making for Autonomous Driving: A Noncooperative Game Theoretic Approach. *arXiv* **2020**, arXiv:2005.11064. [CrossRef]

20. Hang, P.; Lv, C.; Huang, C.; Xing, Y.; Hu, Z. Cooperative Decision Making of Connected Automated Vehicles at Multi-Lane Merging Zone: A Coalitional Game Approach. *arXiv* **2021**, arXiv:210307887.

21. Chen, J.; Tian, S.; Xu, H.; Yue, R.; Sun, Y.; Cui, Y. Architecture of Vehicle Trajectories Extraction with Roadside LiDAR Serving Connected Vehicles. *IEEE Access* **2019**, *7*, 100406–100415. [CrossRef]

22. Yang, W.; Wan, B.; Qu, X. A Forward Collision Warning System Using Driving Intention Recognition of the Front Vehicle and V2V Communication. *IEEE Access* **2020**, *8*, 11268–11278. [CrossRef]

23. Xiang, X.; Qin, W.; Xiang, B. Research on a DSRC-Based Rear-End Collision Warning Model. *IEEE Trans. Intell. Transp. Syst.* **2014**, *15*, 1054–1065. [CrossRef]

24. Yang, T.; Zhang, Y.; Tan, J.; Qiu, T.Z. Research on Forward Collision Warning System Based on Connected Vehicle V2V Communication. In Proceedings of the 2019 5th International Conference on Transportation Information and Safety (ICTIS), Liverpool, UK, 14–17 July 2019; pp. 1174–1181.

25. Patra, S.; Veelaert, P.; Calafate, C.; Cano, J.-C.; Zamora, W.; Manzoni, P.; González, F. A Forward Collision Warning System for Smartphones Using Image Processing and V2V Communication. *Sensors* **2018**, *18*, 2672. [CrossRef] [PubMed]

26. Marano, S.; Gifford, W.; Wymeersch, H.; Win, M. NLOS Identification and Mitigation for Localization Based on UWB Experimental Data. *IEEE J. Sel. Areas Commun.* **2010**, *28*, 1026–1035. [CrossRef]

27. Lu, Y.; Yi, J.; He, L.; Zhu, X.; Liu, P. A Hybrid Fusion Algorithm for Integrated INS/UWB Navigation and Its Application in Vehicle Platoon Formation Control. In *Proceedings of the 2018 International Conference on Computer Science, Electronics and Communication Engineering (CSECE 2018)*; Wuhan, China, 7–8 February 2018, Atlantis Press: Sanya, China, 2018.

28. Wang, M.; Zhou, A.; Chen, X.; Shen, Y.; Li, Z. A Novel Asynchronous UWB Positioning System for Autonomous Trucks in an Automated Container Terminal. *SAE Int. J. Adv. Curr. Pract. Mobil.* **2020**, *2*, 3413–3422. [CrossRef]

29. Sun, S.; Hu, J.; Li, J.; Liu, R.; Shu, M.; Yang, Y. An INS-UWB Based Collision Avoidance System for AGV. *Algorithms* **2019**, *12*, 40. [CrossRef]

30. Liu, X.; Jin, F.; Lv, X.; Zhan, Y.S.; Zhang, D. Design and Development of Vehicle Collision-Avoidance System Based on UWB Wireless Sensor Networks. In *Proceedings of the 7th International Conference on Computer Engineering and Networks—PoS (CENet2017)*; Shanghai, China, 22–23 July 2017, Sissa Medialab: Shanghai, China, 2017; p. 037.

31. Pittokopiti, M.; Grammenos, R. Infrastructureless UWB Based Collision Avoidance System for the Safety of Construction Workers. In Proceedings of the 2019 26th International Conference on Telecommunications (ICT), Hanoi, Vietnam, 8–10 April 2019; pp. 490–495.

32. Kianfar, A.E.; Uth, F.; Baltes, R.; Clausen, E. Development of a Robust Ultra-Wideband Module for Underground Positioning and Collision Avoidance. *Min. Metall. Explor.* **2020**, *37*, 1821–1825. [CrossRef]

33. ISO 8855:2011. Available online: https://www.iso.org/cms/render/live/en/sites/isoorg/contents/data/standard/05/11/51180.html (accessed on 6 April 2021).

34. Forkenbrock, G.J.; O'hara, B. A Forward Collision Warning (FCW) Performance Evaluation. In *Proceedings of the International Technical Conference on the Enhanced Safety of Vehicles*; Stuttgart, Germany, 15–18 June 2009, National Highway Traffic Safety Administration: Washington, DC, USA, 2009; Volume 2009.

35. Transportation Industry Standard of the People's Republic of China. In *JT/T 883-2014: Commercial Vehicle Driving Dangerous Warning System Technical Requirements and Test Procedures*; China Communications Press: Beijing, China, 2014.

Article

Small Object Detection in Traffic Scenes Based on Attention Feature Fusion

Jing Lian, Yuhang Yin, Linhui Li *, Zhenghao Wang and Yafu Zhou

Faculty of Vehicle Engineering and Mechanics, School of Automotive Engineering, Dalian University of Technology, Dalian 116024, China; lianjing@dlut.edu.cn (J.L.); yinyuhang@mail.dlut.edu.cn (Y.Y.); zhwangv@mail.dlut.edu.cn (Z.W.); dlzyf@dlut.edu.cn (Y.Z.)
* Correspondence: lilinhui@dlut.edu.cn

Abstract: There are many small objects in traffic scenes, but due to their low resolution and limited information, their detection is still a challenge. Small object detection is very important for the understanding of traffic scene environments. To improve the detection accuracy of small objects in traffic scenes, we propose a small object detection method in traffic scenes based on attention feature fusion. First, a multi-scale channel attention block (MS-CAB) is designed, which uses local and global scales to aggregate the effective information of the feature maps. Based on this block, an attention feature fusion block (AFFB) is proposed, which can better integrate contextual information from different layers. Finally, the AFFB is used to replace the linear fusion module in the object detection network and obtain the final network structure. The experimental results show that, compared to the benchmark model YOLOv5s, this method has achieved a higher mean Average Precison (mAP) under the premise of ensuring real-time performance. It increases the mAP of all objects by 0.9 percentage points on the validation set of the traffic scene dataset BDD100K, and at the same time, increases the mAP of small objects by 3.5%.

Keywords: traffic scenes; object detection; multi-scale channel attention; attention feature fusion

Citation: Lian, J.; Yin, Y.; Li, L.; Wang, Z.; Zhou, Y. Small Object Detection in Traffic Scenes Based on Attention Feature Fusion. *Sensors* **2021**, *21*, 3031. https://doi.org/10.3390/s21093031

Academic Editor: Chao Huang

Received: 25 March 2021
Accepted: 19 April 2021
Published: 26 April 2021

Publisher's Note: MDPI stays neutral with regard to jurisdictional claims in published maps and institutional affiliations.

Copyright: © 2021 by the authors. Licensee MDPI, Basel, Switzerland. This article is an open access article distributed under the terms and conditions of the Creative Commons Attribution (CC BY) license (https://creativecommons.org/licenses/by/4.0/).

1. Introduction

In traffic scenes, the visual perception technology of intelligent vehicles can help automatic driving systems to perceive complex environments accurately and in time, which is a requirement for avoiding collisions and for safe driving. With the rapid development of computer vision technology, vehicle visual perception is increasingly being adopted in the field of automatic driving. For example, object detection based on deep learning has played a very important role in the field of automatic driving.

Object detection involves the delineation of the bounding box of an object to be detected in the given image, and then the determination of the class that the object in the box belongs to. Due to their large amount of calculations, redundant marker boxes, and poor robustness of manual features, traditional object detection algorithms are currently being replaced by their deep learning counterparts. Lightweight real-time object detection models, such as the "you only look once" (YOLO) algorithm [1–3], the single shot multibox detector (SSD) algorithm [4], Light-Head R-CNN [5], and ThunderNet [6], have already demonstrated good detection effects in actual application scenarios.

At present, the prevailing deep learning-based object detection algorithms, such as YOLOv5 [7], treat each region of the whole feature map equally by default, that is, each region has the same contribution to the final detection result. This means that they do not weigh the convolution features extracted from the network according to their position and importance. However, compared with simple ordinary scenes, there are usually more complex and rich semantic features around the object to be detected in actual traffic scenes. If the features of the object area are weighted according to their importance, the objects to

117

be detected can be better positioned in the feature map and the detection accuracy and generalization ability of the model can be improved.

Furthermore, in traffic scenes, there are many small objects in the distance. These objects offer limited feature information due to their relatively small size, which makes detection more difficult. Research on small object detection includes a deconvolutional single shot detector (DSSD) [8], scale normalization for image pyramids (SNIP) [9], high-resolution detection network (HRDNet) [10], etc. The DSSD algorithm mainly improves the detection performance of the object detector for small objects by using a better feature extraction network and adding context information. The SNIP algorithm uses a novel training scheme, called scale normalization for image pyramids (SNIP), which selectively back-propagates the gradients of object instances of different sizes as a function of the image scale to better detect small objects. The HRDNet algorithm feeds high-resolution input into a shallow network to reserve more positional information while feeding low-resolution input into a deep network to extract more semantics. By extracting various features from high to low resolutions, the algorithm improves the detection performance of small objects as well as maintaining the detection performance of medium and large objects. These algorithms each have their own advantages and limitations. Improving the detection of small objects in traffic scenes as much as possible is also one of the current research hotspots in the field of visual perception for autonomous vehicles. The YOLOv5 model is a milestone object detection method, which achieves a good balance between accuracy and speed, but it still has the possibility for improvement in small object detection problems in traffic scenes.

In response to the above problems, in this paper, we first propose an MS-CAB to alleviate the problems caused by scale changes to small object detection. This block effectively improves the feature inconsistency between objects at different scales, and at the same time, focuses attention on the objects in the area that need to be focused on, which reduces the unnecessary shallow feature information of the background. In other studies [11,12], the attention mechanism also considers the scale, such as by aggregating contextual information through convolution kernels of different sizes or from the feature pyramid inside the attention module. The MS-CAB proposed here aggregates contextual information along the channel dimensions of the feature map. It can not only focus on large objects that are distributed globally, but also deal with small objects that are distributed more locally. This block helps the model to detect and identify objects with extreme size differences.

Second, based on MS-CAB, an AFFB is proposed that is different from linear fusion schemes such as addition and concatenation, which are completely context-independent. The block is non-linear and can better capture the contextual information from different network layers by fusing features that are inconsistent semantically and in terms of scale. By replacing the simple addition or concatenation operation with the AFFB, a network model with fewer parameters and higher detection accuracy can be obtained, and the detection effect of small objects is improved greatly.

The remainder of this paper is organized as follows: Section 2 introduces the related works and existing problems of the three topics of object detection, attention mechanisms, and feature fusion. Section 3 briefly introduces the benchmark model, YOLOv5s, and then elaborates on the principle and structure of the proposed MS-CAB and the AFFB. Section 4 presents the experiments and an analysis of the results. The paper ends with our conclusions and suggestions for future work.

2. Related Works

2.1. Object Detection

Object detection algorithms are mainly divided into one-stage and two-stage methods. Relatively speaking, one-stage object detection algorithms have better real-time performance, but lower accuracy, while two-stage algorithms have better accuracy, but weaker real-time performance. He et al. proposed a two-stage spatial pyramid pooling network

(SPPNet) in 2014 [13]. By introducing a spatial pyramid pooling layer, the convolutional neural network (CNN) can receive inputs of non-fixed size without considering the size of the region of interest. The SPPNet method was ultimately 20 times faster than R-CNN [14], with comparable accuracy. Ren et al. proposed Faster R-CNN [15], and the region proposal network (RPN) candidate box generation algorithm based on Fast R-CNN [16], which greatly improved the speed of object detection. Besides, Lin et al. proposed feature pyramid networks (FPN) [17], which solved the multi-scale problem in object detection. Through a relatively simple network connection change, the detection effect of small objects is greatly improved while maintaining the original model's computational load. The YOLO algorithm [1], which divides the image into multiple regions, formulates the bounding box, and predicts the probability of an object belonging to a class at the same time, was proposed by Redmon et al. It was the first one-stage object detection algorithm based on deep learning and started a new approach towards object detection. The author subsequently proposed the improved versions of YOLOv2 [2] and YOLOv3 [3], which further improved the detection accuracy while maintaining a relatively high detection speed. Then, Liu et al. proposed the SSD algorithm [4], which greatly improved the accuracy of object detection by introducing multi-reference and multi-resolution detection technology, especially for small objects.

To solve the problem of imbalance between positive and negative categories, Lin et al. proposed the RetinaNet algorithm [18], in which the focal loss is derived so that the algorithm can maintain a relatively fast detection speed, while the detection accuracy can be equivalent to that of two-stage object detection algorithms. Zhu et al. proposed the feature selective anchor-free (FSAF) module [19], which can be inserted into a one-stage detector with a feature pyramid structure to enhance the decision feature layer to which each input instance belongs to make full use of the performance of FPN, and this method has a high mAP value and little additional computation. Zhou et al. proposed CenterNet [20], which uses the object center point predicted by the heatmap instead of the anchor mechanism to predict the object and uses a higher-resolution output feature map. This network has strong scalability and simple model design, and thus achieves good results in detection speed and accuracy. Tan et al. proposed EfficientDet [21], which is a weighted bi-directional feature pyramid network (BiFPN) and a composite scale expansion method to refresh the mAP of the MS COCO dataset. In the above works, the detection accuracy of the object detection algorithms was improved to varying degrees. However, it is more important to make full use of the effective information of the input features to improve the detection performance of the model, especially the detection accuracy of small objects in traffic scenes while keeping the number of model parameters and the real-time performance of the model basically unchanged.

2.2. Attention Mechanism

When facing the external environment, the human visual system can quickly identify useful information and ignore irrelevant information. This characteristic is gradually being considered by computer vision researchers. Deep learning's attention mechanism first appeared as an imitation of the human visual attention mechanism [22]. Non-local neural networks were proposed by Wang et al. in one of the important works on attention mechanisms in the field of computer vision [23]. Non-local operations calculate the response at a position as a weighted sum of the features at all positions and establish remote dependencies through self-attention, and they can also be used as general modules for various tasks, which can lead to improvements in the model accuracy. The squeeze-and-excitation network (SENet) [24] proposed by Hu et al. was the first attention mechanism that focused on the channel level dependencies of the model, and could adaptively adjust the characteristic response value of each channel. This network won the ImageNet 2017 classification competition and has been recognized as an important advancement in the field. Woo et al. proposed the convolutional block attention module (CBAM) [25], which contains two modules of channel attention and spatial attention so that the model has better performance and interpretability and pays more attention to foreground objects. The selective kernel network (SKNet) was proposed by Li et al. [26], which utilizes a building block called a selection kernel unit that allows each neuron to adaptively adjust the size of the receptive field, depending on the scale of the input information. Experiments showed that SKNet achieved better detection accuracy through its relatively low model complexity.

Roy et al. proposed spatial and channel squeeze-and-excitation (scSE) [27] for semantic segmentation. They proposed three variants of the squeeze-and-excitation (SE) module, channel squeeze-and-excitation (cSE), spatial squeeze-and-excitation (sSE), and scSE, as improvements of the SE module. Experiments have shown that these modules can enhance useful features and suppress useless ones. Combining the advantages of non-local neural networks and SENet, Cao et al. proposed the global context network (GCNet) [28], which uses a relatively small amount of calculations to optimize the global context modeling capabilities. Huang et al. proposed the criss-cross network (CCNet) [29], which was also based on Non-local Neural Networks. Its special feature is the novel criss-cross attention module, which can obtain contextual information from remote dependencies in a more effective way. The dual attention network (DANet) was proposed by Fu et al. [30], which adds two attention modules to a dilated fully convolutional network to model semantic dependencies in the spatial and the channel dimensions. This model achieved excellent results on the semantic segmentation dataset, Cityscapes.

Most of the above-mentioned attention mechanisms use global channel attention mechanisms, which are more suitable for the detection of large objects with a more global distribution. However, the scale range of objects is very large in actual traffic scenes. If only the contextual information is extracted from the global range, the detection effect of the model is better for large objects with more distribution in the global range, but will be weaker for small objects with more distribution in the local range. Therefore, a simplified multi-scale channel attention block composed of local channel attention and global channel attention is needed to adaptively extract contextual object information to improve the detection effect of small objects.

2.3. Feature Fusion

In many object detection tasks, the fusion of features at multiple scales is an important way to improve detection performance. Low-level object features have high resolution and usually contain more location and detail information, but they lack semantic information and have more noise. High-level features have richer semantic information after the convolution operation, but their resolution is reduced, and the location and detail information are lacking. Efficient integration of low-level and high-level features is key to improving a model's detection performance. Depending on the sequence of feature fusion and prediction, feature fusion can be divided into early fusion and late fusion methods. Early fusion fuses features of different layers first and then trains predictors on the fused features, such as the addition operation in ResNet [31] and the concatenation operation in U-Net [32]. Late fusion improves the detection performance by combining the detection results of different layers, and can be mainly divided into two types. The first separately predicts the features of multiple scales before fusion, and then the obtained prediction results are processed comprehensively, such as in SSD [4], multi-scale CNN [33], etc. The second approach uses the idea of feature pyramid networks for reference, and then predicts after fusing the features, such as in YOLOv3 [3], feature fusion single shot multibox detector (FSSD) [34], etc.

The feature fusion problem is currently a research hotspot in the field of object detection. Chaib et al. improved the effect of feature fusion using a discriminant correlation analysis-based feature fusion strategy [35], which incurred only a small computational cost. The FSSD was proposed by Li et al. [34], which includes a feature fusion module. The module first fuses the features of different layers through concatenation operations to obtain a larger-scale feature, and then a feature pyramid is constructed on this feature map. This significantly improves the detection accuracy of the SSD model, with only a slight speed reduction. Lim et al. proposed the SSD with feature fusion and attention (FA-SSD) [36], which includes a feature fusion module and an attention module. The results showed that the network improved the accuracy of object detection, especially the detection performance of small objects.

Pang et al. proposed Libra R-CNN [37], which integrates features from different layers to obtain more balanced semantic feature information. Compared with [15] and [18], the detection effect on the MS COCO dataset was significantly improved. Ghaisi et al. proposed the neural architecture search feature pyramid network (NAS-FPN) [38], which uses a neural architecture search algorithm to customize a feature pyramid network that merges features across a range. This approach produced significant improvements in many object detection networks. An adaptive spatial feature fusion (ASFF) strategy was proposed by Liu et al. [39], which combines features of different layers by learning weight parameters. Experimental results showed that this method was superior to concatenation and element-wise methods. In addition to the feature fusion using deep learning technology, Gao et al. analyzed the limitations of only using deep learning methods, and proposed a new fusion logic that can effectively combine the advantages of known knowledge used by a traditional method with the self-extracted features learned by a deep learning method [40]. A better detection performance can be achieved by properly designing traditional and deep learning detectors. However, the above methods of feature fusion are biased towards constructing complex paths to combine the features of different network layers or groups. They are all too complicated. Therefore, we propose an AFFB with a simple structure to improve the integration of various object context features in traffic scenes using fewer parameters and smaller models to ultimately improve the network's object detection performance, especially the detection accuracy of small objects.

3. Benchmark Model and Proposed Methods

In this section, we briefly introduce the benchmark model YOLOv5s, then elaborate on the principle and structure of the proposed MS-CAB, and finally present the AFFB based on MS-CAB.

3.1. The YOLOv5s Benchmark Model

The development of the YOLO series ushered in a change in object detection technology through the adoption of deep learning. At present, the YOLO series includes YOLOv1 [1], YOLOv2 [2], YOLOv3 [3], YOLOv4 [41], and YOLOv5 [7]. The YOLOv5 model is the latest iteration of the model, and constitutes an improvement over YOLOv4. The model is faster, more accurate, has fewer model parameters, and can be more easily adapted to various devices embedded in vehicles. The YOLOv5 model refers to four models of different sizes, namely, YOLOv5s, YOLOv5m, YOLOv5l, and YOLOv5x, where smaller models have fewer parameters, lower accuracy, and are faster. To better meet the real-time requirements of object detection in traffic scenes, in this study, we chose the YOLOv5s model as the benchmark model for improvement.

3.2. Multi-Scale Channel Attention Block

Based on the idea of combining local and global features in the convolutional neural networks adopted in ParseNet [42] and multi-scale channel attention [43], we propose MS-CAB, with the main difference being that we use 1×1 convolution rather than kernels of different sizes to control the channel attention scale. Similar to spatial attention, channel attention also has a scale, and the variable that controls that scale is the size of the pooling. Figure 1 shows a diagram of the MS-CAB structure, which is divided into two scales, the local scale and the global scale, where context features are aggregated through both scales. The branch that uses global average pooling is the global scale, while the other is the local scale. This block gathers contextual information along the channel dimension of the feature map, and can simultaneously focus on large objects that are more distributed in the global range and small objects that are distributed more in the local range, which helps the model to detect and identify objects with extreme scale changes in traffic scenes. In the following, we introduce the details of the implementation of the proposed MS-CAB.

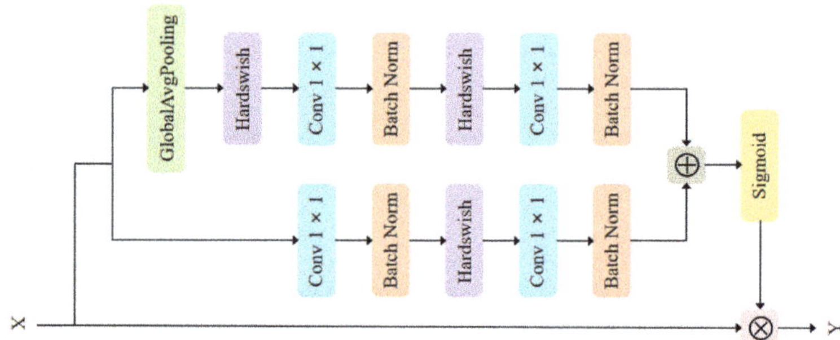

Figure 1. The MS-CAB structure. The global average pooling branch is the global channel attention, while the other is the local channel attention.

Suppose that the output of a certain layer in the middle of the network is X and $X \in R^{C \times H \times W}$, where C is the channel number of the feature map, and H and W are the height and width of the feature map, respectively. Then, X is used as the input of MS-CAB. The global and local channel attention can be obtained by changing the pooling size, and 1×1 convolution is used as the local channel context aggregator to extract the channel interaction at each spatial location. The local channel context $L(X) \in R^{C \times H \times W}$ can be expressed as

$$L(X) = BN(Conv_2(Hs(BN(Conv_1(X))))),\tag{1}$$

where the convolution kernel parameters of $Conv_1$ and $Conv_2$ are $\frac{C}{r} \times C \times 1 \times 1$ and $C \times \frac{C}{r} \times 1 \times 1$, r is the channel reduction ratio, BN stands for batch normalization [44], and Hs stands for the Hardswish activation function [45]. The local channel context $L(X)$ has the same shape as the input feature map X, and retains and highlights the richly detailed information of the low-level features. It focuses more on the small object information present in the local range.

The global channel context $G(X) \in R^{C \times 1 \times 1}$ can be expressed as

$$G(X) = BN(Conv_2(Hs(BN(Conv_1(Hs(g(X))))))),\tag{2}$$

$$g(X) = \frac{1}{H \times W}\sum_{i=1}^{H}\sum_{j=1}^{W}X_{[:,i,j]}\tag{3}$$

where $g(X) \in R^C$ stands for global average pooling. Here, $G(X)$ has the same number of channels as the input feature map X and pays more attention to large object information that is distributed more globally.

Combining the local channel context $L(X)$ and the global channel context $G(X)$, the output $Y \in R^{C \times H \times W}$ of the MS-CAB can be expressed as follows:

$$Y = X \otimes MSCAB(X) = X \otimes \sigma(L(X) \oplus G(X))\tag{4}$$

where $MSCAB(X) \in R^{C \times H \times W}$ represents the output weight of the MS-CAB, σ represents the sigmoid function, $\otimes$ represents element-wise multiplication, and $\oplus$ represents the addition of the broadcast mechanism.

The proposed MS-CAB was embedded in the four Concat operation branches of the YOLOv5s model, and a new network model, MS-CAB_YOLOv5s, was obtained. The network structure diagram is shown in Figure 2. In the diagram, "Input" refers to the network input, and "Prediction" is the prediction result made by the network on the feature map on three scales. "Upsample" represents an upsampling operation, "Concat" denotes a concatenation operation, and "Conv" denotes a convolution operation. The composition of the "Focus" block is shown in Figure 3. It performs a slicing operation on the input red/green/blue (RGB) image, ultimately integrating the width and height information into the channel dimension. Its main function is to reduce floating point operations and improve the running speed of the model. The CBL block is composed of a convolution layer, batch normalization, and the Hardswish activation function, and its composition is shown in Figure 4. The YOLOv5s model contains two cross stage partial (CSP) structures [46], of which the CSP1 structure is used in the backbone of the network, while the CSP2 structure is used in the neck of the network. The composition of CSP1_X is shown in Figure 5. Here, CSP1_X indicates that it contains X residual units; for example, CSP1_1 contains one residual unit, and CSP1_3 contains three residual units. The composition of each residual unit is shown in Figure 6. The composition of CSP2_X is shown in Figure 7. Here, CSP2_X means that, in addition to the first CBL component, there are $2 \times X$ CBL components in the middle. The size of the convolution kernel in the first CBL component is 1×1, while in the second CBL component it is 3×3. For example, in addition to the first CBL component in CSP2_1, there are $2 \times 1 = 2$ CBL components in the middle, and the convolution kernel sizes in the two CBL components are 1×1 and 3×3, respectively. The SPP block uses the maximum pooling method to perform "Concat" operations on feature maps of different scales, and its composition is shown in Figure 8.

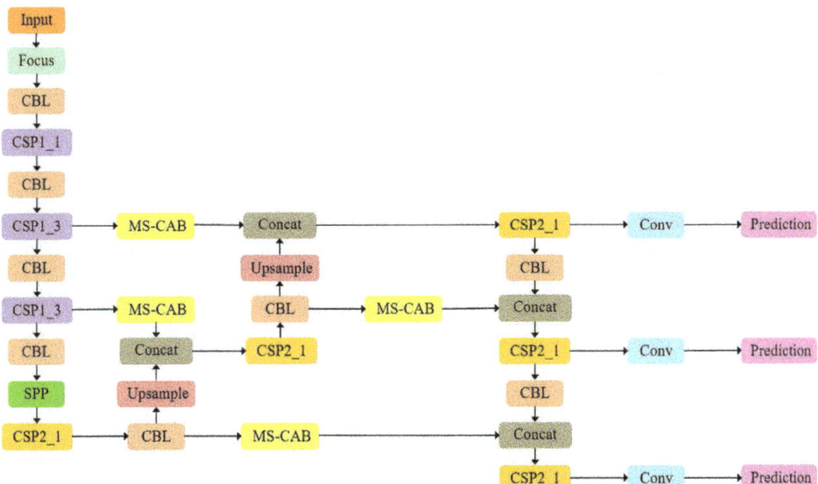

Figure 2. The MS-CAB_YOLOv5s network structure.

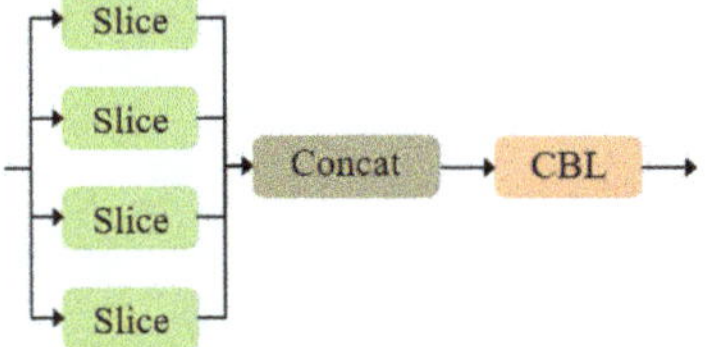

Figure 3. Composition of the "Focus" block.

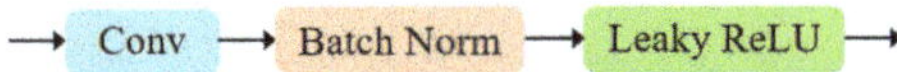

Figure 4. Composition of the CBL block.

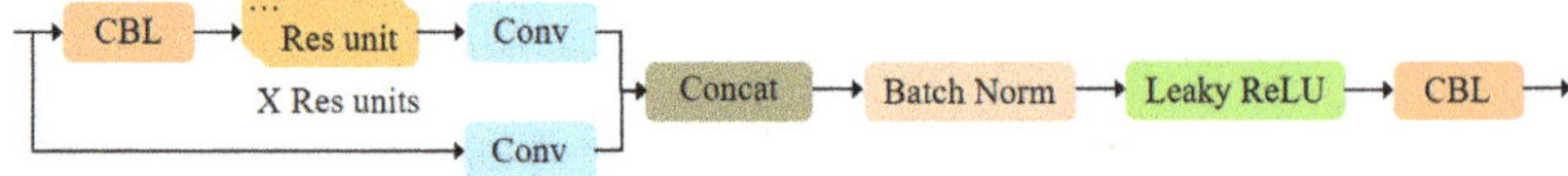

Figure 5. Composition of the CSP1_X block.

Figure 6. Composition of the residual unit block.

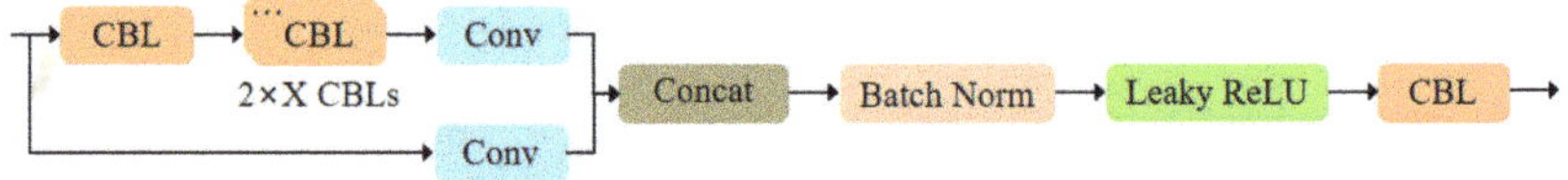

Figure 7. Composition of the CSP2_X block.

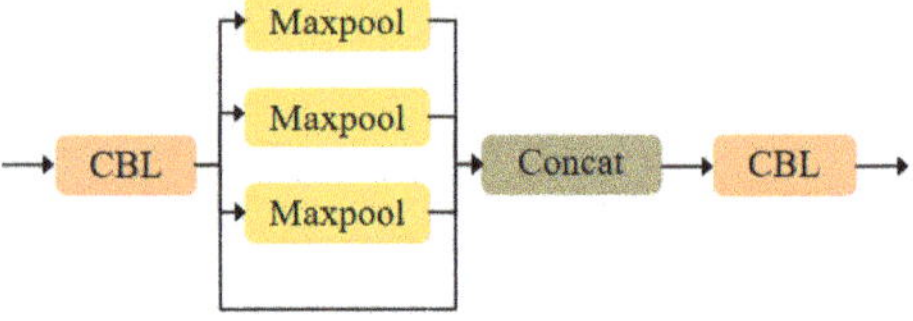

Figure 8. Composition of the SPP block.

3.3. Attention Feature Fusion Block

In combination with the multi-scale channel attention block proposed above, we propose AFFB, which can better capture contextual information from different network layers by fusing semantic and scale-inconsistent features and thus achieve better object detection. Figure 9 is a structure diagram of the AFFB. Due to the presence of the multi-scale channel attention block, the output $Z \in R^{C \times H \times W}$ of the AFFB can be expressed as

$$Z = MSCAB(X_1 \oplus X_2) \otimes X_1 + (1 - MSCAB(X_1 \oplus X_2)) \otimes X_2 \tag{5}$$

where $X_1 \in R^{C \times H \times W}$ and $X_2 \in R^{C \times H \times W}$ are two input feature maps, with X_1 being a low-level semantic feature map and X_2 a high-level semantic feature map. The values of the fusion weights $MSCAB(X_1 \oplus X_2)$ and $1 - MSCAB(X_1 \oplus X_2)$ are both between 0 and 1, which corresponds to a weighted averaging operation between X_1 and X_2.

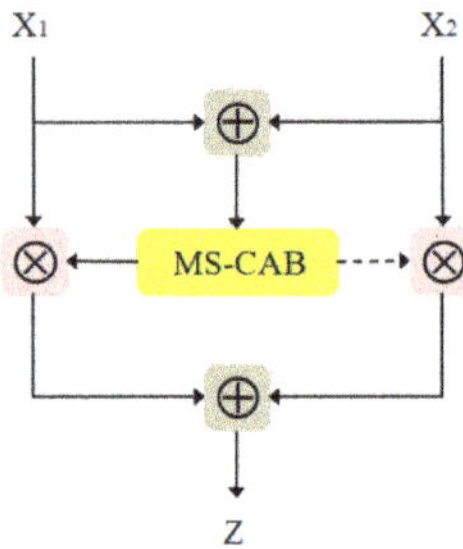

Figure 9. The AFFB structure.

In YOLOv5s, linear feature fusion is performed through concatenation, which only yields a fixed linear aggregation of feature maps, and is not adaptable to the object to be detected. The AFFB has fewer parameters, is non-linear, and can capture the contextual information from different network layers better through the fusion of features that are inconsistent semantically and in terms of scale. The four "Concat" operations are then replaced in the YOLOv5s model with the proposed AFFB to obtain a new network model AFFB_YOLOv5s, as shown in Figure 10.

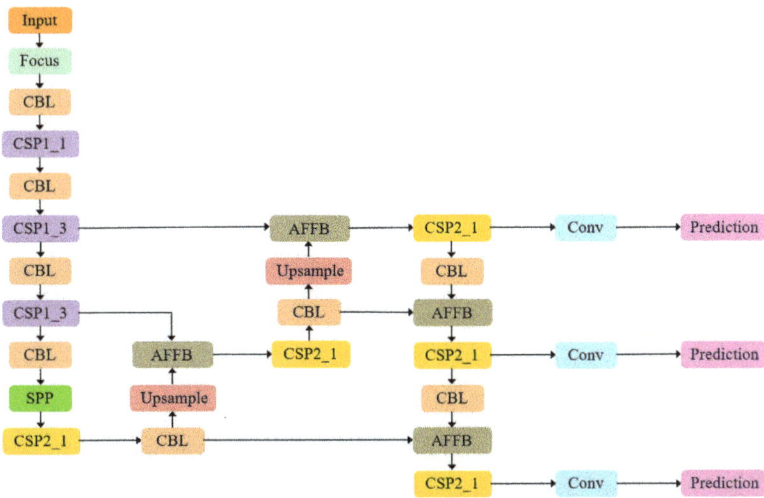

Figure 10. The AFFB_YOLOv5s network structure.

4. Experiments and Result Analysis

4.1. Datasets and Experimental Settings

4.1.1. Datasets

In this paper, the object detection task is oriented towards traffic scenes, and thus the experimental part mainly used the BDD100K dataset [47], while the PASCAL VOC dataset [48] was used as an auxiliary validation dataset.

The BDD100K dataset is the largest open autonomous driving dataset, and includes ten categories of traffic scene objects: car, bus, person, bike, truck, motor, train, rider, traffic sign, and traffic light. It has a very rich diversity of geography, environments, and weather to enable models to recognize a variety of complex traffic scenes and make the models' generalization ability stronger at the same time. The dataset has a total of 100,000 images with a resolution of 1280×720 pixels. The official usage guidelines recommend splitting the dataset into a training set, a validation set, and a test set at a 7:1:2 ratio. As the labels of the test set are not disclosed, we used the validation set to test the model and evaluate the model's detection performance of the model. The final training set consisted of 70,000 images, and the test set consisted of 10,000 images. (The BDD100K dataset is available at https://bdd-data.berkeley.edu, accessed on 25 November 2020).

The PASCAL VOC dataset is a commonly used object detection dataset, and it includes two parts, VOC2007 and VOC2012, with a total of 20 categories: airplane, bicycle, bird, boat, bottle, bus, car, cat, chair, cow, dining table, dog, horse, motorbike, person, potted plant, sheep, sofa, train, and TV monitor. In this paper, 22,136 images of the VOC2007 and VOC2012 training and validation sets were used for model training. The test set of VOC2007 has a total of 4952 images and was used to evaluate the detection performance of the model. (The PASCAL VOC dataset is available at http://host.robots.ox.ac.uk/pascal/VOC/, accessed on 30 November 2020).

4.1.2. Experimental Settings

(a)   Network loss function

The loss function of the network designed in this paper is divided into three parts: bounding box regression loss L_{box}, confidence loss L_{obj}, and classification loss L_{cls}. The total loss of the network is the sum of the three functions. The bounding box regression loss uses the complete intersection over union (CIoU) loss [49], and both the confidence loss and classification loss use the binary cross-entropy (BCE) with logits loss (BCEWithLogitsLoss). The CIoU loss considers three important geometric factors of the bounding box regression loss: the overlap area between the prediction and the ground truth boxes; the center point distance of the prediction and the ground truth boxes; and the aspect ratio between the prediction and the ground truth boxes, which improves the speed and accuracy of bounding box regression. The bounding box regression loss L_{box} can be expressed as follows:

$$L_{\text{box}} = 1 - CIoU = 1 - \left(IoU - \frac{\rho^2}{c^2} - \alpha v \right) \tag{6}$$

where intersection-over-union (IoU) is the ratio of the intersection area to the union area of the prediction box and the ground truth box, ρ is the Euclidean distance between the center points of the prediction and the ground truth boxes, and c is the diagonal length of the smallest enclosing box covering both the prediction box and the ground truth box. Besides, α is the trade-off parameter, which is defined as

$$\alpha = \frac{v}{(1 - IoU) + v} \tag{7}$$

here, v is a parameter that measures the consistency of the aspect ratio between the ground truth box and the prediction box, and it is expressed as follows:

$$v = \frac{4}{\pi^2} \left(arctan \frac{w^{gt}}{h^{gt}} - arctan \frac{w^p}{h^p} \right)^2 \tag{8}$$

where w^{gt} and h^{gt} are the width and height of the ground truth box, while w^p and h^p are the corresponding values of the prediction box.

The BCEWithLogitsLoss mainly measures the binary cross-entropy between the target value and the output value of the model. It can be expressed as

$$L_n = -w_n [y_n log\sigma(x_n) + (1 - y_n)log(1 - \sigma(x_n))] \tag{9}$$

where w_n is the loss weight of each category, y_n is the target value, x_n is the output value of the model, and σ is the sigmoid function.

(b)   Training parameter settings

In this study, we used the stochastic gradient descent algorithm [50] to optimize the loss function. The momentum was set to 0.937, the weight decay coefficient was set to 0.0005, and the initial learning rate was set to 0.01. We used warmup training [51], cosine annealing [52], gradient accumulation, exponential moving average, and other optimization strategies. In terms of data augmentation, in addition to the most advanced mosaic data augmentation method [41], common data augmentation methods, such as random hue, saturation, value transformation, image horizontal and vertical translation, image scaling, and image left and right flip, were also used. The batch size was set to 32, the epochs were set to 300, and the resolution size of the input image was set to 640 × 640. The channel reduction ratio r was set to 4. The k-means clustering algorithm was used to obtain new anchor boxes. Other parameter settings were consistent with the default settings of YOLOv5. The computer configuration used in the experiment is shown in Table 1.

Table 1. Computer configuration.

Project	Content
CPU	Intel Xeon E5-2620 v4
RAM	32GB
GPU	NVIDIA TITAN Xp
Operating System	Ubuntu 18.04.5 LTS
Cuda	Cuda 10.1 with Cudnn 7.5.1
Data Processing	Python 3.8, OpenCV
Deep Learning Framework	Pytorch 1.7.0

(c)   Testing parameter settings

The batch size was set to 1, the resolution size of the input image was set to 640×640, the confidence threshold for the filtering prediction box was set to 0.001, and the IoU threshold for non-maximum suppression was set to 0.6. Other parameter settings were consistent with the default YOLOv5 settings.

4.2. Quantitative Result Analysis

The three models, YOLOv5s, MS-CAB_YOLOv5s, and AFFB_YOLOv5s, were trained on the BDD100K dataset to test the effectiveness of the proposed MS-CAB and AFFB blocks. Five indicators commonly used in the field of object detection, namely, precision, recall, mAP, frames per second (FPS), and the number of parameters, were used to quantitatively evaluate the accuracy of the model [7]. To quantitatively study the impact of the proposed improvements on the detection of small objects, we examined small objects of the size defined by the COCO dataset [53], that is, those with a pixel area smaller than 32×32 pixels. Moreover, to verify the generalization ability of the model on other datasets, we used the same parameter settings as above on the public dataset PASCAL VOC for network training, and then tested to complete the auxiliary validation.

The accuracy evaluation results of the three models on the BDD100K validation set are shown in Table 2. It is evident that under the premise of ensuring the real-time requirements of a vehicle's environment perception, compared with the original YOLOv5s model, the precision, recall, and mAP of the MS-CAB_YOLOv5s and AFFB_YOLOv5s models proposed in this paper were improved to varying degrees. Among them, the mAP of the AFFB_YOLOv5s model increased by 0.9 percentage points, which is a significant improvement given the complexity of the BDD100K traffic scene dataset. The 63 FPS achieved by both improved networks can fully meet the real-time requirements of vehicles' environment perception systems. Furthermore, the parameters of the model were reduced to a certain extent. The size of the model is only 14.7 MB, which makes it quite suitable for embedded vehicle platforms.

Table 2. Model performance comparison on the BDD100K validation set.

Model	Precision (%)	Recall (%)	mAP (%)	FPS	Parameters (M)
YOLOv5s	32.5	57.7	50.6	77	7.28
MS-CAB_YOLOv5s	32.5	58.1	51.0	63	7.45
AFFB_YOLOv5s	33.0	58.3	51.5	63	7.20

The BDD100K dataset is a traffic scene dataset, and thus contains many cars and traffic signs at a distance with a pixel area less than 32×32 pixels. These objects are defined as small objects that need to be detected. Table 3 shows the comparison results of the three models for small object detection performance. Compared with the original YOLOv5s model, the MS-CAB_YOLOv5s and AFFB_YOLOv5s models proposed in this paper had a significantly improved precision of small object detection, while the recall decreased slightly, and the mAP, respectively, improved by 1.6 and 3.5 percentage points.

This shows that the MS-CAB and AFFB significantly improved the model's detection effect on small objects.

Table 3. Comparison of models on small object detection performance.

Model	Precision (%)	Recall (%)	mAP (%)
YOLOv5s	11.7	51.9	21.5
MS-CAB_YOLOv5s	16.4	49.8	23.1
AFFB_YOLOv5s	23.1	48.6	25.0

To verify the generalization ability of the model, the three models were trained and tested on the PASCAL VOC dataset. The performance comparison for each model is shown in Table 4. Under the premise of ensuring real-time performance, the two models, MS-CAB_YOLOv5s and AFFB_YOLOv5s, had improved precision, recall, and mAP. This again verifies the effectiveness of the MS-CAB and AFFB to improve the performance of object detection. At the same time, it shows that our improved model can adapt to different datasets or scenes and has good generalization ability.

Table 4. Performance comparison of models on PASCAL VOC test set.

Model	Precision (%)	Recall (%)	mAP (%)	FPS	Parameters (M)
YOLOv5s	60.3	82.3	79.4	76	7.31
MS-CAB_YOLOv5s	62.0	82.7	80.2	61	7.48
AFFB_YOLOv5s	63.4	82.9	80.8	61	7.23

4.3. Comparative Analysis of Detection Results

Figure 11 shows a visual comparison of the detection results of the YOLOv5s model, the MS-CAB_YOLOv5s model, and the AFFB_YOLOv5s model. To see the differences between the three models more easily, the yellow rectangles in the detection result of column (a) in Figure 11 indicate the objects that were not detected by YOLOv5s. Similarly, the yellow rectangles in the detection result of column (b) indicate the objects that were not detected by MS-CAB_YOLOv5s. The AFFB_YOLOv5s model could detect small objects with small pixel areas, such as cars, people, and traffic signs, at long distances that were not detected by the YOLOv5s model. At the same time, the detection effect was also excellent under dark night conditions. Moreover, compared with the benchmark model YOLOv5s, the detection effect of the MS-CAB_YOLOv5s model was better. It could detect some objects that the YOLOv5s model did not detect, but its effect was not as good as that of AFFB_YOLOv5s. For example, in column (b) of Figure 11, the person on the left side of the figure on the second row and the traffic sign on the right side of the figure on the third row were not detected by the MS-CAB_YOLOv5s model, but they were all accurately detected by the AFFB_YOLOv5s model. Based on these detection results in Figure 11, both the MS-CAB_YOLOv5s model and the AFFB_YOLOv5s model could improve the effect of object detection in traffic scenes, and the AFFB_YOLOv5s model had the best detection effect, especially for small objects that are away from the vehicle, which is of great significance for improving the stability and efficiency of automatic driving systems and preventing traffic accidents.

(a) YOLOv5s   (b) MS-CAB_YOLOv5s   (c) AFFB_YOLOv5s

Figure 11. Comparison of the detection results of YOLOv5s, MS-CAB_YOLOv5s, and AFFB_YOLOv5s.

5. Conclusions and Future Work

The high accuracy and fast real-time performance of object detection algorithms are very important for the safety and real-time control of autonomous vehicles. In this paper, we presented a small object detection method for traffic scenes based on attention feature fusion for autonomous driving systems as an improvement to the YOLOv5s architecture. To aggregate the effective information at the local and global scales, MS-CAB simultaneously focuses on small objects that are more distributed within a local range and large objects that are more distributed on the global range. Using AFFB to fuse contextual information from different network layers, we obtain a model with fewer parameters and higher accuracy. Under the condition of meeting the real-time requirements of vehicles' environment perception systems, compared with the benchmark model YOLOv5s, the model proposed in this paper increased the mAP of all objects on the validation set of the traffic scene dataset BDD100K by 0.9 percentage points. Specifically, small objects' mAP was increased by 3.5%. Therefore, the model achieves a better balance between object detection accuracy and speed in traffic scenes, and can effectively improve the performance of vision-based object detection systems for autonomous vehicles.

Since our proposed method is essentially based on deep learning, there are some general limitations. First, the interpretability of deep learning is poor. It learns the implicit relationship between input and output features, but not the causal relationship. Secondly, the neural network has many parameters, and network training requires a large amount of time and relatively large computing power. Therefore, the deep learning method requires stronger computer hardware equipment. Finally, the accuracy of the model based on the deep learning method greatly relies on the collected data, and the accuracy of the dataset label directly determines the accuracy of the model detection. A traditional method based on manual feature extraction is a beneficial supplement to the deep learning method. In future research, we will try to combine the two methods to further improve object detection performance. We plan to deploy the model proposed in this paper to embedded vehicle devices to develop more convenient portable applications. Moreover, we will explore the extent to which the proposed blocks improve the performance of larger YOLOv5 models.

Author Contributions: Conceptualization, J.L. and Y.Y.; methodology, J.L.; software, J.L., Y.Y., and Z.W.; validation, L.L. and Y.Z.; investigation, L.L.; resources, Y.Z.; writing—original draft preparation, Y.Y.; writing—review and editing, Y.Y., L.L., and Z.W.; visualization, Y.Y.; supervision, Y.Z. All authors have read and agreed to the published version of the manuscript.

Funding: This work was supported by the National Natural Science Foundation of China (Grant Nos. 51775082, 61976039) and the China Fundamental Research Funds for the Central Universities (Grant Nos. DUT19LAB36, DUT20GJ207), and Science and Technology Innovation Fund of Dalian (2018J12GX061).

Institutional Review Board Statement: Not applicable.

Informed Consent Statement: Not applicable.

Data Availability Statement: Not applicable.

Conflicts of Interest: The authors declare no conflict of interest.

References

1. Redmon, J.; Divvala, S.; Girshick, R.; Farhadi, A. You only look once: Unified, real-time object detection. In Proceedings of the 2016 IEEE Conference on Computer Vision and Pattern Recognition (CVPR), Las Vegas, NV, USA, 27–30 June 2016; pp. 779–788.
2. Redmon, J.; Farhadi, A. YOLO9000: Better, faster, stronger. In Proceedings of the 2017 IEEE Conference on Computer Vision and Pattern Recognition (CVPR), Honolulu, HI, USA, 21–26 July 2017; pp. 6517–6525.
3. Redmon, J.; Farhadi, A. YOLOv3: An incremental improvement. *arXiv* **2018**, arXiv:1804.02767.
4. Liu, W.; Anguelov, D.; Erhan, D.; Szegedy, C.; Reed, S.; Fu, C.Y.; Berg, A.C. SSD: Single shot multibox detector. In *Computer Vision-ECCV 2016*; Springer: Cham, Switzerland, 2016.
5. Li, Z.; Peng, C.; Yu, G.; Zhang, X.; Deng, Y.; Sun, J. Light-Head R-CNN: In defense of two-stage object detector. *arXiv* **2017**, arXiv:1711.07264.
6. Qin, Z.; Li, Z.; Zhang, Z.; Bao, Y.; Yu, G.; Peng, Y.; Sun, J. ThunderNet: Towards real-time generic object detection. *arXiv* **2019**, arXiv:1903.11752.
7. Jocher, G.; Stoken, A.; Borovec, J.; Changyu, L.; Hogan, A.; Diaconu, L.; Ingham, F.; Poznanski, J.; Fang, J.; Yu, L.; et al. YOLOv5. Available online: http://doi.org/10.5281/zenodo.4154370 (accessed on 16 November 2020).
8. Fu, C.; Liu, W.; Ranga, A.; Tyagi, A.; Berg, A.C. DSSD: Deconvolutional single shot detector. *arXiv* **2017**, arXiv:1701.06659.
9. Singh, B.; Davis, L.S. An Analysis of Scale Invariance in Object Detection—SNIP. In Proceedings of the 2018 IEEE/CVF Conference on Computer Vision and Pattern Recognition (CVPR), Salt Lake City, UT, USA, 18–23 June 2018; pp. 3578–3587.
10. Liu, Z.; Gao, G.; Sun, L.; Fang, Z. HRDNet: High-resolution detection network for small Objects. *arXiv* **2020**, arXiv:2006.07607.
11. Li, H.; Xiong, P.; An, J.; Wang, L. Pyramid attention network for semantic segmentation. *arXiv* **2018**, arXiv:1805.10180.
12. Wang, W.; Zhao, S.; Shen, J.; Hoi, S.C.H.; Borji, A. Salient Object Detection with Pyramid Attention and Salient Edges. In Proceedings of the 2019 IEEE/CVF Conference on computer vision and pattern recognition (CVPR), Long Beach, CA, USA, 15–20 June 2019; pp. 1448–1457.

13. He, K.; Zhang, X.; Ren, S.; Sun, J. Spatial pyramid pooling in deep convolutional networks for visual recognition. In *Computer Vision-ECCV 2014*; Springer: Cham, Switzerland, 2014; pp. 346–361.
14. Girshick, R.; Donahue, J.; Darrell, T.; Malik, J. Rich feature hierarchies for accurate object detection and semantic segmentation. In Proceedings of the 2014 IEEE Conference on Computer Vision and Pattern Recognition (CVPR), Columbus, OH, USA, 23–28 June 2014; pp. 580–587.
15. Ren, S.; He, K.; Girshick, R.; Sun, J. Faster R-CNN: Towards real-time object detection with region proposal networks. *IEEE Trans. Pattern Anal. Mach. Intell.* **2017**, *39*, 1137–1149. [CrossRef]
16. Girshick, R. Fast R-CNN. In Proceedings of the 2015 IEEE International Conference on Computer Vision (ICCV), Santiago, Chile, 7–13 December 2015; pp. 1440–1448.
17. Lin, T.; Dollár, P.; Girshick, R.; He, K.; Hariharan, B.; Belongie, S. Feature pyramid networks for object detection. In Proceedings of the 2017 IEEE Conference on Computer Vision and Pattern Recognition (CVPR), Honolulu, HI, USA, 21–26 July 2017; pp. 936–944.
18. Lin, T.; Goyal, P.; Girshick, R.; He, K.; Dollár, P. Focal loss for dense object detection. In Proceedings of the 2017 IEEE International Conference on Computer Vision (ICCV), Venice, Italy, 22–29 October 2017; pp. 2999–3007.
19. Zhu, C.; He, Y.; Savvides, M. Feature selective anchor-free module for single-shot object detection. In Proceedings of the 2019 IEEE/CVF Conference on Computer Vision and Pattern Recognition (CVPR), Long Beach, CA, USA, 15–20 June 2019; pp. 840–849.
20. Zhou, X.; Wang, D.; Krhenbühl, P. Objects as Points. *arXiv* **2019**, arXiv:1904.07850.
21. Tan, M.; Pang, R.; Le, Q.V. EfficientDet: Scalable and Efficient Object Detection. In Proceedings of the 2020 IEEE/CVF Conference on Computer Vision and Pattern Recognition (CVPR), Seattle, WA, USA, 13–19 June 2020; pp. 10778–10787.
22. Fan, D.; Wang, W.; Cheng, M.; Shen, J. Shifting more attention to video salient object detection. In Proceedings of the 2019 IEEE/CVF Conference on Computer Vision and Pattern Recognition (CVPR), Long Beach, CA, USA, 15–20 June 2019; pp. 8546–8556.
23. Wang, X.; Girshick, R.; Gupta, A.; He, K. Non-local neural networks. In Proceedings of the 2018 IEEE/CVF Conference on Computer Vision and Pattern Recognition (CVPR), Salt Lake City, UT, USA, 18–23 June 2018; pp. 7794–7803.
24. Hu, J.; Shen, L.; Sun, G. Squeeze-and-excitation networks. In Proceedings of the 2018 IEEE/CVF Conference on Computer Vision and Pattern Recognition (CVPR), Salt Lake City, UT, USA, 18–23 June 2018; pp. 7132–7141.
25. Woo, S.; Park, J.; Lee, J.; Kweon, I.S. CBAM: Convolutional Block Attention Module. In *Computer Vision-ECCV 2018*; Springer: Cham, Switzerland, 2018; pp. 3–19.
26. Li, X.; Wang, W.; Hu, X.; Yang, J. Selective Kernel Networks. In Proceedings of the 2019 IEEE/CVF Conference on Computer Vision and Pattern Recognition (CVPR), Long Beach, CA, USA, 15–20 June 2019; pp. 510–519.
27. Roy, A.G.; Navab, N.; Wachinger, C. Concurrent Spatial and Channel 'Squeeze & Excitation' in Fully Convolutional Networks. *Med Image Comput. Comput. Assist. Interv.* **2018**, *11070*, 421–429.
28. Cao, Y.; Xu, J.; Lin, S.; Wei, F.; Hu, H. GCNet: Non-local networks meet squeeze-excitation networks and beyond. In Proceedings of the 2019 IEEE/CVF International Conference on Computer Vision Workshop (ICCVW), Seoul, Korea, 27–28 October 2019; pp. 1971–1980.
29. Huang, Z.; Wang, X.; Wei, Y.; Huang, L.; Shi, H.; Liu, W.; Huang, T.S. CCNet: Criss-cross attention for semantic segmentation. In Proceedings of the 2019 IEEE/CVF International Conference on Computer Vision (ICCV), Seoul, Korea, 27 October–2 November 2019; pp. 603–612.
30. Fu, J.; Liu, J.; Tian, H.; Li, Y.; Bao, Y.; Fang, Z.; Lu, H. Dual attention network for scene segmentation. In Proceedings of the 2019 IEEE/CVF Conference on Computer Vision and Pattern Recognition (CVPR), Long Beach, CA, USA, 15–20 June 2019; pp. 3141–3149.
31. He, K.; Zhang, X.; Ren, S.; Sun, J. Deep residual learning for image recognition. In Proceedings of the 2016 IEEE Conference on Computer Vision and Pattern Recognition (CVPR), Las Vegas, NV, USA, 27–30 June 2016; pp. 770–778.
32. Ronneberger, O.; Fischer, P.; Brox, T. U-Net: Convolutional Networks for Biomedical Image Segmentation. *Med Image Comput. Comput. Assist. Interv.* **2015**, *9351*, 234–241.
33. Cai, Z.; Fan, Q.; Feris, R.S.; Vasconcelos, N. A Unified multi-scale deep convolutional neural network for fast object detection. In *Computer Vision-ECCV 2016*; Springer: Cham, Switzerland, 2016; pp. 354–370.
34. Li, Z.; Zhou, F. FSSD: Feature Fusion Single Shot Multibox Detector. *arXiv* **2018**, arXiv:1712.00960.
35. Chaib, S.; Liu, H.; Gu, Y.; Yao, H. Deep Feature Fusion for VHR Remote Sensing Scene Classification. *IEEE Trans. Geosci. Remote Sens.* **2017**, *55*, 4775–4784. [CrossRef]
36. Lim, J.; Astrid, M. Small object detection using context and attention. *arXiv* **2019**, arXiv:1912.06319.
37. Pang, J.; Chen, K.; Shi, J.; Feng, H.; Ouyang, W.; Lin, D. Libra R-CNN: Towards balanced learning for object detection. In Proceedings of the 2019 IEEE/CVF Conference on Computer Vision and Pattern Recognition (CVPR), Long Beach, CA, USA, 15–20 June 2019; pp. 821–830.
38. Ghiasi, G.; Lin, T.; Le, Q.V. NAS-FPN: Learning scalable feature pyramid architecture for object detection. In Proceedings of the 2019 IEEE/CVF Conference on Computer Vision and Pattern Recognition (CVPR), Long Beach, CA, USA, 15–20 June 2019; pp. 7029–7038.
39. Liu, S.; Huang, D.; Wang, Y. Learning spatial fusion for single-shot object detection. *arXiv* **2019**, arXiv:1911.09516.
40. Gao, F.; Wang, C.; Li, C. A combined object detection method with application to pedestrian detection. *IEEE Access* **2020**, *8*, 194457–194465. [CrossRef]

41.	Bochkovskiy, A.; Wang, C.Y.; Liao, H.Y.M. YOLOv4: Optimal speed and accuracy of object detection. *arXiv* **2020**, arXiv:2004.10934.
42.	Liu, W.; Rabinovich, A.; Berg, A.C. ParseNet: Looking wider to see better. *arXiv* **2015**, arXiv:1506.04579, 2015.
43.	Dai, Y.; Gieseke, F.; Oehmcke, S.; Wu, Y.; Barnard, K. Attentional Feature Fusion. In Proceedings of the IEEE Winter Conference on Applications of Computer Vision (WACV), 5–9 January 2021; pp. 3560–3569.
44.	Ioffe, S.; Szegedy, C. Batch normalization: Accelerating deep network training by reducing internal covariate shift. In Proceedings of the 32nd International Conference on Machine Learning (ICML), 6–11 July 2015; pp. 448–456.
45.	Howard, A.; Sandler, M.; Chu, G.; Chen, L.C.; Chen, B.; Tan, M.; Wang, W.; Zhu, Y.; Pang, R.; Vasudevan, V.; et al. Searching for MobileNetV3. In Proceedings of the 2019 IEEE/CVF International Conference on Computer Vision (ICCV), Seoul, Korea, 27 October–2 November 2019; pp. 1314–1324.
46.	Wang, C.Y.; Liao, H.Y.M.; Wu, Y.H.; Chen, P.Y.; Hsieh, J.W.; Yeh, I.H. CSPNet: A new backbone that can enhance learning capability of CNN. In Proceedings of the 2020 IEEE/CVF Conference on Computer Vision and Pattern Recognition Workshops (CVPRW), Seattle, WA, USA, 14–19 June 2020; pp. 1571–1580.
47.	Yu, F.; Chen, H.; Wang, X.; Xian, W.; Chen, Y.; Liu, F.; Madhavan, V.; Darrell, T. BDD100K: A diverse driving dataset for heterogeneous multitask learning. In Proceedings of the 2020 IEEE/CVF Conference on Computer Vision and Pattern Recognition (CVPR), Seattle, WA, USA, 13–19 June 2020; pp. 2633–2642.
48.	Everingham, M.; Zisserman, A.; Williams, C.; Gool, L.V.; Allan, M.; Bishop, C.M.; Chapelle, O.; Dalal, N.; Deselaers, T.; Dorkó, G.; et al. The pascal visual object classes (voc) challenge. *Int. J. Comput. Vis.* **2010**, *88*, 303–338. [CrossRef]
49.	Zheng, Z.; Wang, P.; Liu, W.; Li, J.; Ye, R.; Ren, D. Distance-IoU Loss: Faster and better learning for bounding box regression. In Proceedings of the AAAI Conference on Artificial Intelligence, New York, NY, USA, 7–12 February 2020; pp. 12993–13000.
50.	Ruder, S. An overview of gradient descent optimization algorithms. *arXiv* **2016**, arXiv:1609.04747.
51.	Goyal, P.; Dollár, P.; Girshick, R.; Noordhuis, P.; Wesolowski, L.; Kyrola, A.; Tulloch, A.; Jia, Y.; He, K. Accurate, large minibatch SGD: Training ImageNet in 1 hour. *arXiv* **2017**, arXiv:1706.02677.
52.	Loshchilov, I.; Hutter, F. SGDR: Stochastic gradient descent with warm restarts. *arXiv* **2016**, arXiv:1608.03983.
53.	Lin, T.; Maire, M.; Belongie, S.; Hays, J.; Perona, P.; Ramanan, D.; Dollár, P.; Zitnick, C.L. Microsoft COCO: Common objects in context. In *Computer Vision-ECCV 2014*; Springer: Cham, Switzerland, 2014; pp. 740–755.

 *sensors*

Article

Distributed Urban Platooning towards High Flexibility, Adaptability, and Stability

Sangsoo Jeong [1], Youngmi Baek [2,*] and Sang H. Son [1]

[1] Department of Information and Communication Engineering, DGIST, Daegu 42988, Korea; 88jeongss@dgist.ac.kr (S.J.); son@dgist.ac.kr (S.H.S.)

[2] Department of Computer Software Engineering, Changshin University, Changwon 51352, Korea

* Correspondence: ymbaek@cs.ac.kr; Tel.: +82-55-250-1314

Abstract: Vehicle platooning reduces the safety distance between vehicles and the travel time of vehicles so that it leads to an increase in road capacity and to saving fuel consumption. In Europe, many projects for vehicle platooning are being actively developed, but mostly focus on truck platooning on the highway with a simpler topology than that of the urban road. When an existing vehicle platoon is applied to urban roads, many challenges are more complicated to address than highways. They include complex topology, various routes, traffic signals, intersections, frequent lane change, and communication interference depending on a higher vehicle density. To address these challenges, we propose a distributed urban platooning protocol (DUPP) that enables high mobility and maximizes flexibility for driving vehicles to conduct urban platooning in a decentralized manner. DUPP has simple procedures to perform platooning maneuvers and does not require explicit conforming for the completion of platooning maneuvers. Since DUPP mainly operates on a service channel, it does not cause negative side effects on the exchange of basic safety messages on a control channel. Moreover, DUPP does not generate any data propagation delay due to contention-based channel access since it guarantees sequential data transmission opportunities for urban platooning vehicles. Finally, to address a problem of the broadcast storm while vehicles notify detected road events, DUPP performs forwarder selection using an analytic hierarchy process. The performance of the proposed DUPP is compared with that of ENSEMBLE which is the latest European platooning project in terms of the travel time of vehicles, the lifetime of an urban platoon, the success ratio of a designed maneuver, the external cost and the periodicity of the urban platooning-related transmissions, the adaptability of an urban platoon, and the forwarder selection ratio for each vehicle. The results of the performance evaluation demonstrate that the proposed DUPP is well suited to dynamic urban environments by maintaining a vehicle platoon as stable as possible after DUPP flexibly and quickly forms a vehicle platoon without the support of a centralized node.

Keywords: urban platooning; vehicle-to-vehicle communication; in-vehicle network; analytic hierarchy architecture

Citation: Jeong, S.; Baek, Y.; Son, S.H. Distributed Urban Platooning towards High Flexibility, Adaptability, and Stability. *Sensors* **2021**, *21*, 2684. https://doi.org/10.3390/s21082684

Academic Editors: Chao Huang, Haiping Du, Wanzhong Zhao, Yifan Zhao, Fuwu Yan and Checn Lv

Received: 14 February 2021
Accepted: 6 April 2021
Published: 10 April 2021

Publisher's Note: MDPI stays neutral with regard to jurisdictional claims in published maps and institutional affiliations.

Copyright: © 2021 by the authors. Licensee MDPI, Basel, Switzerland. This article is an open access article distributed under the terms and conditions of the Creative Commons Attribution (CC BY) license (https:// creativecommons.org/licenses/by/ 4.0/).

1. Introduction

Recently, sensors, advanced data processing techniques, and wireless networking technology have enabled vehicles to generate a variety of information and share road infrastructure, and share it with others, contributing to ensuring safety and efficiency while driving. Besides, automated vehicle technology is accelerating the further development of self-driving towards the highest level of automation. This is expected to bring significant changes to our car-centric lifestyle linked with cooperative intelligent transportation systems (C-ITS). In a mixed traffic environment with human-driven vehicles and partially-automated vehicles under level 3 automation, smart transportation applications already begin to appear in the form of sensing driving (e.g., intersection collision warning and cooperative adaptive cruise control), awareness driving (e.g., emergency vehicle warning, traffic jam warning, and intersection collision warning), and cooperative driving (e.g., vehicle platooning, cooperative overtaking, and cooperative lane change) [1]. In automated vehicles, to support these applications, sensing information is mainly obtained from equipped sensors such as the radar, lidar, or vision sensors. Sensor-centric applications such as lane detection, lane-keeping, and obstacle detection, therefore, utilize either these sensors individually or sensor fusion technology [2–4]. However, there are times when it is more difficult to obtain an accurate measurement of the driving environment due to an increase in uncertainty derived from dynamic driving environments and sensor measuring errors. In other words, only using sensor readings could not guarantee the safe driving of automated vehicles. To overcome the limitations inherent in physical sensors and adapt quickly to unexpected driving conditions, wireless communication technology is very useful. Moreover, to support vehicle platooning, it is useful that any sensing information related to the same physical variable is combined with the control information received from surrounding vehicles. From the safety point of view, the communication functionality is necessary to monitor any behavior of surrounding vehicles and increase the stability of a given platoon by maintaining its velocity and safety distance.

Until now, the vehicle platooning technique is mainly investigated in Europe and is referred to as truck platooning. For instance, it is studied through various projects such as KONVOI, SARTRE (Safe Road Trains for the Environment), and ENSEMBLE [5–7]. It is noted that their truck platooning techniques are proposed for a particular type of road because it is expected that the effectiveness of the vehicle platooning increases on a highway with a simple topology. In other words, since a highway is the main road with a few interchanges connecting towns or cities, it may last a long time on the highway once a vehicle platoon is formed. For this reason, they focus on platooning for trucks to reduce logistics costs. With complex topologies and dynamic vehicle movements, there are challenges in applying the highway platooning techniques directly to an urban road. For instance, vehicle routes may become diverse due to many road sections divided by intersections. When traffic flow is temporarily blocked by intersections and traffic signal lights, the number of vehicles in the local road section may increase. Crosswalks, pedestrians, or vehicles parked on the side of the road also become a source of bottlenecks of traffic flow. They tend to interfere with the movement of other vehicles and may lead to change in the vehicle route frequently. In this urban environment, the duration time of the vehicle platoon may be shorter than that on the highway. Besides, an urban area has already many existing transportation applications and communication infrastructures that may cause interference with V2V (vehicle-to-vehicle) communication required for platooning maneuvers.

Vehicle platooning enables vehicles to respond to speed changes occurring upstream with a faster reaction than that of drivers [8]. According to the European Commission, the majority of fatalities occur on rural roads and urban roads, with 55% of a road accident fatality occurring on rural roads and 37% on urban roads [9]. In this regard, urban platooning can contribute to the reduction of accidents on urban roads. Using vehicle platooning, the following vehicles can reduce 14% of fuel consumption because two or more vehicles form a chain in the same lane and a lead vehicle in front reduces the air resistance of the following vehicles [10]. In addition, it is possible to reduce the inter-vehicle distance, thereby increasing road capacity. The travel time taken from the vehicle's source to its destination may decrease as well.

To fully take these advantages of the vehicle platooning on urban roads, in this paper, we investigate how to provide stable and reliable urban platooning for vehicles with V2V communication functionality. In addition, we focus on the development of the method in which urban platooning vehicles are aimed at quickly participating in a local platoon on urban roads and sharing control information required for urban platooning with low delay. As the protocol suite for vehicular ad-hoc networks, IEEE WAVE (Wireless Access in Vehicular Environment) is specified in the IEEE 1609 family of standards [11]. Its medium access control (MAC) and physical layers are based on the IEEE 802.11p. It is designed to broadcast particular messages, i.e., basic safety messages (BSMs), based on the collision-sense multiple access with collision avoidance (CSMA/CA) mechanism [12]. Therefore, it is difficult to guarantee road safety as either the traffic density of roads, the amount of data to be sent, or the number of flows increases [13]. To develop safety-critical applications based on V2V communication, a critical issue is to deliver data within a given time. Due to the uncertainty of the wireless channel, its link delay might be as high as hundreds of milliseconds [14]. Especially, as the vehicle density on the road increases, the transmission delay could become much higher. Increasing delay may influence the stability of a given platoon [15]. Meanwhile, to achieve the automation and reliable control of self-driving, sensor reading data of every vehicle should be transmitted within a given transmission cycle, i.e., every minimum 5 to maximum 100 milliseconds, in in-vehicle networks [16–19]. Therefore, we jointly consider the requirements of intra-vehicular control and inter-vehicular communication to guarantee reliable urban platooning. It is indicated that control data of each vehicle in a given platoon should be shared until the next transmission cycle to ensure the stability of a given platoon when the V2V communication-based data is exploited. Another issue is to maximize the duration of the formed urban platoon. It may depend on the flexible and autonomous formation of an urban platoon [20]. To clarity the terms flexible and autonomous platooning, we specify the requirements of flexible and autonomous platooning as follows. A vehicle should join a given platoon at the rear when it participates in a given platoon without regard to its location. When a vehicle intends to disjoin the given platoon, the given platoon allows it to leave regardless of its location. Therefore, this formation should be performed either in the lane in which they are driving or through lane changes. If a given platoon encounters a new platoon, they can merge with each other autonomously, depending on a given condition. The separated platoons with the existing formation should be maintained as much as possible even though the existing platoon is separated by unexpected situations. The other issue is related to the nature of broadcasting communication. During driving, both vehicles in an urban platoon and vehicles under normal-driving conditions could encounter unexpected traffic conditions. In IEEE WAVE, every vehicle should inform surrounding vehicles of the events as soon as events that negatively affect driving are detected. It may result in a broadcast storm of redundant information in a local road section. To increase the efficiency of data propagation, it is necessary to perform notifications of the emergency messages required for safety-critical applications by minimizing the number of forwarders to notify this information.

In this paper, we propose a distributed urban platooning protocol (DUPP) that is designed to maximize flexibility for vehicles, considering their high mobility. It conducts urban platooning in a decentralized manner. We adopt four distinct approaches to eliminate unnecessary competition in contention-based medium access. First, in DUPP, a distributed medium access method as a layer is designed and added on top of IEEE 802.11p to ensure that all members of an urban platoon transmit messages fairly and sequentially. Second, although urban platooning is one of the safety-critical applications, DUPP operates mainly on a service channel rather than a control channel designated for safety-critical applications. Third, to adapt quickly to the complex urban topology and the changing traffic conditions, DUPP is designed with very simple operations for the urban platooning maneuvers consisting of creation, joining, leaving, merging, and splitting. In addition, it never requires explicit acknowledgments. Fourth, during urban platooning, only one forwarder transmits the related information when unexpected events on roads occur. To enhance the propagation efficiency, DUPP determines one forwarder by an analytic hierarchy process (AHP) using vehicle status indicators. Therefore, not only the upkeep cost of urban platooning under DUPP is minimized but also DUPP-enabled vehicles are capable of quickly responding to the dynamic driving environment in urban roads.

It is important to validate the effectiveness of the proposed urban platooning. We exploit the PLEXE simulator with a microscopic vehicle control model. To make a virtual traffic environment mimicking a real city's road, we use the NYC public traffic data collected for 24 h and a part of the NYC road network. The DUPP's performance is evaluated by examining (1) the vehicle travel time to see how much our DUPP increases road capacity and efficiency, (2) the lifetime of an urban platoon in order to see how quickly an existing urban platoon responds on the urban road, (3) the success ratio of each urban platooning maneuver, (4) the drop ratio of BSMs on the control channel in order to show how much the operations of our DUPP affect the performance of other vehicles not involved in urban platooning, (5) the transmission periodicity of urban platooning-related messages in order to show both the stability of an urban platooning and satisfying the requirement for the control frames that should be transmitted through in-vehicle networks for reliable driving control, (6) the maintenance of safety distances between the vehicles to show the adaptability to unexpected situations through forwarder selection, and (7) the selection ratio for each vehicle in an urban platoon to demonstrate the performance of the designed AHP-based forwarder selection.

The contribution of this paper is as follows. First, to quickly adapt to the dynamic traffic flow and complex topology of an urban road network, we propose a novel urban platooning protocol enabling distributed coordination and decentralized autonomous maneuvers. It is well suited to zero infrastructure communications and to support adaptive, flexible, and simple platooning. Especially, DUPP does not require any control message to explicitly confirm the completion of each maneuver and is distributed to each vehicle interested in a given platoon. Although all transmissions in DUPP are based on the IEEE 802.11p that employs contention-based access, the delay of the data propagation and the maneuver conduction are minimized due to our distributed coordination. In addition, we address the problem of the broadcast storm by employing the approach of the analytic hierarchy process to regulate the number of forwarders even if vehicles simultaneously detect event occurrence. To demonstrate the effectiveness of the proposed urban platooning, we construct a new environment to which actual traffic public data is applied.

The remainder of this paper is organized as follows. In Section 2, we review the recent related research and discuss the requirements for urban platooning. Section 3 provides a detailed description of the proposed DUPP. We evaluate the performance of DUPP, comparing it with that of ENSEMBLE in Section 4. Finally, the conclusions and future work are provided in Section 5.

2. Related Work

In this section, we mainly review previous studies for vehicle platooning performed in a distributed manner and discuss the requirements imposed on urban platooning. Many protocols for vehicle platooning on highways have been designed, which adopt either a centralized approach or a distributed approach to vehicle platoon formation [7,21–25]. The centralized approach requires a central system that is responsible for determining which vehicle platoon a given vehicle will join. To determine a certain platoon to be joined, the central system uses status information collected from all of the driving vehicles on roads. In the distributed approach, a driving vehicle determines joining platoon by itself, based on status information collected from neighboring vehicles. Once a vehicle platoon is formed, regardless of which approach is used for the platoon formation, a leader vehicle driving in front of a given platoon arranges how to control vehicle movement during vehicle platooning. The operation of vehicle platooning, therefore, is generally performed in a centralized manner.

Heinovski and Dressler have proposed the distributed formation method where a vehicle independently determines a specific platoon with its preceding vehicle to be joined [21]. To select one preceding vehicle to be joined, a vehicle exploits status information collected from preceding vehicles, which are driving in front of the vehicle, through beacon messages based on the IEEE 802.11p. A vehicle joins behind the selected preceding vehicle after investigating which of all preceding vehicles can drive with it for a long time. If there are no vehicle platoons around, a vehicle creates a vehicle platoon for itself. To start a joining maneuver for a specific platoon, a vehicle explicitly sends a request message to the selected preceding vehicle, and then receives a response message from it. On a highway, this procedure is effective to reliably form the platoon. However, once a joining maneuver is started by the given vehicle, it is not allowed to change the selected preceding vehicle to another preceding vehicle even if traffic conditions change. They assume that once a joining maneuver is successful, any vehicle does not leave its platoon until it reaches its destination. In addition, if the given vehicle is in its joining maneuver, other following vehicles cannot join the given vehicle and just wait until it finishes its joining maneuver. It is noted that in urban roads, the driving speed and the driving route may frequently change depending on vehicle density and intersections with traffic lights. Therefore, it is not efficient if the formed platoon does not allow vehicles to change or the selection of the platoon to join cannot be changed. Although explicitly sending and receiving control messages for the joining maneuver contribute to improving the reliability of the joining maneuver, it also imposes an additional processing delay during the joining maneuver depending on the driving environment.

There is a study that enables the joining of a platoon regardless of the driving position of a given vehicle [22]. Vehicles on the road exchange information of their speeds and positions with the surrounding vehicles to support both lateral and longitudinal control models. To perform a joining maneuver, this protocol adopts a three-way handshake process with explicit control messages by sending a request message to a leader vehicle, receiving a response message from it, and finally reporting back to it using an acknowledge message after a vehicle approaches the selected preceding vehicle to be joined by speeding up. In this process, although a platoon can be stably formed by using explicit control messages, a leader vehicle is involved in the whole procedure for the platoon formation. That results in a relatively long time for the completion of the maneuver, depending on the vehicle density. Furthermore, it allows a vehicle to join a platoon at the side of a platoon by changing the driving lane. It means that a joining maneuver needs to operate more sophisticatedly. It might be useful in an environment with a simple topology and few negative factors affecting its completion. However, it is not suitable for urban platooning where driving route changes frequently occur since an urban environment requires a quick response to more complicated situations than those of highways. When an urban platooning protocol is designed, it needs to consider whether to improve the flexibility by allowing vehicles to join at the side of a platoon or improve stability by allowing vehicles to join only at the back. It is noted that the former requires more sophisticated and complicated maneuvers than those of the latter. In this paper, we are interested in improving the adaptability and stability of urban platooning at the same time.

One platooning protocol is to mainly use a radar sensor to maintain the inter-vehicle distance and only occasionally use wireless communication to share control information for unexpected driving situations [23]. A vehicle is allowed to join a platoon regardless of its driving position. When a joining maneuver occurs at the side of the platoon, it is necessary to change the distance between two vehicles that exist backward and forwards in the position to be joined at the side of a platoon. To achieve this, the vehicle preceding the two vehicles located at the front and back of a new joining position sends a control message to notify those two vehicles of an increase or a decrease in speed. The preceding vehicle transmits the control message through ultra-short distance communication so that only the following vehicle could receive it. For every 100 ms, a leader vehicle allocates a different channel used for communication to each vehicle belonging to its platoon. Using different channels for communication leads to a decrease in communication interference among many vehicles. However, since urgent information must be transmitted to rear vehicles across multiple hops, using different channels increases communication delay.

Satisfying the periodicity of control message transmission during vehicle platooning is a critical issue in terms of control. In this regard, Böhm and Kunert have proposed a fair communication method in which a leader vehicle allocates a time slot of transmission to each vehicle belonging to its platoon within a given superframe [24]. To ensure the reliability of vehicle platooning, a leader vehicle explicitly acknowledges individual vehicles whenever member vehicles transmit their messages. In addition, every vehicle is also required to acknowledge receipt of the leader vehicle's message. In this approach, a given superframe is divided into two phases. The first phase is designed for a leader vehicle to collect the information of the member vehicles as each vehicle transmits its status information in the time slot assigned to it. In the second phase, a leader vehicle individually informs each vehicle of control information for vehicle platooning. During the first phase, one or more vehicles can fail to transmit their message in the assigned time slots. A leader vehicle continuously allocates a new time slot to all failed vehicles until either allfailed vehicles succeed the transmission of its control message or all the time slots assigned in the first phase are exhausted. In the worst case, thus, the second phase is postponed until a maximum period defined for the first phase is consumed. Even in the second phase, if a leader vehicle fails to transmit a control message for each member vehicle, it should also retransmit control messages after its scheduled transmissions for every vehicle are finished. In this protocol, there should exist an explicit phase to collect status information so that a leader vehicle controls all vehicles in a platoon by transmitting control messages required for vehicle platooning. All platooning vehicles are responsible for explicit acknowledgments for all data transmission within the assigned time slot. Depending on the vehicle density on an urban road, contentions for transmission can be significantly intensified so that it may be difficult to guarantee a stable platoon. Furthermore, when a leader node leaves its platoon, its platoon is bound to be destroyed and a new platoon should be constructed again. In a dynamic driving environment such as an urban road, there is a desire to maintain a platoon no matter which vehicle leaves.

In addition, one study also points out the difficulty of guaranteeing transmission reliability [25]. Especially, under high channel load, there can occur many packet collisions due to the nature of IEEE 802.11p. To ensure transmission reliability, they propose a token-based MAC protocol. In this protocol, one member of the local platoon with the best connectivity among the neighboring local platoons plays a role of a token manager as a central controller. After the token manager determines a receiver, called a token holder, for its frame, it transmits its data with a token. A token holder is responsible for sending its frame as its token is delivered to a new token holder. If the token holder cannot transmit its frame with a token for a new token holder, this is excluded from a given local platoon. At the same time, the token manager creates a new token that indicates retransmission starting at the beginning. When this token-based MAC protocol is applied to urban platooning, there may be tricky issues. For instance, if a node cannot transmit data due to a communication failure, according to their token management mechanism, the token manager creates a new token. While ensuring the reliability of the transmission, the periodicity of control message transmission required for urban platooning may not be satisfied, which may lead to a dangerous platoon driving situation.

ENSEMBLE is designed as a truck platooning protocol with flexibility regardless of the type of trucks [7]. This protocol operates on a control channel and uses the extension of the cooperative awareness message (CAM) specified as standard in EN 302 637-2 [26]. It focuses on specifying the creation, joining, and leaving maneuvers for efficient vehicle platooning. During driving, platooning vehicles should periodically transmit not only their CAM extensions once every 100 ms but also platooning-related control messages once every 50 ms. To join a platoon, a vehicle that does not participate in a platoon uses the information of the CAM extension received from surrounding vehicles belonging to the platoon. If a vehicle intends to join a specific platoon, it should send a request message to the rear vehicle at the end of the platoon. The rear vehicle allows it to join by explicitly responding with a response message. If a vehicle intends to leave a platoon, it is allowed to leave the platoon after it transmits a leaving-related message to all vehicles in the platoon ten times. It is noted that the ENSEMBLE-enabled platoon is destroyed even if one vehicle leaves the platoon. In the urban road where many vehicles frequently join and leave an existing platoon, it is hard to maintain vehicle platoons constructed by ENSEMBLE. Furthermore, since a single control channel is used for vehicle platooning, performance degradation might be severe in urban environments with high vehicle density. While urgent information related to the detected events is shared with surrounding vehicles during vehicle platooning, this may hinder the transmissions of platooning-related control messages by imposing the additional load to the control channel in a dense area.

As described above, there have been many attempts to perform vehicle platooning in a distributed manner. However, they are not suitable for urban driving environments with complex topology and high uncertainty. From the literature review, the requirements for urban platooning are derived. First, a vehicle should determine a vehicle platoon to join by itself by minimizing the amount of information collected from surrounding vehicles. Second, even without the explicit acknowledgment from the vehicles of the platoon, the platoon should be reliably maintained. Third, once a vehicle platoon is created, various maneuvers suitable for driving environments need to be specified for a long lifetime. In addition, specified maneuvers should provide the adaptability and flexibility of the platoon formation. Fourth, information for urgent events should be shared quickly with platooning vehicles to ensure their stability by actively controlling the vehicle movement. We should consider minimizing the additional load on the platooning-related transmission. Considering the derived requirements for urban platooning, we exploit a fully distributed coordination based on a beaconing mechanism to maximize flexibility and adaptability, which has been actively studied for ultra-wideband [27,28]. It is capable of quickly organizing a new network connection and rapidly joining a new network [29].

3. Distributed Urban Platooning Protocol

In this section, considering the mobility of vehicles, we introduce a novel distributed urban platooning protocol that is designed to maximize flexibility, adaptability, and stability. DUPP is developed to support fair and reliable data propagation by suppressing unnecessary transmission among surrounding vehicles on the urban road and allowing an urban platoon to adapt rapidly to a frequently changing driving environment (i.e., changeable routes of vehicles, irregular traffic flow, and a signalized intersection). The functionality of DUPP is distributed among vehicles involved in urban platooning and is composed of the distributed control for medium access and urban platooning maneuvers. First, for distributed control for medium access, we design a distributed coordination method added on top of IEEE 1609.4 and IEEE 802.11p, which is described in Section 3.1. Second, for distributed control for urban platooning maneuvers, flexible and autonomous platooning (FAP) maneuvers are proposed which are discussed in Section 3.2.

An urban platoon consists of two or more vehicles driving on the same lane. The length of an urban platooning is limited to η in order to maintain efficiency and stability under an urban driving environment with complex topologies and dynamic vehicle movements, where η can be defined by experiments. From the literature, η would be set to 7 [7]. In this paper, an urban platoon with a certain length below η is referred to as a local platoon. A local platoon is defined as $P = \{L_{v_i}, F_{v_i}, R_{v_i}\}$, where L_{v_i}, F_{v_i}, and R_{v_i} are individual subsets of $S = \{v_1, v_2, \ldots, v_\eta\}$ which represents a set of vehicles joining a local platoon with the identification of $1 \leq i \leq \eta$. Furthermore, this set P becomes a partition set of the set S when the number of vehicles in a local platoon is over three. During driving, a vehicle that intends to perform urban platooning is referred to as a candidate node $v_k \in V$, where set V is a set with the elements of a driving vehicle on the road. When a new local platoon is initially formed, it might have one vehicle that is an owner and creator of the new local platoon. This vehicle is referred to as a leader node $L_{v_i}(v_i \in S, |L_{v_i}| = 1)$ which is a node driving at the very front of the local platoon. After that, many vehicles as candidate nodes can participate in the local platoon of this leader node, which are referred to as member nodes $M_{v_i}(v_i \in S - \{v_k | v_k = L_{v_k}\}, |M_{v_i}| = \eta - 1)$. Among the member nodes M_{v_i}, one vehicle driving at the tail end of the local platoon is called a rear node $R_{v_i}(v_i \in S, |R_{v_i}| = 1)$ and the other vehicles are referred to as follower nodes $F_{v_i}(v_i \in S, |F_{v_i}| = \eta - 1 - 1)$ except for a rear node. According to this definition, in DUPP, a vehicle should take at least one of the four roles (i.e., candidate, leader, follower, and rear nodes) of vehicles related to a local platoon corresponding to four defined vehicle sets (i.e., $\{v_k\} \cap \{L_{v_i}, F_{v_i}, R_{v_i}\}$). Member nodes may consist of one rear node and one and more follower nodes while $|S| \geq 3$. If the number of vehicles in a local platoon is below two, the member node is a rear node.

3.1. A Distributed Coordination for Urban Platooning

In this subsection, we introduce a distributed coordination method by which vehicles can form a local platoon, sequentially participating in urban platooning, and obtaining the control information from member nodes and a leader node within a given time. It is well suited to zero infrastructure communications and to support FAP maneuvers.

IEEE WAVE is referred to as the suite of IEEE 1609.x standards including IEEE 802.11p [11]. In WAVE, IEEE 1609.4 that describes multi-channel operations works on top of the IEEE 802.11p MAC [30]. DUPP exploits alternative channel access among the four methods for multi-channel operations. The alternative channel access divides the channel access time into a synchronization interval with a fixed length of 100 ms. The synchronization interval consists of a control channel interval (CCH-I) with a fixed length of 50 ms and a service channel interval (SCH-I) with a fixed length of 50 ms. During CCH-I, vehicles should reside in one designated control channel and share BSMs for supporting safety-critical applications. During SCH-I, one of four service channels (SCHs) is used by vehicles to support non-safety applications.

In DUPP, vehicles of a local platoon should share platooning-related information in the form of a platoon control message (PCM) on a given SCH during SCH-I. Regardless of an urban platooning, it is compulsory for driving vehicles to broadcast BSMs on the designated CCH during CCH-I. A leader node is required to transmit a WAVE service advertisement (WSA) message notifying the availability of an urban platooning. The multi-channel operation of DUPP works regardless of the number of transceivers of a vehicle. To coordinate PCM transmissions among vehicles during SCH-I, the DUPP's superframe, indicating periodic time interval as shown in Figure 1, consists of two major parts: a beacon period (BP) followed by a data period. The data period for sharing platooning control information may be divided into a contention-free period (CFP) and an event-based period (EBP). The EBP is an optional period activated depending on whether a certain event occurs or not. The BP is defined for vehicles to send only beacon messages designated to get scheduled access to the medium. The CFP aims to share PCMs within the existing local platoon to conduct the urban platooning by controlling vehicles' acceleration. During driving, the nodes of a local platoon should respond to four events that may have the potential negative effects on the stability and safety of the urban platooning against various events predefined for ETSI's DENM (Decentralized Environmental Notification Message) [31]. They include emergency brake lights, hazard locations of dangerous curves, obstacles and road construction, and road conditions such as heavy rain, snow, and slippery road. All nodes detecting one of the events should be responsible for transmitting a platoon event message (PEM) within the EBP of a given superframe. In order to exclude the possibility of all nodes struggling with intensified competition, DUPP selects one transmitting node among them in a distributed manner.

Each superframe starts with the BP where the leader node of a local platoon transmits a beacon message. Each BP has only two beacon messages transmitted independently by two vehicles. After the first one is transmitted by a leader node L_{v_i}, the second one is sent by a new candidate node $v_k \in V$ based on IEEE 802.11p. A new candidate node attempts the transmission of its beacon message only after receiving the beacon message of the leader node. It is noted that only one candidate node succeeds in this competition even though there may exist several candidate nodes accessing to medium to send their beacon messages. In other words, DUPP allows only one vehicle for each superframe to participate in the existing local platoon.

After finishing the BP, the CFP starts with the PCM transmission of a leader node. An individual member node sequentially transmits its PCM upon receiving the PCM from its preceding member node. Since the wireless channel is shared with contention-based medium access and is easily influenced by various driving environments, nodes' transmissions in the local platoon might be delayed or failed. DUPP is responsible for periodically transmitting accurate control information through in-vehicle networks. Therefore, if one of them does not succeed in transmitting the PCM, all nodes behind the failed node are regulated from transmitting their PCM. It means that the nodes behind the delayed or failed node cannot send their PCMs during the superframe and should be separated from the existing local platoon. Those nodes need to perform either a creation, joining, splitting, or merging maneuver.

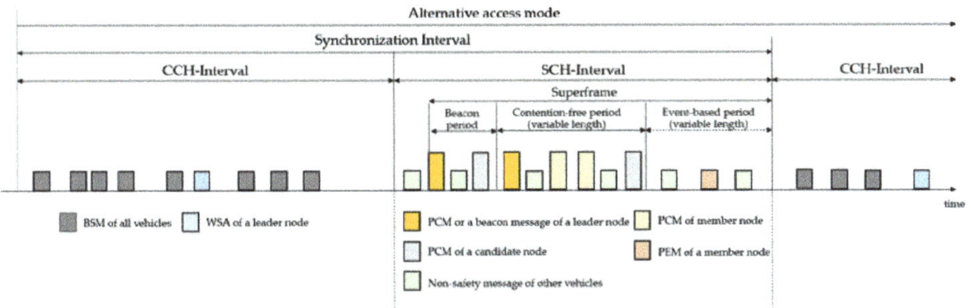

Figure 1. A channel access method defined for a distributed coordination.

During EBP, DUPP allows only one vehicle (i.e., a forwarder node) to broadcast a PEM as a representative of a local platoon in order to give a warning against the detected event. It aims to reduce the redundant messages generated from many nodes detecting a particular event simultaneously. To determine a forwarder node to broadcast its PEM, an individual node of the local platoon makes a decision by itself, based on an analytic hierarchy process. We describe this AHP-based decision in Section 3.3.

A WSA message defined in IEEE 1609.3 is composed of a header and a series of WAVE elements [32]. The header information includes the current WAVE version and extension fields. The WAVE elements may include three segments (i.e., a series of variable-length Service Info, a series of variable-length Channel Info, and a WAVE Routing Advertisement). Since the WAVE Routing Advertisement segment is to provide information about infrastructure internetwork connectivity, it is not necessary for DUPP designed for distributed coordination using zero-infrastructure communication.

In DUPP, three segments (Service Info, Channel Info, and Platooning Info) are used for the leader node's WSA message as shown in Figure 2, which illustrates a new WSA message used for DUPP designed for urban platooning. To notify the availability of urban platooning, the Platooning Info segment is newly added. The size of the WSA message used for DUPP is a total of 48 bytes. We discuss each segment in detail in the following.

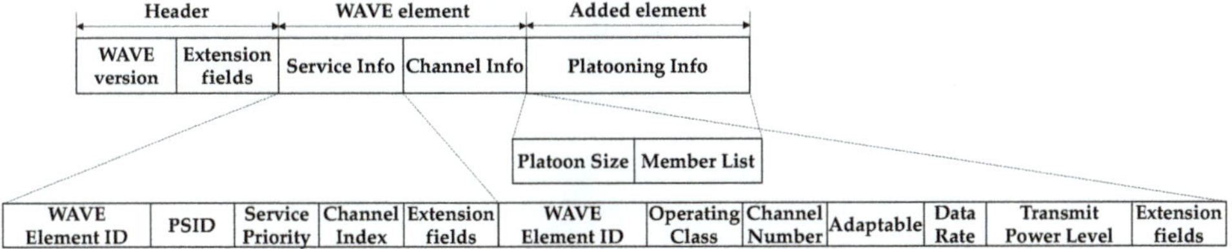

Figure 2. A WAVE service advertisement message for urban platooning.

First, the Service Info segment contains information about a supported service. The first field of the Service Info segment is a WAVE Element ID field with a value of 0x01. To distinguish the type of the supported services, a unique identifier (ID) is allocated to each service. This service ID is used to the value of the PSID (Provider Service Identifier) field. The Service Priority field determines access to service channels to respond to the service request of this specified priority. In this regard, we use the value having the highest priority among given services to occupy the service channel. The Channel Index field indicates the service channel where the advertised urban platooning service operates. Normally, in one WSA message, there may be information on one or more channels to support various services at the same time. The n value of the Channel Index field of the Service Info segment connects to the n-th Channel Info segment. However, it is assumed that the nodes for DUPP concentrate on this urban platooning service. Therefore, the WSA message of DUPP has only one Channel Info segment and the Channel Index field which is set as the value of 0x01.

Second, the Channel Info segment aims to provide the information on the wireless channel used for a defined service. In the Channel Info segment, the WAVE Element ID is specified by the value of 0x02. The Operating Class field allows the Channel Number to identify a specific channel uniquely in the context of a country. The Channel Number field indicates the number of the channel for urban platooning to operate. To communicate with each other, the nodes of a given local platoon should specify how fast and how strong they transmit after they move to the channel where urban platooning operates. It is related to the two fields of Data Rate and Transmit Power Level. The Adaptable field enables them to operate in a more flexible way. If the Adaptable field is set to zero, they are required to communicate by complying with the values specified in the Data Rate and Transmit Power Level fields. If not, the value of the Data Rate field is used as a minimum value for transmission, and the transmission power of the nodes cannot exceed the value specified in the Transmit Power Level field. Finally, the Platooning Info segment contains two fields of the Platoon Size with the number of nodes and Member List with the list of IDs of all nodes belonging to a local platoon. These fields are used to assist platooning maneuvers.

In DUPP, the three messages supporting a local platoon are designed: a beacon message, PCM, and PEM as discussed above. To maintain stable urban platooning, DUPP exploits a part of the existing SAE J2735-based BSM part 1 to share vehicle information and extends it to contain additional information regarding a local platoon. As shown in Figure 3, this extension illustrates the basic form of three messages. The basic form consists of two elements of vehicle information and platoon information.

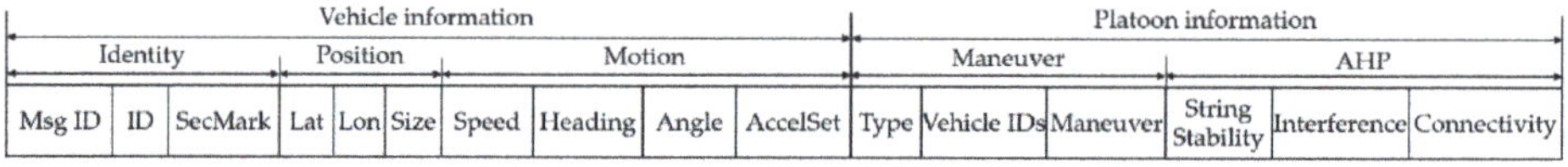

Vehicle information										Platoon information					
Identity			Position			Motion				Maneuver				AHP	
Msg ID	ID	SecMark	Lat	Lon	Size	Speed	Heading	Angle	AccelSet	Type	Vehicle IDs	Maneuver	String Stability	Interference	Connectivity

Figure 3. The basic format of three messages for urban platooning.

First, the vehicle information element is categorized into three segments. As shown in Figure 3, the Identity segment has three fields related to basic information about a given message.

The first field of Msg ID is a unique ID of a message, the second ID field is the vehicle's ID of a sender node, and the third SecMark field is a generated time of the message. To maintain shorter inter-vehicle space, it is necessary to exchange dynamics information of all DUPP nodes. In this regard, the Position segment has information consisting of Latitude, Longitude, and Size fields of the vehicle. The Motion segment contains vehicle dynamics information defined as Speed, Heading, Angle, and AccelSet fields. In detail, the AccelSet field consists of four acceleration values (i.e., longitude acceleration, latitude acceleration, vertical acceleration, and a yaw rate).

Second, the platoon information element consists of maneuver and AHP segments. In the maneuver segment, the value of the Type field of each message is given as a different value to distinguish three messages. The beacon message, PCM, and PEM have 0x01, 0x02, and 0x03 respectively. To specify an existing local platoon, the Vehicle IDs field contains a list with the IDs of all nodes in sequence in this local platoon. Since DUPP does not perform explicit confirmation of nodes' maneuvers, this Vehicle IDs field is designed to ensure its reliability while vehicles are conducting distributed coordination. When a leader node generates a beacon message, the leader node specifies the Vehicle IDs field with a list of all nodes belonging to its local platoon. The Vehicle IDs field of the leader node is used to notify candidate nodes of information on all nodes belonging to its local platoon and can be used to identify this local platoon from others on the same channel. A candidate node adds its own vehicle ID to a leader node's list and puts the extended list in this field to notify the leader node of its intention to attend. It indicates that the candidate node designates a specific local platoon to participate. In the case of the PCM, as it does in the beacon message, the leader node also constructs the list for all nodes belong to its local platoon. When there is a candidate node succeeding to the transmission of a beacon message during this superframe and the current number of all nodes is below η, the list is extended by adding the vehicle ID of that candidate node. This extended list indicates implicitly that the leader node accepts a new joining attempt presented by the beacon message of the candidate node. During the CFP, the members receiving the PCM of the leader node also extend the list by adding the vehicle ID of that candidate node. The Maneuver field is designed to describe nodes' maneuvers defined for a distributed urban platooning. In the case of a beacon message, the value of the Maneuver field can be either 0x00 or 0x01. The value of 0x00 means the message of the leader node and the value of 0×01 means the message of the newly joining node (i.e., one of the candidate nodes). When the Type field has the value representing PCM, the Maneuver field has one of four different values: 0x00 representing a normal driving node within a given local platoon, 0x01 representing a newly joining node, 0x02 representing a node intending for leaving, 0x03 representing a newly merging node, and 0x04 representing a node to conduct splitting. When the value of the Type field is PEM, the Maneuver field indicates the ID of an event. The total size of each of the DUPP messages is fixed as 86 bytes.

Three fields of the AHP segment are used to determine a forwarder for a local platoon. In DUPP, each node of the local platoon only uses the information of PCMs when a forwarder is determined in a distributed manner. Therefore, the beacon message and PEM do not specify the AHP segment. The string stability field contains the difference between expected and real position, indicating how well the node keeps the safety distance required by the leader node of the local platoon. The interference field specifies the number of vehicles coexisting outside the local platoon in order to convey information about how much data transmission of this PCM's owner affects the data transmission outside the local platoon. Since the connectivity represents a capability of data dissemination of a given node, the connectivity field has the number of nodes behind this PCM's owner.

3.2. Flexible and Autonomous Platooning

To maximize the flexibility and adaptability of urban platooning, flexible and autonomous platooning is designed for DUPP. FAP defines the vehicle's maneuvers with five platooning maneuvers: creation, joining, merging, leaving, and splitting maneuvers. For ease of understanding, the flow diagram for each FAP is illustrated in Appendix A.

3.2.1. A Creation Maneuver

A creation maneuver is a series of processes in which a candidate node becomes a leader node to create a new local platoon. To create a local platoon, it is assumed that there is no platoon in the transmission range of a candidate node. After a candidate node determines creating a new local platoon under those conditions, a candidate node sends a new WSA message on CCH during CCH-I and becomes a leader node. At the end of CCH-I, it switches to SCH specified in the WSA message. When a leader node transmits a beacon message to start the BP, a new local platoon is generated on SCH. While there is one vehicle belonging to a local platoon, the vehicle is not only a leader node but also a rear node. Therefore, a leader node completes creating a local platoon within one superframe on SCH.

There may be situations that violate the assumptions above. If there is an existing local platoon, a candidate node can receive a WSA message broadcast by a certain vehicle.

If a candidate node is interested in this local platoon and satisfies four join requirements, it does not generate a new WSA message but performs a joining maneuver. The four join requirements are described in Section 3.2.2 in detail. If a candidate node intends not to participate in this local platoon, it transmits its own WSA message to create a local platoon.

3.2.2. A Joining Maneuver

A joining maneuver is a series of processes for a candidate node to join an existing platoon. To join a local platoon, a candidate node should satisfy four join requirements: (1) there is at least one local platoon in the transmission range of the candidate node regardless of the driving lane, (2) the candidate node follows the given local platoon, (3) the total length of the local platoon should be less than η before the candidate node participates in the existing local platoon, and (4) the existing local platoon and the candidate node should not be blocked by obstacles such as vehicles of no interest.

A candidate node that decides on participating in the leader node's local platoon performs its speed control and channel switching simultaneously. To inform a new vehicle of a joining position, the WSA message of a leader node includes the ID of the rear node in the existing local platoon. By using the ID of the rear node, a candidate node obtains position information from the BSM of the rear node and approaches within 20 m of the rear node to perform the joining maneuver. If the candidate node and the rear node are not driving in the same lane, the candidate node changes the lanes and approaches the rear node.

Simultaneously, at the beginning of SCH-I, the candidate node moves to the SCH specified in the WSA of the leader node to perform a joining maneuver. On SCH, the candidate node starts to perform the joining maneuver by transmitting its beacon message during the BP. It is allowed to transmit a beacon message after the leader node transmits its beacon message. The beacon message of the candidate node should specify the value of 0x01 in the Maneuver field and have the Vehicle IDs field with the leader node's list to specify the local platoon to join. Since the candidate node can obtain the leader node's list from the WSA message on CCH, it can confirm the leader node's beacon message transmitted on SCH. After the candidate node successfully transmits the beacon message, the leader node of the local platoon expands the list of the existing nodes belonging to this local platoon to include the new candidate node. The leader node allows the candidate node to join this local platoon by transmitting the PCM with the extended list in the given superframe during the SCH-I. While the existing local platoon is maintained, the leader node is responsible for managing the list of the vehicles' IDs including a new rear node and providing the list periodically through the beacon message, the WSA message, and the PCM.

Assuming no contention, the candidate node (i.e., winner) finishes the joining maneuver by transmitting its PCM and can complete it within one synchronization interval. There might be one and more new candidate nodes to attempt to join the existing local platoon. However, since one candidate node wins in contention-based access, it transmits its beacon message after a leader node succeeds to broadcast its beacon message during the BP. This winner becomes a new member of the local platoon and becomes a rear node at the same time. Accordingly, the rear node is changed into a follower node. The candidate nodes that were not successful to join the local platoon in the first attempt can try in the next superframe. In this regard, if there are several candidate nodes to join the local platoon at the same time on SCH, to complete the joining maneuver, it takes as many superframes as the number of candidate nodes attempting at the same time. Nevertheless, it is noted that there might exist two cases for a candidate node not to completely join an existing local platoon over time: (1) the case that the rear of the vehicle is blocked by another vehicle that is not interested in this platoon and (2) the case that the number of the member and leader nodes exceeds the designated size η of the local platoon after a winner among new candidate nodes conducts the joining maneuver. In these cases, the candidate nodes except for the winner should perform the creation maneuver to generate a new local platoon in the next CCH-I.

3.2.3. A Merging Maneuver

A merging maneuver is conducted when two consecutive local platoons meet in the same lane. To merge local platoons, it is assumed that (1) two local platoons should be in the same lane, (2) the leader node of each local platoon should be in the transmission range of each other, (3) the sum of the length of two platoons should be below η, and (4) two local platoons should not be blocked by obstacles such as vehicles that are not interested in the urban platooning.

When driving on the road, the length of a local platoon will increase until the number of nodes is below η. However, this local platoon can be split by traffic lights and communication failure. It is possible that there are many small local platoons, reducing the efficiency of the urban platooning. To improve the performance of the DUPP, we design a merge maneuver allowing small local platoons to merge to form one large local platoon. For instance, if there are two small local platoons nearby, two leader nodes can hear each other's WSA message during CCH-I. According to the distributed coordination of DUPP, the following local platoon can merge into the preceding local platoon. To perform the merging maneuver, all nodes of the following local platoon change the speed to approach the preceding local platoon. To do this, the leader node of the following local platoon should send a PCM with the desired speed. After that, if the leader node of the following local platoon approaches within 20 m of the rear node of the preceding local platoon, the merging maneuver will be performed by hearing each other's WSA message during CCH-I. In the next SCH-I, the nodes in the following local platoon switch the service channel designating in the WSA message of the preceding local platoon.

During the BP in the next SCH-I, two leader nodes of the preceding and following local platoons should transmit individual beacon messages. First, the leader node of the preceding local platoon sends a beacon message with its managed node IDs. Second, the leader node of the following local platoon responds with a beacon message containing all the node IDs of the preceding and following local platoons. In the CFP, the leader node of the preceding local platoon sends a PCM containing the list of all node IDs which is obtained from the beacon message of the leader node of the following local platoon. It indicates that the leader node of the preceding local platoon allows the following local platoon to be merged to its local platoon. As a confirmation of newly merged members, all the nodes in the preceding local platoon transmit PCM which contains not only the nodes in the preceding local platoon but also the nodes in the following local platoon. The PCMs of the following local platoon nodes are the same as that of the preceding local platoon nodes except for the value of a particular field. The Maneuver field in the PCM has the value of 0x04 that indicates that they are newly merged nodes. As the nodes of the following local platoon merge into the preceding local platoon, the value of the Maneuver field will be the value of 0x00 in the next superframe. As the merging maneuver is completed when the nodes in the following local platoon transmit a PCM containing the Maneuver field as the value of 0x04, the merging maneuver is finished within one and a half synchronization interval if two platoons are close enough before the merging maneuver is performed.

3.2.4. A Leaving Maneuver

A leaving maneuver represents the behavior of the vehicle to leave without participating in the local platoon anymore regardless of its position. There are two cases that the vehicle leaves from a platoon: the expected and the unexpected. The former indicates that the vehicle has a plan to change its route in advance due to the road traffic conditions (e.g., congestion, illegal parking, and traffic disruption due to a signalized intersection). The latter indicates that a vehicle is considered as leaving a local platoon by other members that cannot send PCMs within a deadline of one synchronization interval.

As a node of the local platoon changes its route due to road traffic conditions, the node is separated from the local platoon. Before this separation, a node should inform the other nodes of its intention of leaving the local platoon by transmitting its PCM with the Maneuver field of the value of 0x02. However, the leaving maneuver is different depending on whether a leaving node is a leader node or not. If a leaving node is the rear node of the local platoon, the follower node just in front of it becomes a new rear node. In this case, the leader node removes the leaving node's ID from the managed list in this local platoon. In the next CCH-I, the leader node broadcasts the changed information through its WSA message, providing tacit approval for a leaving maneuver. If a leaving node is one of the follower nodes, it should conduct a lane change to leave after it transmits its PCM. Due to this leaving node, there occurs a large distance gap in the middle of the local platoon. To maintain the safety distance specified by its leader node, the follower node behind the leaving node controls its speed, using the distance gap from its new preceding node and the position and acceleration of the PCM received from the leader node in a given superframe. If a leaving node is the leader node, the leaving maneuver enables the member node just behind it to inherit the leader node's role. Therefore, the member node just behind the leader node should become a new leader node at the next CCH-I. The new leader node transmits its WSA message after it constructs its node list excluding the ID of the leaving leader node. The expected case requires one synchronization interval for completing it.

The unexpected leaving maneuver is more complex than the expected leaving maneuver. An unexpected case may occur when a node cannot transmit a PCM within a given superframe due to a communication failure. A leader node is responsible for detecting an unexpected case within a given superframe. As soon as detecting it, the leader node determines ruling out this failure node and removes this node ID from the managed list. According to the distributed coordination in DUPP, all member nodes behind the failure node are also removed from the local platoon since they are not allowed to transmit their PCM. After that, they can perform creation, joining, or merging maneuvers after they are eliminated against the local platoon. In other words, performing these maneuvers is preferred than accepting a potential risk that may occur if they drive during a certain grace period without the latest PCMs. The leaving maneuver of the unexpected case is completed when the leader node transmits the updated WSA message in the next CCH-I. Accordingly, the leaving maneuver is also finished within one synchronization interval.

On the road, the eliminated nodes should perform one among the creation, joining, splitting, and merging maneuvers. Determining an appropriate maneuver is based on the number of the eliminated nodes. We define the length of the local platoon as ρ and the driving order number of a given node in the local platoon as i. The leader node (i.e., $i = 1$) and the rear node (i.e., $i = \rho$) are not related to the decision depending on the number of eliminated nodes since they comply with the expected leaving maneuver. Depending on driving conditions as discussed above, it can perform either creation or joining maneuver. When the unexpected leaving node is identified by the driving order number i ($i \neq 1$), the i-th node divides the given local platoon into a front group and a rear group. The front group is part of the given local platoon and is maintained as a local platoon by the leader node (i.e., $i = 1$). However, the rear group consists of the eliminated nodes and is not the local platoon because the leader node does not exist in that. If the number of the eliminated nodes in the rear group is less than or equal to three, the eliminated nodes independently and sequentially perform joining maneuvers.

When the number of the eliminated nodes in the rear group is greater than three, they perform a splitting maneuver. To improve the efficiency of DUPP, the split local platoon is designed to merge into a small local platoon. Among the eliminated nodes, the first driving nodes start the splitting maneuver for eliminated nodes. After the splitting maneuver, a new rear local platoon should merge into the front local platoon according to the assumptions of the merging maneuver (as described in Section 3.2.2). This is because it is more efficient to perform the joining maneuver than performing the splitting maneuver when the maximum number of the eliminated nodes is three. The join maneuver requires at least one to three synchronization intervals depending on the number of the eliminated nodes. However, the splitting maneuver accompanied by the merging maneuver requires the minimum of three synchronization intervals: an interval to confirm the elimination by the WSA message of the leader node in the front local platoon, an interval to create a new rear local platoon by sending a WSA message, and an interval to perform the merging maneuver.

If the failed node recovers its communication capability after performing any maneuver of the eliminated nodes, this node can perform either the creation or joining maneuver according to the creation and join conditions.

3.2.5. A Splitting Maneuver

A local platoon might be physically split when all nodes of the local platoon cannot completely cross the intersection because of the traffic signal after the leader node of the local platoon goes into an intersection. Furthermore, a certain vehicle might interrupt the smooth flow of a local platoon when the local platoon is unstable. It leads to the separation of the local platoon in the DUPP. As discussed above, a group consisting of the nodes eliminated from the existing local platoon can determine performing the splitting maneuver according to the local platoon condition. Among the nodes that will be separated physically, the member node driving at the front is responsible for the splitting maneuver and becomes a new leader node. Therefore, the new leader node creates a new local platoon (i.e., a rear local platoon) separated from the given local platoon, and the new platoon consists of the existing separated nodes. The new leader node should transmit its WSA message in the next CCH-I for starting splitting maneuver which is the same as the creation maneuver. In contrast to the creation maneuver, however, they do not transmit their beacon messages during BP since the member nodes already belong to the new local platoon. The new leader node transmits a PCM with the value of 0x03 in the Maneuver field and the list of the existing member nodes that are separated from the previous leader node's list. All member nodes of the new local platoon sequentially transmit their PCM with the new list of nodes and the value of 0x03 in the Maneuver field. In this regard, the splitting maneuver is completed within one synchronization interval.

3.3. Analytic Hierarchy Process-Based Forwarder Selection

The EBP is an optional period activated depending on whether a certain event occurs or not. In DUPP, certain events with a negative effect on urban platooning are pre-defined. They include emergency brake lights, hazard locations of dangerous curves, road construction and obstacles on the road, and road conditions such as heavy rain, snow, and slippery road. When nodes in the local platoon detect any such pre-defined event, they may transmit redundant messages using the wireless channel. To address this problem, we regulate the number of nodes for event notification by selecting only one forwarder among vehicles. As soon as CFP is finished by receiving PCMs of all nodes in the local platoon, each node performs AHP-based selection to determine the one-time forwarder using information distributed in PCMs.

The AHP approach is useful in systematically solving the problem of decision-making that may be differentiated depending on the degree of influence of the interrelated and complex criteria required for decision making [33]. In the AHP approach, the definition of criteria and the calculation of their weight are critical to assess the alternatives. The structure of the proposed AHP-based forwarder selection of DUPP is shown in Figure 4. The fundamentals of AHP consist of the definition of criteria, the pairwise comparison between criteria, pairwise comparison between alternatives, and the priority calculation for achieving an objective [34–37]. The proposed AHP-based forwarder selection of DUPP follows the AHP's fundamental processes, focusing on the development of the following four items: the definition of three criteria, the definition of grades for each criterion and for each node, the calculation of the decision weight, and the priority decision. The AHP methodology allows DUPP to determine which alternative is the most consistent with our three criteria and the degree of importance.

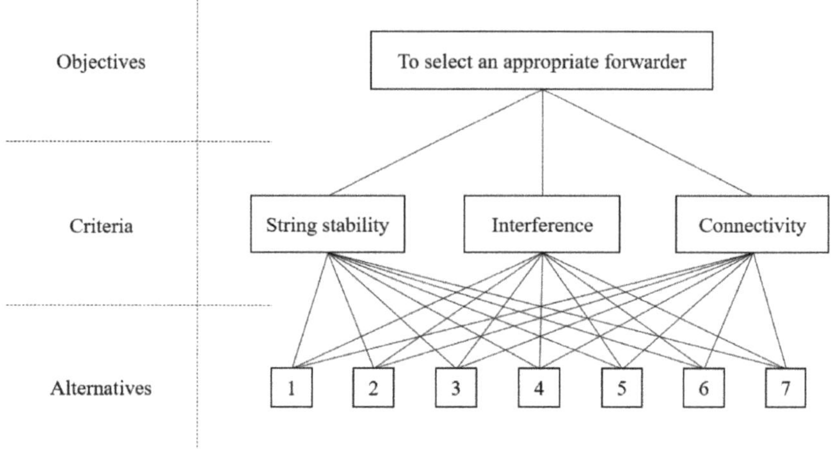

Figure 4. A structure for the AHP-based selection method.

3.3.1. Definition of Three Criteria

In the first step, to select an appropriate forwarder (i.e., objective in AHP) among all nodes (i.e., alternatives in AHP) in a local platoon, the DUPP defines three criteria consisting of string stability, interference, and connectivity. In AHP-based forwarder selection, DUPP intends to select a node heavily affected by the detected event to become a forwarder. DUPP also considers the individual communication conditions of the nodes in the local platoon. The string stability is related to the degree of the detected event's effect and both interference and connectivity represent the communication capability of a node. After the priority for each node is calculated using three criteria, the node with the highest priority is determined as a forwarder in a given superframe.

The string stability refers to the stability of a driving node in a local platoon and indicates how well the node keeps the safety distance required by the leader node in the local platoon. The string stability of node i at time t, denoted as $S_i(t)$, is given as:

$$S_i(t) = \left| \frac{ep_i(t) - \{rp_i(t') + v_i(t - t') \times (t - t')\}}{d_{safe}} \right|, \tag{1}$$

where $ep_i(t)$ is the expected position of node i at time t, $rp_i(t')$ indicates i node's position given at time t', $v_i(\Delta)$ is the average speed of node i during Δ that is the amount of time elapsed from the time t', and d_{safe} indicates the safety distance the nodes comply with depending on a given headway time between nodes. The time t' is the time at which node i has received a leader node's PCM in a given superframe. Therefore, the given time t refers to the time Δ after time t'. $ep_i(t)$ is estimated with the position of node i at time $t - 1$, the acceleration value in the longitude acceleration field of the PCM received from a leader node at time $t' - 1$, and the amount of time elapsed from time $t - 1$ to time t. Therefore, the closer the value of S_i is to 0, the better the distance between nodes is maintained and the higher the stability is. The node with a high value in the stability has a high probability to be selected as a forwarder.

The interference criterion represents the degree of interference resulting from the data transmission of a given node on other nodes outside the local platoon for their transmission. The interference of node i at time t, denoted as $I_i(t)$, is given as:

$$I_i(t) = \frac{m_i(t)}{M_i},$$

(2)

where m_i is the number of all nodes on the road that node i can interfere with, regardless of whether they belong to the local platoon, and M_i indicates the maximum number of nodes that can exist in the transmission range of node i. The maximum number of nodes in the transmission range of node i is defined by the vehicle's length, safety distance, road units, and lanes. It is assumed that one road segment is divided into small road units of a certain size (in this paper, it is given as 20 m). When a node tries to transmit data, many road units can be included within its transmission range. Meanwhile, a vehicle occupies a certain size of space on the road. It is reasonable to assume that a vehicle needs space equal to the length of the vehicle plus the safety distance. For instance, when the length of a vehicle is 4 m, it needs about 17.3 m of space on the road when its speed is given as 60 km/h and the time headway is given as 0.8 s in urban platooning. This length may vary depending on the type of vehicle. Considering the situation of a small road unit filled with vehicles for each lane, the total number of vehicles, n, within a small road unit can be calculated. When a road unit consists of multiple lanes, it is the total number of vehicles in a road unit multiplied by the number of lanes. In this regard, M_i is defined as:

$$M_i = n * \sum_{r=1}^{u} L_r,$$

(3)

where n is the total number of a vehicle within road unit r, u is the number of road units within the transmission range of node i, and L_r is the number of lanes of the road unit r. Therefore, the closer the value of I_i is to 0, the less the interference with other nodes is. In terms of forwarder selection, the node with a low interference level has a high probability to be selected as a forwarder.

The connectivity criterion represents the capability of data dissemination of a given node. The connectivity of node i at time t, denoted as $C_i(t)$, is given as:

$$C_i(t) = \frac{n_i(t)}{\rho},$$

(4)

where $n_i(t)$ is the number of nodes behind node i in the local platoon at time t and the current number of nodes in the local platoon is denoted as ρ. Therefore, the closer the value of C_i is to 1, the better the communication capability is. In terms of forwarder selection, the node with high connectivity level has a high probability to be selected as a forwarder.

3.3.2. Pairwise Comparison between Criteria

The AHP determines the relative superiority of the alternatives after the relative importance of the criterion is determined through a pairwise comparison [35]. DUPP performs the first pairwise comparison in each criterion during the second step and the second pairwise comparison to all nodes for each criterion during the third step. The first one aims to compare the importance of criteria through pairwise comparisons, two at a time. The second one aims to compare the importance of nodes, two at a time, through pairwise comparisons. These pairwise comparisons are used to generate a decision weight set for each criterion and each node, respectively, as a decision weight vector. To perform the first pairwise comparison of each criterion, a general grade is defined as shown in Table 1. Using the grade in AHP provides a way to include experience and knowledge of the DUPP in an intuitive way for DUPP [35]. We divide the importance into three grades and give each grade a score from one to five increasing by two points to widen the difference in grade.

We weight the criteria using the general grade. In other words, this general grade enables DUPP to represent the preference between criteria by assessing them.

Table 1. A general grade used for pairwise comparison to criteria.

Grade	Description
1	Equally important
3	A little important
5	Very important

When each node performs a pairwise comparison to criteria, DUPP requires constructing an n-by-n decision matrix (i.e., $n = 3$) which is a square matrix [38]. In the AHP-based forwarder selection, the decision matrix is generated using the general grade in order to represent relative importance between defined criteria. As discussed above, in DUPP, a node that is heavily affected by the detected event should transmit the related information to many nodes (the more, the better) without affecting other nodes in terms of communication. In this regard, the string stability is considered to have higher importance than the interference. The connectivity is designed to have the lowest importance. In detail, depending on the general grade of Table 1, the relative importance indicates that the importance of the string stability to the interference is five-point and the importance of the interference to the string stability gets the reciprocal of this five-point. Note that the decision matrix may vary according to the definition of relative importance with different grades [33].

The results of the relative importance between the two criteria are presented in the n-by-n decision matrix. Table 2 shows how the criteria are rated against each other. In Table 2, the result of the decision matrix is denoted as $\mathbf{M}_p(S, I, C) = [a_{ij}]$, $1 \le i, j \le n$ for string stability denoted as S, interference denoted as I, and connectivity denoted as C. From the n-by-n decision matrix, a decision weight vector (i.e., n-by-1 matrix) is calculated as the normalized eigenvector corresponding to the largest eigenvalue of a pairwise comparison matrix as follows. A pairwise comparison matrix denoted as $\mathbf{P}_p(S, I, C) = [b_{ij}]$, $1 \le i, j \le n$, for defined criteria is constructed through normalization dividing each element of the decision matrix by the total value of the corresponding columns. For instance, b_{11} is obtained by dividing a_{11} by 1.533 that is the total sum of the values in the first column as shown in Table 2. After that, weight values to each criterion are calculated by dividing the sum of the values of all columns of each row by the number of rows. The vector of decision weights, denoted as $\mathbf{W}_p(S, I, C) = [w_k]$, $1 \le k \le n$, is generated for three criteria and should satisfy the conditions $0 \le w_k \le 1$ and $\sum_1^n w_k = 1$. The results of the decision matrix, the pairwise comparison matrix, and the decision weight vector are summarized in Table 2. Each value of the calculated eigenvector $\mathbf{W}_p(S, I, C)$ is considered as the predetermined decision weight for each criterion, respectively.

Table 2. The results of the decision, pairwise comparison, and decision weight matrices.

Criteria	$\mathbf{M}_p(S, I, C)$			$\mathbf{P}_p(S, I, C)$				$\mathbf{W}_p(S, I, C)$
	S	I	C	S [1]	I [1]	C [1]	Subtotal [1]	Decision Weight [2]
S	1	3	5	0.652	0.692	0.556	1.900	0.637
I	1/3	1	3	0.217	0.231	0.333	0.781	0.258
C	1/5	1/3	1	0.130	0.077	0.111	0.318	0.105

[1] All values are rounded to three decimal places. [2] Its consistency rate to assess the consistency of the comparison matrix is 4%.

3.3.3. Pairwise Comparison between Alternatives

DUPP performs the second pairwise comparison for all nodes in the same way as performing the first pairwise comparison to the criteria. The second pairwise comparison for all nodes is independently performed in terms of the string stability ($S_i(t)$, $1 \leq i \leq \rho$), the interference ($I_i(t)$, $1 \leq i \leq \rho$), and connectivity ($C_i(t)$, $1 \leq i \leq \rho$) at time t when the number of the nodes belonging to the local platoon is given as ρ. In other words, DUPP compares the ρ node alternatives to the criteria. For each criterion, determining the relative grade between nodes is based on the difference of the given criterion of nodes. The difference between node i and j for a given criterion χ which is defined as Equation (5).

$$\Delta_{ij}(\chi) = |\chi_i - \chi_j| \tag{5}$$

If the difference $\Delta_{ij}(\chi)$ ranges from a given low-boundary denoted as $\beta_{low}(\chi)$ to a given high-boundary denoted as $\beta_{high}(\chi)$, node i is given as the defined grade and the grade of node j is given as the reciprocal of the grade of node i. Hence, the relative grade for pairwise comparison can be described by the following equation:

$$\beta_{low}(\chi) < \Delta_{ij}(\chi) \leq \beta_{high}(\chi) \tag{6}$$

We divide the different values into five grades and give each grade a score from one to nine increasing by two points. Hence, we provide the relative grade of nodes' pairwise-comparison according to the range defined by experimental values as shown in Table 3 [33]. When giving the relative grade between two nodes, it depends on the difference between two nodes for each criterion. For interference, in the comparison with node i and node j, node j obtains one from the relative grade but node i has the reciprocal of the relative grade obtained when the difference between node i and node j is positive (i.e., $\chi_i - \chi_j > 0$). For string stability and the connectivity between node i and node j, node i obtains one among the relative grade but node j has the reciprocal of the obtained relative grade when the difference between node i and node j is positive (i.e., $\chi_i - \chi_j > 0$).

Table 3. A relative grade used for the pairwise comparison of nodes.

$\Delta_{ij}(\chi)$		$\beta(\chi)$				
$\Delta_{ij}(S_i(t))$	$\beta_{low}(S_i(t))$	0	0.3	0.5	0.7	0.9
	$\beta_{high}(S_i(t))$	0.1	0.5	0.7	0.7	1
$\Delta_{ij}(I_i(t))$	$\beta_{low}(I_i(t))$	0	0.3	0.5	0.7	0.9
	$\beta_{high}(I_i(t))$	0.1	0.5	0.7	0.7	1
$\Delta_{ij}(C_i(t))$	$\beta_{low}(C_i(t))$	0	0.1714	0.3428	0.5142	0.6857
	$\beta_{high}(C_i(t))$	0.1714	0.3428	0.5142	0.6857	0.8571
Grade		1	3	5	7	9

Based on Table 3, the decision matrix (i.e., $n \times n$ matrix, $1 \leq n \leq \rho$) for all nodes is constructed by the same method as for constructing the decision matrix $\mathbf{M}_p(S, I, C)$ for each criterion above. Three matrices of the pairwise comparison to all nodes are constructed for each criterion by using the defined relative grade of Table 3. The priority weight matrix of a node for all criteria, denoted as $\mathbf{W}_{i(t)} = \begin{bmatrix} W_{S(t)} & W_{I(t)} & W_{C(t)} \end{bmatrix}$ (i.e., 1×3 matrix at time t), consists of the priority weight value for each criterion. To help understand the AHP-based forwarder selection technique, we provide an example of the priority weight matrix at time t, which is summarized in Table 4. The values of this priority weight matrix are changed whenever a one-time forwarder is determined. This is because the information for AHP-based selection is synchronized through PCMs in the local platoon.

Table 4. A priority weight matrix for all nodes at a given time.

Node ID	$W_{S(t)}$ [1]	$W_{I(t)}$ [2]	$W_{C(t)}$ [3]
1	0.076532	0.205409	0.350396
2	0.076532	0.205409	0.237473
3	0.076532	0.179435	0.158966
4	0.31917	0.157457	0.105558
5	0.298168	0.100314	0.069645
6	0.076532	0.083507	0.046163
7	0.076532	0.06847	0.031798

[1] Its consistency ratio to assess the consistency of the comparison matrix is 5.9%. [2] Its consistency ratio to assess the consistency of the comparison matrix is 3.6%. [3] Its consistency ratio to assess the consistency of the comparison matrix is 3.5%.

3.3.4. Priority Calculation for Achieving an Objectives

In the final step on the AHP-based forwarder selection, using the priority weight matrix of all nodes, each node selects the appropriate node with the highest priority as a forwarder. The priority of a node for a forwarder selection, denoted as $P_i(t)$, is a weighted sum of the priority weight for node i ($\mathbf{W}_{i(t)} = \begin{bmatrix} W_{S(t)} & W_{I(t)} & W_{C(t)} \end{bmatrix}$) based on the decision weight of each criterion (i.e., $\mathbf{W}_p(S, I, C) = [w_k]$, $1 \leq k \leq n$) and is given as follows:

$$P_i(t) = w_1 \times W_{S(t)} + w_2 \times W_{I(t)} + w_3 \times W_{C(t)} \tag{7}$$

From Equation (7) using the decision weight vector of Table 2 and the priority weight matrix of the example at time t in Table 4, we derive that Node 4 is determined as a forwarder using Table 5 which shows the calculated priorities of each node.

Table 5. The priority of nodes.

Node	Priority
1	0.138643
2	0.126786
3	0.111789
4	**0.254057** [1]
5	0.222135
6	0.075004
7	0.069586

[1] The highest priority

4. Performance Evaluation

For performance evaluation in terms of flexibility, adaptability, and stability, we use the PLEXE simulator that integrates the traffic simulator Sumo with the networking simulator OMNet++ [39]. Our experiment uses real traffic data in the particular area of New York City in which there exist vehicles without communication capability. However, to construct a mixed traffic environment in which vehicles with or without communication capability coexist, the ratio of vehicle generation with communication capability is adjusted from the actual traffic volume. We assume that in the experiment, all vehicles with communication capability participate in urban platooning.

4.1. Experimental Environment

In this subsection, we describe the experimental environment to evaluate the performance of DUPP. The road network used for the experiment corresponds to the map extracted from Open Street Map (OSM), which is shown in Figure 5.

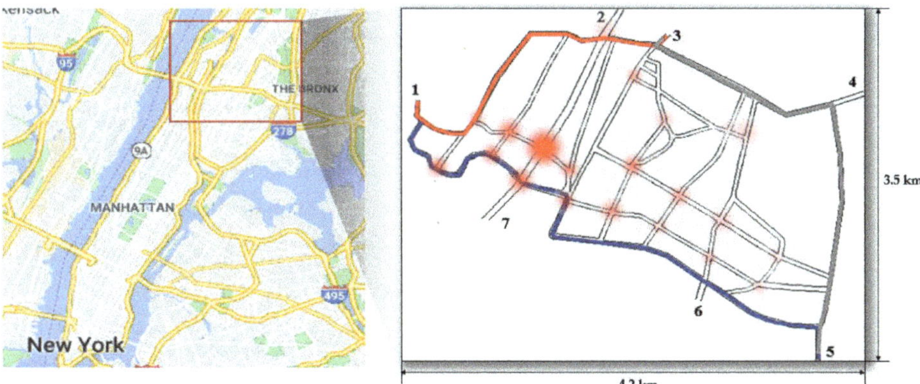

Figure 5. The road network used for the experiment.

Our experiment is performed with the actual traffic data of New York City, collected for 24 h, using a particular area of the road network of New York City with a size of 4.2 × 3.5 km [40]. In this paper, the actual traffic data indicates traffic volume (i.e., the total amount of vehicles passing through a given route) measured for a certain period of time in a given road network. From measured actual traffic data, we get a vehicle generation ratio using the number of vehicles entering the road network by time. To obtain the generation rate per entry point, the vehicle generation ratio is divided by the number of entry points in the road network. In this experiment, vehicles are generated at seven entry points which are numbered from 1 to 7 as shown in Figure 5. For each of the entry points, a vehicle is generated by a Poisson distribution with a different generation ratio depending on the time, denoted as λ (vehicles per second). The red circle represents the vehicle distribution and the clearer the circle, the more vehicles there are. Table 6 shows the generation ratio of vehicles for time. For 24 h, a total of 41,698 vehicles enter the road network. The peak time in which vehicles are generated the most for 24 h is from 4 to 5 p.m. The amount of traffic flow generated during this time period will affect the traffic conditions of the next time unit (i.e., 5 to 6 p.m.). The maximum speed of the vehicle is regulated as 60 km/h. Vehicles are randomly assigned the entry and exit points, and travel along a determined path in the road network.

Table 6. The generation ratio of vehicles for time.

Time		0	1	2	3	4	5	6	7	8	9	10	11
	From	0	1	2	3	4	5	6	7	8	9	10	11
	To	1	2	3	4	5	6	7	8	9	10	11	12
λ	a.m.	0.061	0.046	0.036	0.028	0.025	0.022	0.024	0.031	0.039	0.046	0.059	0.067
	p.m.	0.084	0.101	0.12	0.122	0.132	0.122	0.118	0.103	0.091	0.072	0.059	0.046

Every vehicle should periodically broadcast a BSM on every 100 ms during the CCH-I. When vehicles perform another non-safety application regardless of the urban platooning, during SCH-I, they periodically broadcast the application-related message with the size of 100 bytes. The vehicles of the local platoon should periodically exchange PCMs with all nodes in a local platoon. All vehicles exchange messages based on the CSMA/CA mechanism of IEEE WAVE. In Table 7, we summarize the experimental environment.

Table 7. Parameters defined for the experimental environment.

Parameters	Values
BSM size	182 bytes
Physical and MAC layers	IEEE 802.11p
Channel coordination	IEEE 1604.4
Bitrate	27 Mbps
Tx range	1 km
Propagation model	Two-ray interference model [41]
Maximum length of a local platoon	7 vehicles
Speed limit	60 km/h
Experiment time	24 h
Map size	4200×3500 (m)

In this environment, we consider the effect of the number of vehicles not interested in an urban platooning on the performance of the DUPP. Therefore, three different scenarios are created by varying the generation ratio of the vehicles with communication capability to 100%, 60%, and 30%, respectively. These scenarios are used in both DUPP and ENSEMBLE. We denote three scenarios for DUPP as DUPP-100, DUPP-60, and DUPP-30, respectively and for ENSEMBLE, three scenarios are expressed as ESB-100, ESB-60, and ESB-30, respectively. It is referred to as NONE when there are only normal driving vehicles that have no communication functionality and never participate in urban platooning in this experiment.

4.2. Experimental Results

We evaluate the effectiveness of the proposed DUPP by comparing its performance with the performance of ENSEMBLE in terms of the travel time of vehicles, the lifetime of an existing local platoon, the success ratio of FAP maneuvers, the external cost of PCM transmission, the periodicity of PCM transmission, the adaptability to unexpected situations, and the forwarder selection ratio in a local platoon.

4.2.1. Vehicle Travel Time

The travel time of a vehicle is one of the significant indicators representing the level of improvement by the proposed DUPP in terms of road capacity and efficiency. It is measured as the time to take a vehicle to drive from an entry point to an exit point. In this experiment, to effectively show the general tendency of the travel time over the vehicle density by time zone, the day is divided into 24 time zones. For instance, time zone 1 indicates a period of 1 to 2 a.m. Figure 6 shows the average travel time of vehicles in each time zone for each protocol (i.e., DUPP, ENSEMBLE, and NONE). Figure 7 presents the results for each distance of given routes for each protocol. In Figure 6a, the tendency of the individual results of DUPP-30 and DUPP-60 is similar to the tendency of the result of DUPP-100. In addition, it is shown that they are also superior in terms of average travel time compared with NONE and ENSEMBLE-30 and 60. Therefore, to further highlight the difference among their results in terms of performance more carefully, all results presented in Figures 6b and 7 are generated under the scenarios of DUPP-100 and ESB-100. As shown in Figure 6b, for 24 h, DUPP shows the best performance in terms of the average travel time. When there are few vehicles such as time zone 3 to 6, there is little difference in the average travel time among the DUPP-100, ESB-100, and NONE scenarios. However, it is clearly shown that the difference in the average travel time gets larger as the vehicle density increases gradually at time zone 6 to 16. This difference becomes prominent after the generation ratio of vehicles is the highest (i.e., time zone 17). In that time zone, the average travel time of DUPP-100 is 28.3% less than that of NONE (i.e., 1071 s) and 20.4% less than that of ESB-100 (i.e., 965 s). Moreover, as the number of DUPP-enabled vehicles increases, it causes a reduction in the average travel time. On the other hand, the performance of the ENSEMBLE-enabled vehicles is better than that of NONE, but its performance is worse than that of DUPP. The difference in the performance between DUPP and ENSEMBLE is derived from the efficient and adaptable FAP maneuvers in a dynamic road environment. It becomes clear by the number of the local platoons as shown in Figure 8c. The experimental result showing that DUPP maintains more urban platoons on the road than ENSEMBLE indicates that DUPP is more adaptable to urban roads.

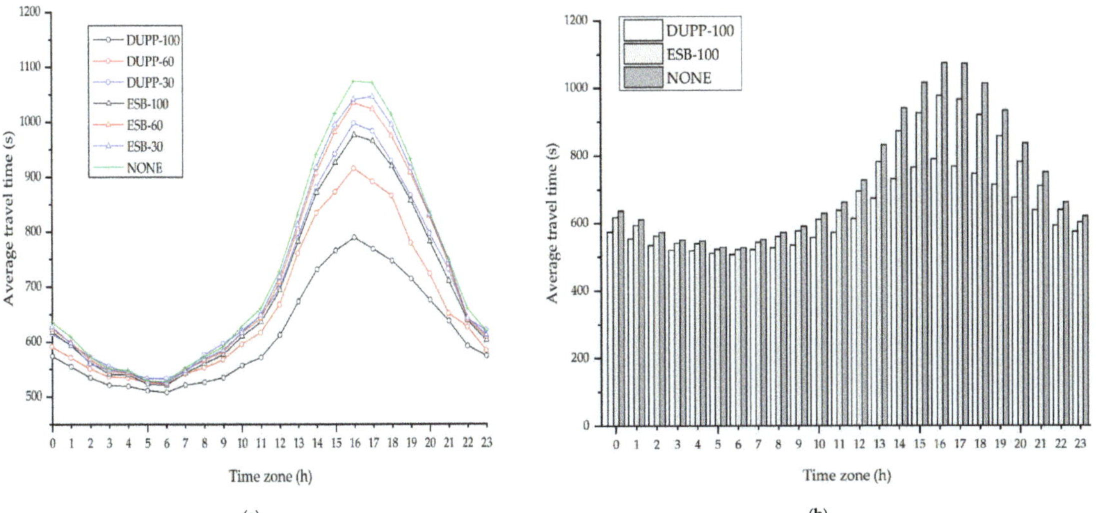

Figure 6. Average travel time according to the time zone for each protocol: (**a**) average travel time; (**b**) average travel time for DUPP-100, ESB-100, and NONE.

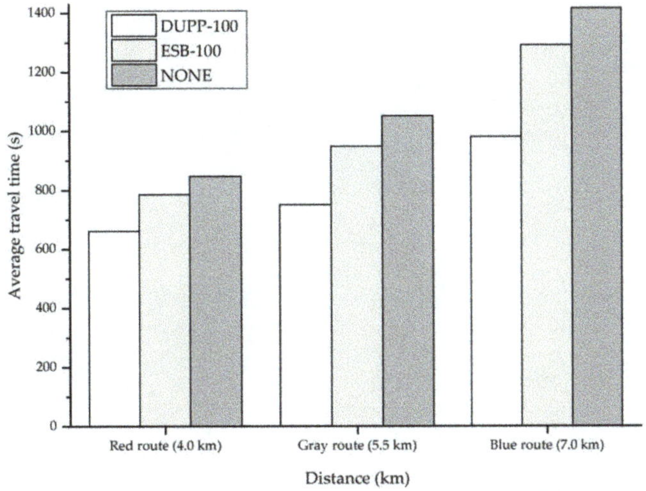

Figure 7. Average travel time over the driving distance at peak time zone 16.

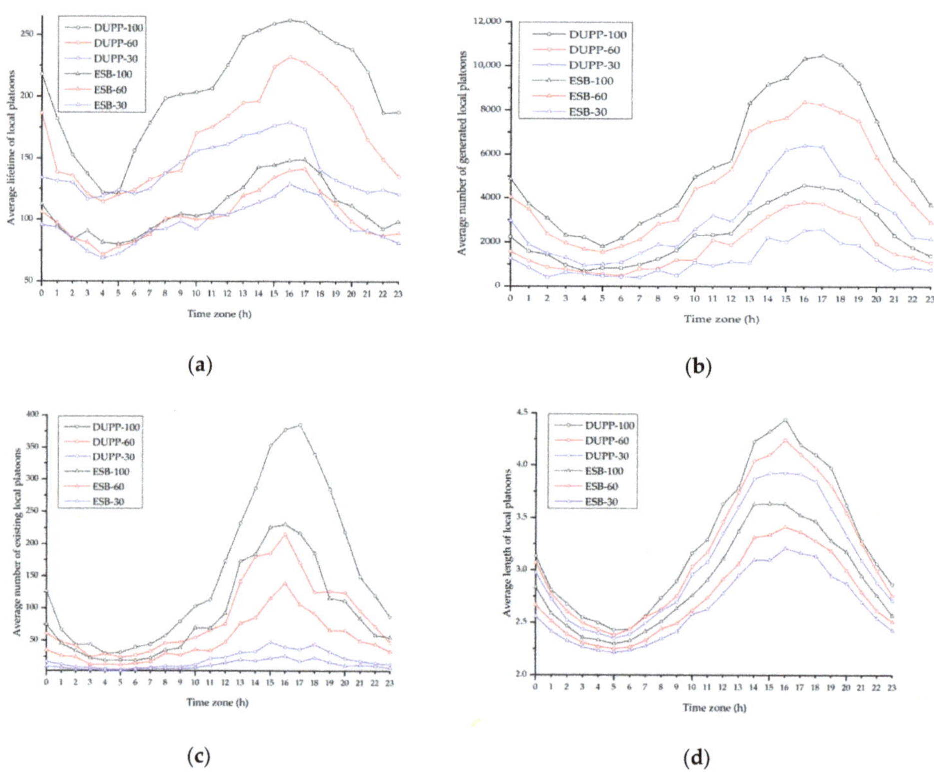

Figure 8. Lifetime of local platoons: (**a**) average lifetime of local platoons, representing the continuity of them; (**b**) average number of generated local platoons, representing the stability of them; (**c**) average number of existing local platoons, representing the stability of them; (**d**) average length of local platoons, representing the durability of them.

To examine whether DUPP affects the travel time as the driving distance varies, we present the average travel time in the peak time zone 16 (i.e., 4 to 5 p.m.) for various distances in Figure 7. Although there exist many routes in the road network, only three of all routes are selected to effectively analyze the performance difference among DUPP, ENSEMBLE, and NONE. The three routes are presented in Figure 5: a red route with a length of 4 km, a gray route with a length of 5.5 km, and a blue route with a length of 7 km. These routes are chosen based on several aspects, including more than 150 vehicles per hour, at least five intersections, and a total distance of at least 3 km. For instance, in Figure 5, we exclude the routes (e.g., entry point 1 to exit point 7) that do not have enough vehicles and the routes (e.g., entry point 2 to exit point 3) that are too short to generate a sufficient amount of measured data.

In the case of the red route with 4.0 km, the average travel time for DUPP-100 corresponds to 84% of that of ESB-100 and 78% of that of NONE, respectively. In the case of the blue route with 7 km, the average travel time for DUPP-100 corresponds to 76% of that of ESB-100 and 69% of that of NONE, respectively. As the route is lengthened, not only the difference between DUPP-100 and ESB-100 but also the difference between DUPP-100 and NONE increases. Furthermore, the average travel time of DUPP-100 does not increase significantly even when comparing the red route with the blue route. DUPP-100 increases by only 319 s from the red route to the blue route while ESB-100 increases by 506 s. Especially, an increase in the distance of the route indicates that the road's uncertainty increases. In other words, due to many signalized intersections and unexpected traffic conditions, vehicles might perform different urban platooning maneuvers frequently. Hence, under high uncertainty, it is difficult for ENSEMBLE-enabled urban platooning to maintain the existing local platoons, since an existing local platoon is destroyed even when one of the vehicles leaves its local platoon. As a result, as the driving distance increases, ENSEMBLE's performance may deteriorate.

4.2.2. Platoon Lifetime

Another performance indicator is the lifetime of an existing local platoon. This performance metric can illustrate if the protocol has the capability to respond quickly to the dynamic topology of the urban environment. In addition, it is significantly associated with the durability, continuity, and stability of existing local platoons. For each protocol, we measure the average lifetime until the local platoon is destroyed after a local platoon is created, the average number of the local platoons generated and maintained, and the average length of the generated local platoons for 24 time zones. The measured results are illustrated in Figure 8. In Figure 8a, DUPP has a tendency that the average lifetime of the local platoons increases as the vehicle density on the road increases. An increase in the vehicle density can be explained by the three scenarios in which the number of vehicles participating in the urban platooning increases (i.e., DUPP-30, DUPP-60, and DUPP-100) and the flow of time towards the peak time zone 16. In DUPP, the inflow and outflow of other nodes to the existing local platoons are performed immediately according to the traffic conditions. Since local platoons are maintained as long as possible even though the nodes are leaving or splitting from it, DUPP shows better performance as vehicle density increases.

In DUPP, the local platoons may be separated into smaller local platoons as the road uncertainty increases but they merge soon. As shown in Figure 8b,c, DUPP-enabled local platoons are maintained relatively long after they are generated, while ENSEMBLE-enabled local platoons are mostly generated, instead of maintaining the existing local platoons. Note that even if the vehicles also perform the merging and joining maneuvers in ENSEMBLE, existing local platoons are easily destroyed as the road uncertainty increases. Although the merging maneuver is likely to occur frequently when there are a lot of small local platoons, it is difficult for the local platoons to merge even at the peak time zone in ENSEMBLE. A large number of generated local platoons as shown in Figure 8b indicates that only creation and joining maneuvers, and the destruction of the local platoons are repeated in ENSEMBLE as the vehicle density increases.

The lifetime of the local platoon is also examined in more detail in connection to the average length of the local platoons of Figure 8d. It shows that the local platoons in DUPP-100 have the longest average length. In DUPP, merging and joining maneuvers occur more actively as the vehicle density increases, positively affecting the length and the lifetime. In contrast to DUPP, ESB-100 shows a shorter length and a lower number of the existing local platoons than those in DUPP-30 since ENSEMBLE does not maintain the existing local platoon whenever nodes perform leaving or splitting maneuvers from it. Therefore, in ENSEMBLE, the average length of the existing local platoons is bound to be small as shown in Figure 8d.

4.2.3. Success Ratio of Maneuvers

To demonstrate the high performance of DUPP-enabled urban platooning, we present the success ratio of FAP maneuvers including creation, joining, leaving, splitting, and merging. In this paper, we present only the result of the joining maneuver in Figure 9 since the success ratios from all maneuvers have a similar tendency. The success ratio of urban platooning is affected by external factors such as communication capability and road environments. The dynamic road environment including unexpected obstacles and signalized intersections may require vehicles to perform different platooning maneuvers. If an obstacle exists between a rear node and a candidate node, communication interference might also occur. Furthermore, a high density of vehicles with communication capability may lead to communication interference. Hence, the failure of the joining maneuver is analyzed in terms of communication capability and road environments in Figure 9b–d.

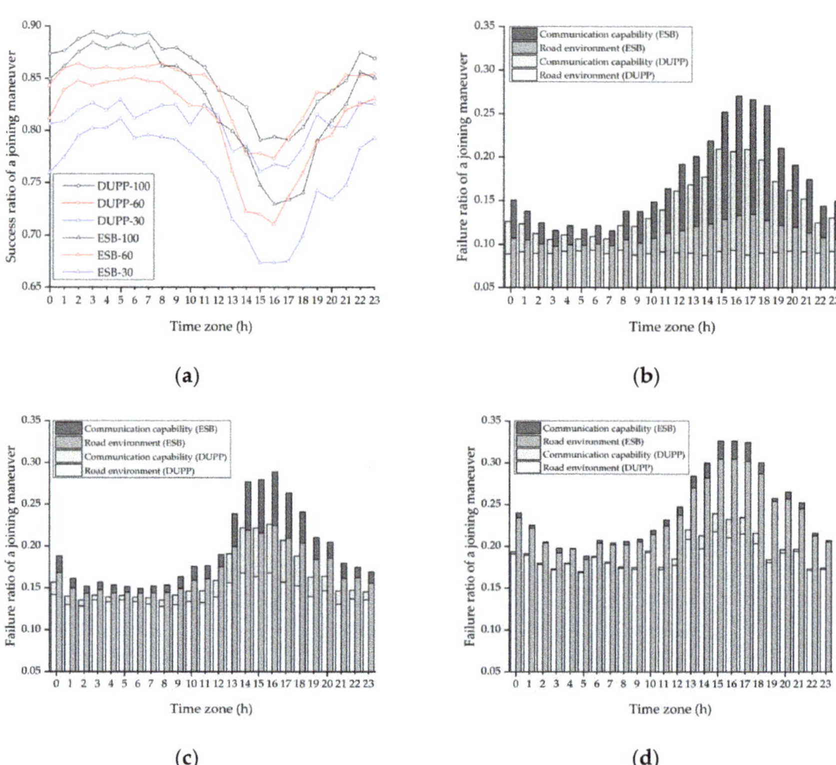

Figure 9. Success ratio of joining maneuver: (**a**) success ratio of a joining maneuver; (**b**) failure ratio of a joining maneuver under DUPP-100 and ESB-100; (**c**) failure ratio of a joining maneuver under DUPP-60 and ESB-60; (**d**) failure ratio of a joining maneuver under DUPP-30 and ESB-30.

For DUPP and ENSEMBLE, Figure 9a shows that the success ratio is getting lower when the number of participating nodes becomes smaller and the vehicle density becomes higher. The number of vehicles participating in the urban platooning (i.e., DUPP-30, DUPP-60, and DUPP-100) represents the degree of communication capability of vehicles on the road. When the vehicles perform the joining maneuver under the scenarios of DUPP-100 and ESB-100, the vehicles in ESB-100 fail much more than those in DUPP-100 as the vehicle density increases. When the number of vehicles participating in urban platooning is the smallest among our scenarios (i.e., DUPP-30 and ESB-30), we can also see that ENSEMBLE's performance is lower than that of the DUPP-30 for quite a long time (12 to 8 p.m.). Specifically, based on the peak time zone, the failure ratio of DUPP-100 is 0.21 and that of ESB-100 is only 0.31 as shown in Figure 9b. As the number of vehicles with communication capability decreases, the difference between their failure ratios widens. In other words, as shown in Figure 9c,d, the failure ratio of the ESB-60 and ESB-30 increase to 0.37 and 0.48, respectively, while the failure ratios of DUPP-60 and DUPP-30 have increased very slightly to 0.222 and 0.24, respectively.

In the case of the fewest vehicles with communication capability (i.e., DUPP-30), due to road environments, it is possible for the joining maneuver to be blocked more frequently by other vehicles when vehicles attempt to join an existing local platoon. The results of Figure 9b–d also show that the DUPP's success ratio is affected more by the road environment as the number of vehicles participating in the urban platooning decreases. Nevertheless, the joining failure in DUPP-30 due to the road environment tends to rarely increase even as it approaches peak time zone 16. Moreover, for all vehicles with communication capability (i.e., DUPP-100), the failure ratio derived from the road environment in DUPP is almost constant as 0.09 as shown in Figure 9b. In other words, in DUPP-100, the impact of the road environment on the performance is negligible. This indicates that its performance is only affected by communication interference since a local platoon is likely in the fully saturated condition while performing the joining maneuver.

4.2.4. Average Drop Ratio

The external cost of PCM transmission is measured as the drop ratio of BSMs on CCH to show the extent of the operations of DUPP affecting the performance of other vehicles regardless of given urban platooning. The average drop ratio of BSMs is measured by dividing the number of received BSMs by the number of transmitted BSMS for all nodes. Figure 10 shows the results of the average drop ratio of DUPP and ENSEMBLE as bar graphs. To fairly compare the performance of the two protocols, we first define the normal scenario, which is a certain driving condition where there exist only vehicles with communication capability and not to participate in urban platooning. The average drop ratio of the normal scenario is shown as a line graph between the upper and lower dashed lines in Figure 10. The upper and lower dashed lines indicate the maximum and minimum drop ratios in the normal scenario, respectively. The average drop ratios of DUPP-100, DUPP-60, and DUPP-30 are 0.063, 0.036, and 0.009, respectively. The average drop ratio of DUPP is usually within the range of variation of the result of the defined normal scenario and the slight difference between DUPP and the normal scenario is derived from the WSA transmission in the operation of DUPP. On the other hand, since ENSEMBLE operates only on CCH, its drop ratio increases significantly. In addition, BSM transmissions cannot be guaranteed while the vehicle density increases due to the nature of CSMA/CA mechanism. In detail, ESB-100 has an average drop ratio of 0.1, and ESB-60 and ESB-30 have those of 0.068 and 0.042, respectively. In the worst case, ENSEMBLE (i.e., ESB-100) has a very high average drop ratio of 0.225.

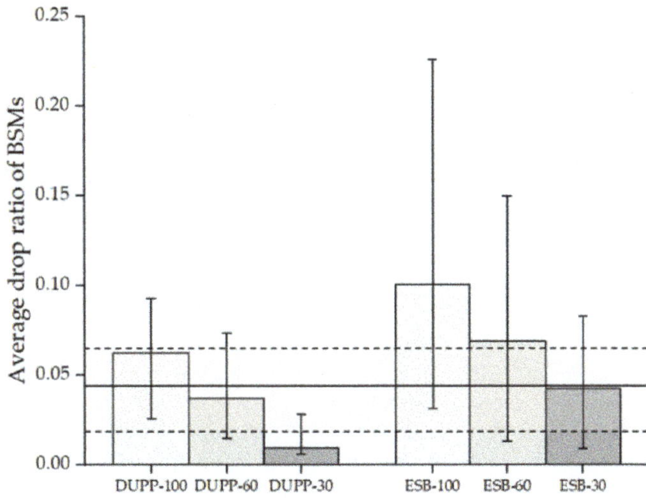

Figure 10. Drop ratio of basic safety messages.

4.2.5. Stability

To show the stability of the urban platooning, the transmission periodicity of urban platooning-related messages is shown in Figure 11. It relates to whether it meets the requirement for the control frames that should be transmitted over in-vehicle networks for reliable driving control. It is known that vehicles in a local platoon cannot maintain constant inter-vehicle space without periodical message transmissions [42,43]. When the urban platooning-related messages are not shared in time, the time headway gradually increases depending on the delayed transmission time. If the transmission is not successful within the maximum 0.8 s, the local platoon can no longer be stable [44]. Moreover, since it also affects the transmission of control messages through the in-vehicle network, providing accurate control information cannot be guaranteed. Therefore, it is imperative to send and receive messages periodically, and especially the transmission should be completely performed within 100 ms which is the time of one synchronization interval, considering the periodicity of the control information. The transmission periodicity is measured as the average number of PCMs transmitted within 100 ms and is presented as transmission success ratio in Figure 11a. Figure 11b,c presents the distance gap between vehicles belonging to a specific local platoon for 60 s at the peak time zone in the blue route as shown in Figure 5.

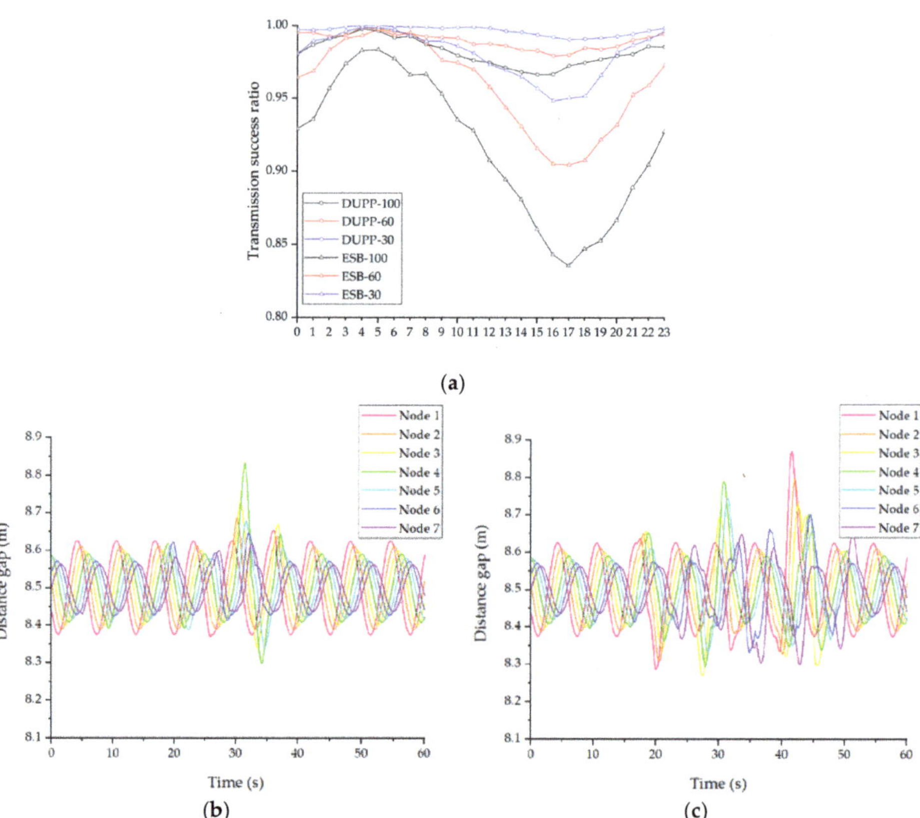

Figure 11. Transmission periodicity of the urban platooning-related messages: (**a**) transmission success ratios of DUPP and ENSEMBLE; (**b**) inter-vehicle distance under DUPP-100 with the transmission periodicity of 94.27%; (**c**) inter-vehicle distance under ESB-100 with the transmission periodicity of 92.19%.

As shown in Figure 11a, the result shows that the periodicity of DUPP-100 is better than that of ESB-30. As interference increases, the transmission success ratio typically decreases. Figure 11a shows there is a tendency that all results related to the performance deteriorate as the time approaches the peak time zone and the number of vehicles participating in urban platooning increases. However, although the worst value of the average transmission success ratio of DUPP is 0.969, the DUPP's performance does not significantly decrease over the entire time zone. Moreover, even in DUPP-100, Figure 11b shows that the DUPP-enabled vehicles stably drive with complying with the velocity specified by a leader node. The inter-vehicle distance under DUPP-100 presented in Figure 11b is during the time DUPP-enabled vehicles have a transmission success ratio of 0.9427. Figure 11c shows the result of ESB-100 having a transmission success ratio of 0.9219. The more its periodicity is destroyed, the more difficult it becomes to maintain stable driving. Due to severe traffic congestion and a signalized intersection, while driving for 60 s, vehicles belonging to that specific local platoon experience the first stable driving from 0 to 10 s, the second slightly unstable driving from 10 to 30 s, the third greatly unstable driving, and the last stable driving from to 60 s. Figure 11b shows that in DUPP-100, the second slightly unstable driving starts at 10 s but it is stabilized soon. After the third greatly unstable driving lasts for about 10 s after 30 s, the last stable driving starts at about 40 s. Figure 11c shows that in ESB-100, the second slightly unstable driving starts at 10 s and lasts until 30 s, the third greatly unstable driving lasts to 45 s. During this period, DUPP-100 quickly recovers from unstable driving such that the distance gap between the intermediate nodes fluctuates for a while, maintaining its safety distance.

4.2.6. Maintenance of Safety Distances

To evaluate the adaptability to unexpected situations through forwarder selection, the stability is examined by showing the maintenance of safety distances between the vehicles for a given local platoon in Figure 12. The number of PEMs transmitted after detecting a certain event is also shown in Figure 12. The unexpected event naturally occurs if vehicles step on the brake when either they enter a congested road section or a vehicle makes a lane change in front of a given local platoon in this experiment. Before a certain event is detected by vehicles of one and more local platoons, all vehicles belonging to a given local platoon drive at a speed of 36 km/h and maintain a safety distance of 8 m. After they enter the congested area from 15 s in Figure 12, a leader node marked as Node 1 in a given platoon starts to apply the brake, and from 25 s, they are in the most congested road section where there are many local platoons. After 40 s, they start to leave this area. Therefore, after 12 s, the PEMs related to the braking event start to be generated and transmitted.

As shown in Figure 12a, in the scenario of ESB-100, the ENSEMBLE-enabled vehicles can no longer maintain the local platoon due to transmissions of PEMs. Although ENSEMBLE does not use PEMs for vehicle platooning, vehicles on the road might receive and transmit the PEMs for detected events [45]. That might lead to placing a large load on the operating wireless channel. Figure 12a shows that it is difficult for ENSEMBLE-enabled vehicles to perform safe driving control since PCMs are not successfully transmitted after a given local platoon enters a congested road section. In other words, since the vehicle density is very high from 15 s, it is difficult to successfully transmit PCMs and this negatively affects urban platooning. Furthermore, in the contested road section, it can be also seen that a maximum of 38 PEMs transmitted on CCH results in hindering the successful transmission of PCMs required for urban platooning. Especially, Node 3 determines that it is difficult to maintain the safety distance anymore because it cannot receive the PCMs of the preceding nodes including Node 1 and Node 2, and then, sends a message with the intention to leave this local platoon. As a result, ENSEMBLE-enabled vehicles can no longer participate in a given platoon because the given local platoon is destroyed after 37 s.

In contrast to the ENSEMBLE's operation for the use of PEMs, DUPP immediately adjusts the driving speed using the information of PEM received. In this regard, Figure 12b shows that DUPP-100 rapidly changes its speed to maintain a safety distance between nodes without destroying the local platoon. Until 15 s, a given local platoon becomes stable. However, we can see that the significant fluctuation starting from 25 s is maintained for about 12 s, and an individual vehicle belonging to a local platoon experiences the difficulty of maintaining the distance gap to the preceding vehicle. This is because the increase in vehicle density triggers a problem in driving control. Therefore, we can see that from 15 s, the state of the given local platoon begins to change to a very gradually unstable state due to the increase in the PCM's drop ratio. Nevertheless, within a short period of time, since the transmissions of the PEMs and PCMs gradually succeed and at least the minimum number of messages required for maintenance is transmitted, the given local platoon is gradually recovered to a stable state. During this period, there is a maximum of 14 PEMS generated by many local platoons within the transmission range of the given local platoon, all of which contribute to a stable platoon state.

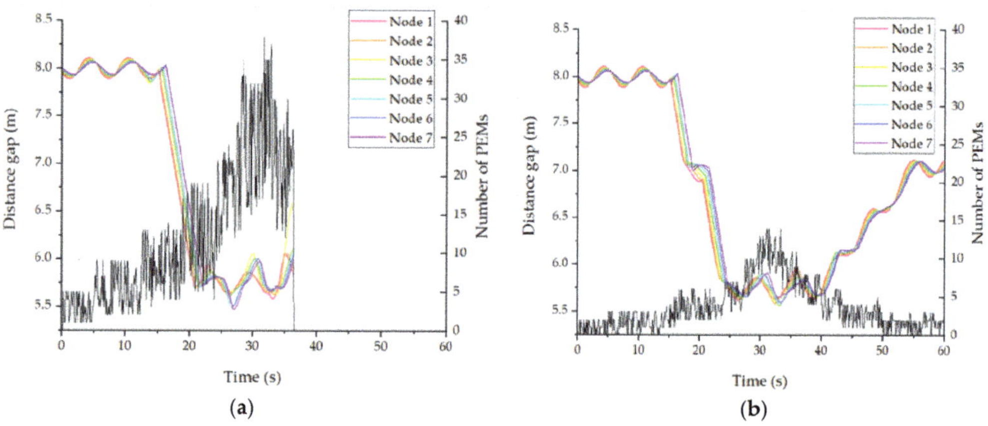

Figure 12. The effect of forwarder selection in distance gap: (**a**) inter-vehicle distance under ESB-100 with PEMs; (**b**) inter-vehicle distance under DUPP-100 with PEMs.

4.2.7. Forwarder Selection Ratio

Finally, we examined the forwarder selection ratio for each vehicle in vehicle platoons to demonstrate the performance of the designed AHP-based forwarder selection using all criteria (i.e., string stability, interference, and connectivity). All results in Figure 13 are derived only from local platoons where the platoon length does not change for at least 60 s to produce valid results. For each road section, chosen randomly among many road sections as shown in Figure 5, an event is generated by a Poisson distribution with a different generation ratio depending on the time, using $\lambda = 10$ (events per second). In Section 3.3, certain events with a negative effect on urban platooning are defined and divided into two types: an event requiring urgent control and an event requiring a warning alarm.

In DUPP, it is noted that each node in a local platoon keeps the stable string stability close to zero except when an event requiring urgent control occurs, such as urgent braking. A node that suddenly brakes is highly likely to become a forwarder by our AHP-based selection since its string stability changes more drastically than connectivity and interference. In this regard, all forwarders as shown in Figure 13a are selected as nodes that have detected the event first. While a selected forwarder shares its PEM with other nodes, the string stabilities of the nodes behind an event-detected node are sequentially affected due to an emergency brake light of the preceding node. Therefore, while the event is detected, a new forwarder is continuously selected for the changed driving conditions.

When an event requiring a warning alarm such as dangerous curves, road construction, obstacles, heavy rain, and slippery road occurs, the criteria of interference and connectivity might affect the AHP-based forwarder selection more than the criterion of string stability. Due to intersections, a vehicle can experience a change in vehicle density. A vehicle can also face it when a vehicle driving in a road section with a low vehicle density approaches a road section with a higher vehicle density or when it exits from a road section with a high vehicle density. A change in vehicle density causes the fluctuation in interference. Especially, while each node in a local platoon sequentially approaches and enters a road section of a high vehicle density, the effect of interference of the node which enters first increases as soon as it enters. Hence, following nodes behind the entering node has a relatively lower effect of interference than that of the entering node. Moreover, among the following nodes, the front node is likely to be selected as a forwarder since it has higher connectivity than that of the other following nodes. In other words, since nodes in a local platoon drive sequentially in a road section of a higher vehicle density along its specified route, a node just before entering the high vehicle density road section is more likely to be selected. This trend is shown in Figure 13b regardless of the platoon length. We can see that Node 4 in a platoon length of 7 and Nodes 2, 3, and 6 in the platoon length of 6 are selected more than the other nodes in Figure 13b. This is because there is no change in road conditions affecting the forwarder selection when a local platoon waits for a green signal at an intersection. Consequently, the forwarder selection is highly affected by interference when an event requiring a warning alarm occurs and there is little change in string stability.

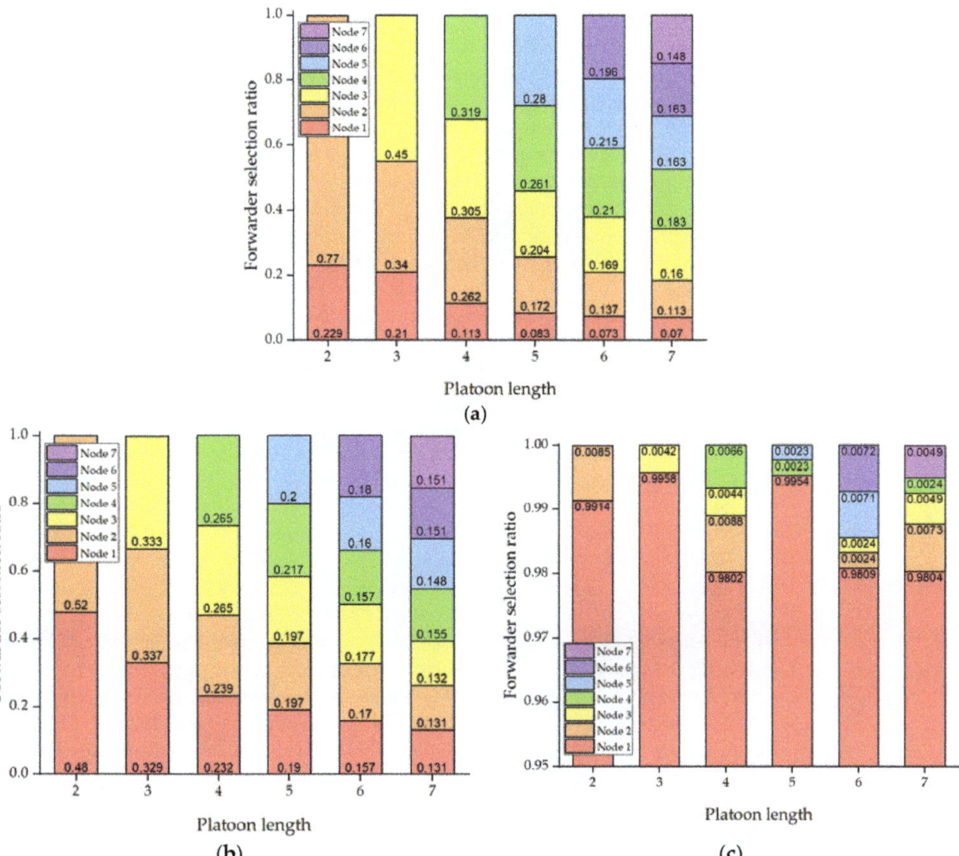

Figure 13. Forwarder selection ratio for each platoon length under DUPP-100: (**a**) the effect of the string stability on AHP-based forwarder selection; (**b**) the effect of the interference on AHP-based forwarder selection; (**c**) the effect of the connectivity on AHP-based forwarder selection.

Excluding the above two situations with an event requiring urgent control and the significant change in vehicle density, forwarder selection is affected by a criterion of connectivity. Figure 13c shows the forwarder selection ratio when a warning event occurs in a low-density road section. We can see that Node 1 in a local platoon is the most selected node due to the highest connectivity. The other nodes except for Node 1 are occasionally selected. This is because their string stabilities fluctuate when the red light of the intersection causes a sudden stop of nodes in a local platoon.

5. Conclusions

Vehicle platooning is a technology that allows multiple vehicles to move as one group on roads, sharing control information through wireless communication and control their movements under the same condition. It reduces the time headway of vehicles, improves fuel efficiency and the driver's convenience, and contributes to the safety of the driving vehicle by responding immediately to the movement of preceding vehicles. Although there are several vehicles platooning studies, they have limitations in flexibility, adaptability, and stability, because they assume only a simple vehicle topology. Urban platooning is characterized by a dynamic road condition depending on signalized intersections, changeable routes, unexpected obstacles, and various vehicle densities. DUPP is an urban platooning protocol to maximize flexibility for vehicles operating in a decentralized manner, considering high mobility. With the compatibility with entities using the existing IEEE 802.11p, the proposed DUPP advertises the existence of a local platoon using WSA during CCH-I and performs an urban platooning during SCH-I. DUPP guarantees that vehicles complete urban platooning maneuvers quickly since they perform FAP maneuvers in a distributed manner without the help of a leader node. AHP-based forwarder selection supports reducing redundant message transmissions.

We assess the applicability of DUPP in urban environments with several aspects and compare the performance of DUPP with that of ENSEMBLE. DUPP reduces the average travel time by 20% compared with that of ENSEMBLE. We have shown that the urban platoons in DUPP-100 last 1.76 times longer and exist 1.64 times larger than those of ESB-100 even at the peak time zone. In addition, the join success rate is 79.3% for DUPP, compared to 72.9% for ENSEMBLE. They indicate that DUPP adapts quickly to urban roads and has its flexibility when performing FAP maneuvers. According to the result in external cost, in the worst cases, DUPP has received 90.7% of BSMs while ENSEMBLE has received only 77.4% of BSMs. DUPP hardly affects road safety even when the vehicle density increases. It is critical to satisfy the requirement for the transmission periodicity that affects the stability of a given local platoon. In this regard, it is demonstrated that DUPP is more stable than ENSEMBLE. Even in the worst, DUPP succeeds 96.5% of transmissions within 100 ms while ENSEMBLE is only 83.9% successful in its transmissions within 100 ms. In DUPP, when a forwarder selected by AHP transmits PEM, the total amount of transmitted PEMs corresponds to 37.8% of ENSEMBLE's event messages. Since redundant transmissions of event messages adversely affect the exchange of platooning control messages in ENSEMBLE, we have seen that its platooning is not stable. By regulating the number of PEMs and quickly sharing them, a local platoon has been stably maintained in DUPP. Finally, we examined how a vehicle is selected as a forwarder as the three criteria of string stability, interference, and connectivity are varied. In this regard, it is demonstrated that DUPP determines an appropriate vehicle as a forwarder for each occurrence of various events. We demonstrate the effectiveness of the proposed urban platooning in terms of flexibility, adaptability, and stability. Consequently, DUPP enables the distributed coordination and autonomous maneuvering to quickly adapt to dynamic traffic flows and complex topologies of urban road networks.

In future work, we plan to implement DUPP in the real vehicle such as an unmanned ground vehicle (UGV) and will compare the performance of an AHP-based forwarder selection with a greedy selection method. Furthermore, the length of an urban platooning (η) represents a trade-off between urban platooning adaptability and effectiveness. A local platoon of a smaller length can be difficult to show good performance in terms of the average travel time but it can adapt well to dynamic traffic environments. Although a local platoon of a longer length can be easy to show good performance, it is not easy to maintain the local platoon under the complex urban roads. Therefore, when the length of an urban platooning is varied, we can examine how well the vehicles adapt to the road condition with high uncertainty when the vehicles drive and how much the performance is improved. We will conduct this validation in the future.

Author Contributions: Conceptualization, Y.B. and S.J.; methodology, S.H.S. and Y.B.; software, Y.B. and S.J.; validation, Y.B.; formal analysis, Y.B.; investigation, Y.B. and S.J.; resources, Y.B. and S.J.; data curation, Y.B.; writing—original draft preparation, Y.B. and S.J.; writing—review and editing, S.H.S. and Y.B.; visualization, Y.B. and S.J.; supervision, S.H.S. and Y.B.; project administration, S.H.S. and Y.B.; funding acquisition, S.H.S. All authors have read and agreed to the published version of the manuscript.

Funding: This work was supported by Changshin University Research Fund of Changshin-2020-046 and Institute for Information and communications Technology Promotion (IITP) grant funded by the Korean government (MSIT) (No. 2014-0-00065, Resilient Cyber-Physical Systems Research).

Conflicts of Interest: The authors declare no conflict of interest.

Appendix A

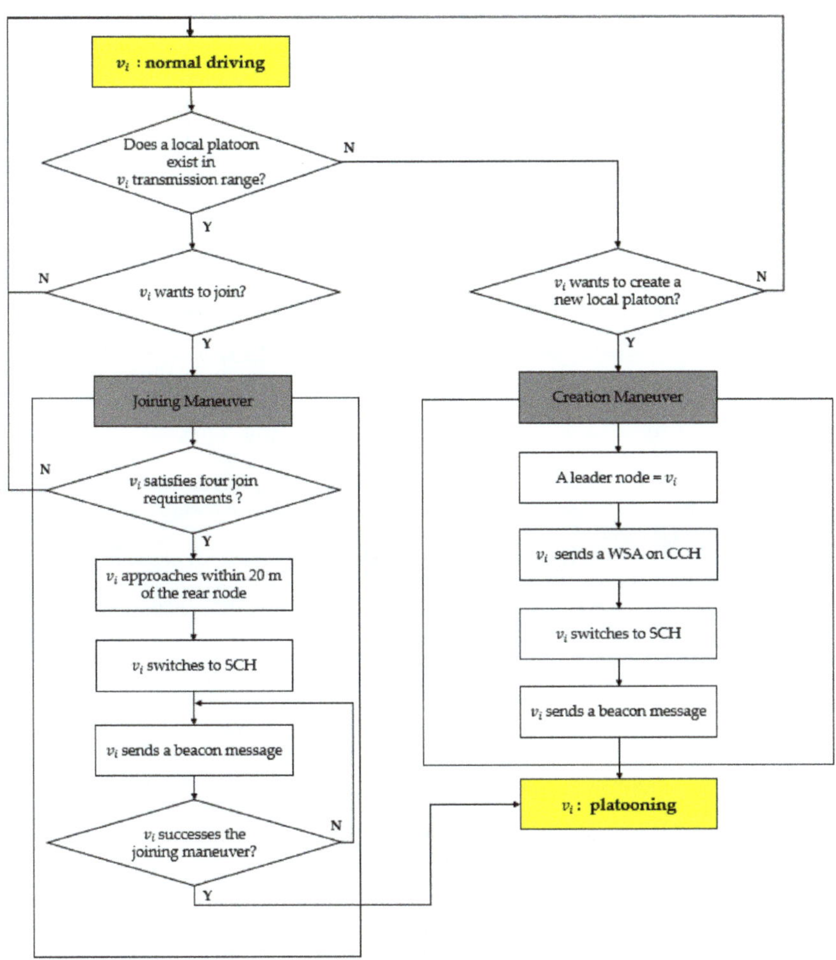

Figure A1. Normal driving in urban platooning.

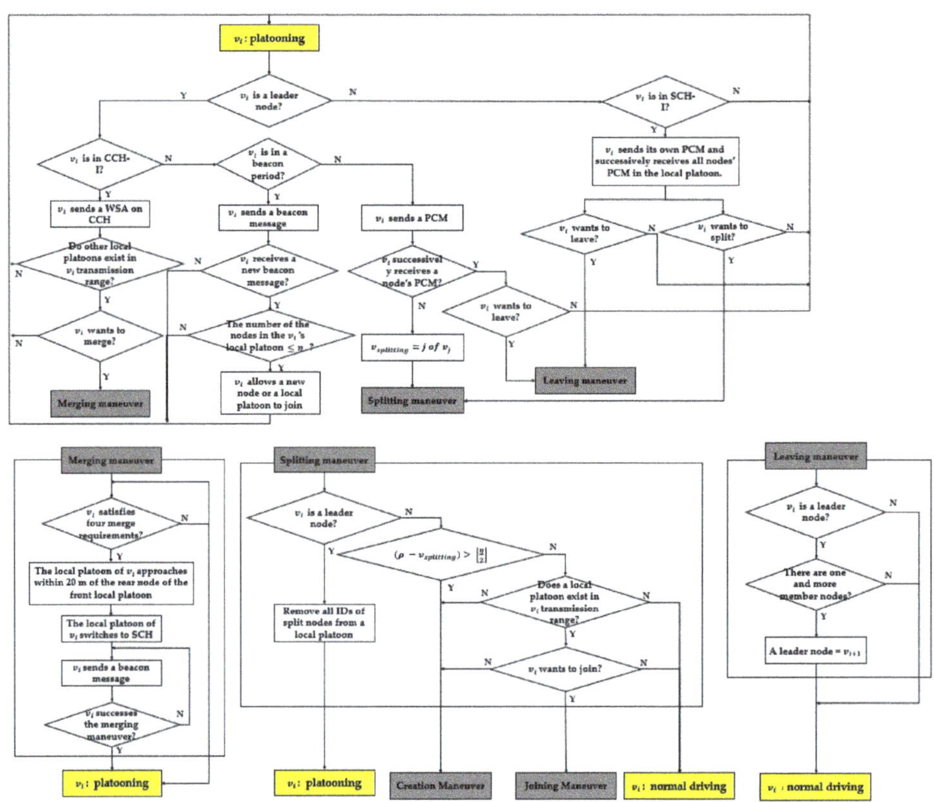

Figure A2. Merging, splitting, and leaving maneuvers in urban platooning to normal driving.

References

1. C-ITS: Cooperative Intelligent Transport Systems and Services. Available online: https://www.car-2-car.org/about-c-its/ (accessed on 7 February 2021).
2. Kim, J.; Emeršič, Ž.; Han, D.S. Vehicle path prediction based on radar and vision sensor fusion for safe lane changing. In Proceedings of the 2019 International Conference on Artificial Intelligence in Information and Communication (ICAIIC), Okinawa, Japan, 11–13 February 2019; pp. 267–271.
3. Lee, H.; Chae, H.; Yi, K.J.I.-P. A geometric model based 2D LiDAR/radar sensor fusion for tracking surrounding vehicles. *FAC-PapersOnLine* **2019**, *52*, 130–135. [CrossRef]
4. Zhong, Z.; Liu, S.; Mathew, M.; Dubey, A.J.E.I. Camera Radar Fusion for Increased Reliability in ADAS Applications. *Electron. Imaging* **2018**, *2018*, 251–254. [CrossRef]
5. Kunze, R.; Ramakers, R.; Henning, K.; Jeschke, S. Organization and Operation of Electronically Coupled Truck Platoons on German Motorways. In *Automation, Communication and Cybernetics in Science and Engineering 2009/2010*; Springer: Berlin/Heidelberg, Germany, 2011; pp. 427–439.
6. Chan, E. SARTRE automated platooning vehicles. Towards Innovative Freight and Logistics. *Towards Innov. Freight Logist.* **2016**, *2*, 137–150.
7. Willemsen, D. V2 Platooning Use Cases, Scenario Definition and Platooning Levels (Version A). Available online: https://platooningensemble.eu/ (accessed on 7 February 2021).
8. Kalbitz, T. A Comparison of Approaches for Platooning Management. 2017. Available online: https://madoc.bib.uni-mannheim.de/42305?rs=true& (accessed on 9 April 2021).
9. EU Commission. Road Safety in the European Union: Trends, Statistics and Main Challenges. Internal Working Material EU DG Mobility Transport 2015. *Tech. Rep.* **2015**. Available online: https://ec.europa.eu/transport/ (accessed on 9 April 2021).
10. Alam, A.; Mårtensson, J.; Johansson, K.H. Look-ahead cruise control for heavy duty vehicle platooning. In Proceedings of the 16th International IEEE Conference on Intelligent Transportation Systems (ITSC 2013), The Hague, The Netherlands, 6–9 October 2013; pp. 928–935.

11. IEEE Standard Association. *IEEE Guide for Wireless Access in Vehicular Environments (WAVE) Architecture*; IEEE: Piscataway, NJ, USA, 2013; p. 1609.
12. SAE. SAE J2735 Dedicated Short Range Communications (DSRC) Message Set Dictionary. *SAE Std.* **2015**, 2735-201. Available online: https://www.sae.org/standards/content/j2735_201603/ (accessed on 9 April 2021).
13. Eichler, S. Performance evaluation of the IEEE 802.11 p WAVE communication standard. In Proceedings of the 2007 IEEE 66th Vehicular Technology Conference, Baltimore, MD, USA, 30 September–3 October 2007; pp. 2199–2203.
14. Xu, Z.; Li, X.; Zhao, X.; Zhang, M.H.; Wang, Z. DSRC versus 4G-LTE for connected vehicle applications: A study on field experiments of vehicular communication performance. *J. Adv. Transp.* **2017**, *2017*, 1–10. [CrossRef]
15. Zeng, T.; Semiari, O.; Saad, W.; Bennis, M. Joint communication and control for wireless autonomous vehicular platoon systems. *IEEE Trans. Commun.* **2019**, *67*, 7907–7922. [CrossRef]
16. Kang, K.-D.; Baek, Y.; Lee, S.; Son, S.H. An attack-resilient source authentication protocol in controller area network. In Proceedings of the 2017 ACM/IEEE Symposium on Architectures for Networking and Communications Systems (ANCS), Beijing, China, 18–19 May 2017; pp. 109–118.
17. SAE. *Class C Application Requirement Considerations*; Technical Report; Society of Automotive Engineers: Warrendale, PA, USA, 1993.
18. Tindell, K.; Burns, A. Guaranteed Message Latencies for Distributed Safety-Critical Hard Real-Time Control Networks. University of York. 1994. Available online: http://citeseerx.ist.psu.edu/viewdoc/download?doi=10.1.1.49.9976&rep=rep1&type=pdf (accessed on 9 April 2021).
19. NetcarBench 3.4. Available online: http://www.netcarbench.org/ (accessed on 7 February 2021).
20. Schindler, J.; Dariani, R.; Rondinone, M.; Walter, T. Dynamic and flexible platooning in urban areas. In Proceedings of the AAET Automatisiertes und Vernetztes Fahren, Braunschweig, Germany, 14–15 March 2018.
21. Heinovski, J.; Dressler, F. Platoon formation: Optimized car to platoon assignment strategies and protocols. In Proceedings of the 2018 IEEE Vehicular Networking Conference (VNC), Taipei, Taiwan, 5–7 December 2018; pp. 1–8.
22. Farag, A.; Hussein, A.; Shehata, O.M.; Garcia, F.; Tadjine, H.H.; Matthes, E. Dynamics platooning model and protocols for self-driving vehicles. In Proceedings of the 2019 IEEE Intelligent Vehicles Symposium (IV), Paris, France, 9–12 June 2019; pp. 1974–1980.
23. Sarker, A.; Qiu, C.; Shen, H. Connectivity maintenance for next-generation decentralized vehicle platoon networks. *IEEE/ACM Trans. Netw.* **2020**, *28*, 1449–1462. [CrossRef]
24. Böhm, A.; Kunert, K. Data age based retransmission scheme for reliable control data exchange in platooning applications. In Proceedings of the 2015 IEEE International Conference on Communication Workshop (ICCW), London, UK, 8–12 June 2015; pp. 2412–2418.
25. Balador, A.; Bohm, A.; Uhlemann, E.; Calafate, C.T.; Cano, J. A Reliable Token-Based MAC Protocol for Delay Sensitive Platooning Applications. In Proceedings of the 2015 IEEE 82nd Vehicular Technology Conference (VTC2015-Fall), Boston, MA, USA, 6–9 September 2015; pp. 1–5.
26. ETSI EN 302 637–2 v1. 3.1-Intelligent Transport Systems (Its); Vehicular Communications; Basic Set of Applications; Part 2: Specification of Cooperative Awareness Basic Service. *ETSI Sept.* **2014**. Available online: https://www.etsi.org/deliver/etsi_ts/102600_102699/10263702/01.02.01_60/ts_10263702v010201p.pdf (accessed on 9 April 2021).
27. Yun, J.; Baek, Y.; Lee, B.; Li, J.; Han, J.; Han, K. An efficient network merging mechanism for WiMedia UWB MAC protocol. In Proceedings of the 2010 12th International Conference on Advanced Communication Technology (ICACT), Gangwon-Do, South Korea, 7–10 February 2010; pp. 425–429.
28. Standard ECMA. ECMA-368: High Rate Ultra Wideband PHY and MAC Standard. Geneva, Switzerland. December 2005. Available online: https://www.ecma-international.org/wp-content/uploads/ECMA-368_1st_edition_december_2005.pdf (accessed on 9 April 2021).
29. Baek, Y.; Kim, Y.; Lee, B.; Lee, J.; Yun, J.; Shu, Q.; Oh, S.; Han, K. Network Initialization for Wireless Distributed Beaconing Networks. In Proceedings of the 2010 International Conference on Cyber-Enabled Distributed Computing and Knowledge Discovery, Huangshan, China, 10–12 October 2010; pp. 125–128.
30. IEEE Working Group. IEEE Standard for Wireless Access in Vehicular Environments (WAVE)-Multi-Channel Operation. IEEE Std 2016, 1609.1604-2016. Available online: https://ieeexplore.ieee.org/document/7435228 (accessed on 9 April 2021).
31. ETSI EN 302 637-3 v1. 2.2 (2014-11), Intelligent Transport Systems (ITS); Vehicular Communications; Basic Set of Applications; Part 3: Specifications of Decentralized Environmental Notification Basic Service. Available online: https://standards.iteh.ai/catalog/standards/etsi/09b1747b-42f8-4b11-be40-760852ca4216/etsi-en-302-637-3-v1.2.2-2014-11 (accessed on 9 April 2021).
32. Intelligent Transportation Systems Committee. IEEE trial-use standard for wireless access in vehicular environments (wave)-networking services. *IEEE Std.* **2007**, *1609*, 1603–2007.
33. Saaty, R. The analytic hierarchy process—what it is and how it is used. *Math. Model.* **1987**, *9*, 161–176. [CrossRef]
34. Mu, E.; Pereyra-Rojas, M. Understanding the analytic hierarchy process. In *Practical Decision Making*; Springer: Berlin/Heidelberg, Germany, 2017; pp. 7–22.
35. Bahurmoz, A.M. The analytic hierarchy process: A methodology for win-win management. *Econ. Adm.* **2006**, *20*. [CrossRef]
36. Saaty, T.L. *Decision Making for Leaders: The Analytic Hierarchy Process for Decisions in a Complex World*, 3rd ed.; RWS Publications: Pittsburgh, PA, USA, 2012.
37. Brunelli, M. *Introduction to the Analytic Hierarchy Process*; Springer: Berlin/Heidelberg, Germany, 2014.

38. Baek, Y.; Son, S.H. Load-aware association with AP for internet-connected vehicles. In Proceedings of the 2016 Eighth International Conference on Ubiquitous and Future Networks (ICUFN), Vienna, Austria, 5–8 July 2016; pp. 173–177.
39. Segata, M.; Joerer, S.; Bloessl, B.; Sommer, C.; Dressler, F.; Cigno, R.L. Plexe: A platooning extension for Veins. In Proceedings of the 2014 IEEE Vehicular Networking Conference (VNC), Paderborn, Germany, 3–5 December 2014; pp. 53–60.
40. NYC Open Data. Available online: https://opendata.cityofnewyork.us/ (accessed on 7 February 2021).
41. Sommer, C.; Dressler, F. Using the Right Two-Ray Model? A Measurement Based Evaluation of PHY Models in VANETs. In Proceedings of the ACM MobiCom, Las Vegas, NV, USA, 19–23 September 2011; pp. 1–3.
42. Ucar, S.; Ergen, S.C.; Ozkasap, O. IEEE 802.11 p and visible light hybrid communication based secure autonomous platoon. *IEEE Trans. Veh. Technol.* **2018**, *67*, 8667–8681. [CrossRef]
43. Ploeg, J.; Semsar-Kazerooni, E.; Lijster, G.; van de Wouw, N.; Nijmeijer, H. Graceful degradation of CACC performance subject to unreliable wireless communication. In Proceedings of the 16th International IEEE Conference on Intelligent Transportation Systems (ITSC 2013), The Hague, The Netherlands, 6–9 October 2013; pp. 1210–1216.
44. Naus, G.J.; Vugts, R.P.; Ploeg, J.; van De Molengraft, M.J.; Steinbuch, M. String-stable CACC design and experimental validation: A frequency-domain approach. *IEEE Trans. Veh. Technol.* **2010**, *59*, 4268–4279. [CrossRef]
45. ENSEMBLE Deliverable D2.8. (December 2018). Platooning Protocol Definition and Communication Strategy. Available online: https://platooningensemble.eu/ (accessed on 7 February 2021).

Article

Data-Driven Object Vehicle Estimation by Radar Accuracy Modeling with Weighted Interpolation [†]

Woo Young Choi [1], Jin Ho Yang [1] and Chung Choo Chung [2],*

[1] Departerment of Electrical Engineering, Hanyang University, Seoul 04763, Korea; wooyoungchoi@hanyang.ac.kr (W.Y.C.); jjz0426@hanyang.ac.kr (J.H.Y.)

[2] Division of Electrical and Biomedical Engineering, Hanyang University, Seoul 04763, Korea

* Correspondence: cchung@hanyang.ac.kr

[†] This paper is an extended version of our paper published in Choi, W.Y.; Yang, J.H.; Lee, S.H.; Chung, C.C. Object Vehicle Tracking by Convex Interpolation with Radar Accuracy. In Proceedings of the 19th International Conference on Control, Automation and Systems (ICCAS), Jeju, Korea, 15–18 October 2019.

Abstract: For accurate object vehicle estimation using radar, there are two fundamental problems: measurement uncertainties in calculating an object's position with a virtual polygon box and latency due to commercial radar tracking algorithms. We present a data-driven object vehicle estimation scheme to solve measurement uncertainty and latency problems in radar systems. A radar accuracy model and latency coordination are proposed to reduce the tracking error. We first design data-driven radar accuracy models to improve the accuracy of estimation determined by the object vehicle's position. The proposed model solves the measurement uncertainty problem within a feasible set for error covariance. The latency coordination is developed by analyzing the position error according to the relative velocity. The position error by latency is stored in a feasible set for relative velocity, and the solution is calculated from the given relative velocity. Removing the measurement uncertainty and latency of the radar system allows for a weighted interpolation to be applied to estimate the position of the object vehicle. Our method is tested by a scenario-based estimation experiment to validate the usefulness of the proposed data-driven object vehicle estimation scheme. We confirm that the proposed estimation method produces improved performance over the conventional radar estimation and previous methods.

Keywords: object vehicle estimation; radar accuracy; data-driven; radar latency; weighted interpolation; autonomous vehicle

Citation: Choi, W.Y.; Yang, J.H.; Chung, C.C. Data-Driven Object Vehicle Estimation by Radar Accuracy Modeling with Weighted Interpolation. *Sensors* **2021**, *21*, 2317. https://doi.org/10.3390/s21072317

Academic Editor: Chao Huang

Received: 25 February 2021
Accepted: 22 March 2021
Published: 26 March 2021

Publisher's Note: MDPI stays neutral with regard to jurisdictional claims in published maps and institutional affiliations.

Copyright: © 2021 by the authors. Licensee MDPI, Basel, Switzerland. This article is an open access article distributed under the terms and conditions of the Creative Commons Attribution (CC BY) license (https://creativecommons.org/licenses/by/4.0/).

1. Introduction

Autonomous driving technologies such as collision risk decision, path planning with collision avoidance, lane change systems, and advanced driver assistance systems (ADASs) are attracting attention [1–4]. These research areas are becoming critical not only for research but also to bring autonomous vehicles to public roads. To improve active safety systems for autonomous driving, it is necessary to accurately estimate the relative position of surrounding vehicles [5,6]. Object vehicle estimation research incorporates various types of sensors, such as radio detecting and ranging (radar), light detection and ranging (LiDAR), and cameras. Among the various sensors, radar is a reliable vehicle sensor that measures the motion of surrounding vehicles. Its advantages lie in its commercial availability and robustness against environmental variation. Radar sensors have been applied in ADASs functions such as blind-spot detection (BSD) and adaptive cruise control (ACC).

However, radar has intrinsic measurement uncertainties in calculating an object vehicle's position and velocity as it uses a virtual polygon box with only partial information [7–9]. To address this limitation, various filters have been applied to improve radar accuracy. In radar applications, the Kalman filter (KF) and the interacting multiple model (IMM) were compared in [10]. A particle filter [11] and an unscented Kalman filter (UKF) [12]

169

for nonlinear systems have been proposed for target tracking using a radar sensor. For reasonable object tracking of a radar system, it has been found that the multiple model approach provides better filtering performance than a single model [13]. Radar tracking performance is improved through IMM [14] and convex interpolation [15] by using different radar accuracies depending on the object vehicle position [16]. In [9], the authors proposed an IMM algorithm using extended Kalman filters (EKF) for multi-target state estimation. In [17], the performances of the IMM and Viterbi algorithm were investigated and compared through radar tracking and detection. A self-adapting variable structure multiple model (VS-IMM) estimation approach combined with an assignment algorithm was presented in [18] for tracking ground targets with constrained motion. Motion uncertainties due to variable dynamic driving situations were handled using the VS-IMM. In [19], the authors presented a data-driven object tracking approach by training a deep neural network to learn situation-dependent sensor measurement models.

Another approach to accurate object tracking using radar adds sensors such as a camera and LiDAR. Research fusing radar and camera sensors is described in [20]. In [21], the authors used visual recognition information to improve tracking model selection, data association, and movement classification. An algorithm to estimate the location, size, pose, and motion information of a threat vehicle was implemented by fusing the information from a stereo-camera and from millimeter-wave radar sensors in [22]. In [23], the authors proposed a fusion architecture using radar, LiDAR, and camera for accurate detection and classification of moving objects. In [24], heuristic fusion with adaptive gating and track to track fusion were applied to a forwarding vehicle tracking system using camera and radar sensors, and the two algorithms were compared. In [25], the authors presented an EKF that reflects the distance characteristics of LiDAR and radar sensors. In [26], the fusion of radar and camera sensor data with a neural network was studied to improve object detection accuracy. In [27], the object was identified and detected using vision and radar sensor data, and YOLOv3 architecture. However, the sensor fusion approach requires a larger number of sensors. Although the estimation performance can be improved through multi-sensor applications, it increases the vehicle's cost. In addition, latency occurs due to the increase in computational cost for sensor fusion [28].

As stated above, by applying a filter without an additional sensor, accurate tracking is possible without increasing the cost. However, radar latency (processing delay) increases with the use of a filter [16,29,30]. This latency increases further depending on the tracking algorithm used (e.g., point cloud clustering, segmentation, single sensor tracking, multi-lateration, classification, and filtering) in vehicle applications [7,14,31,32]. In this regard, the radar sensor was evaluated for the effect of processing latency on the efficiency of detecting, acquiring, and tracking a target [29]. In [33], the authors noted that it is important for delays in the measurement (i.e., the time elapsed since a physical event occurs until it is output to the application) and accurate data on the position of other vehicles in future driver assistance systems. In [34], the authors proposed a classification method based on deep neural networks using automotive radar sensors in consideration of latency. Eventually, this processing latency causes a tracking error depending on the relative speed in autonomous driving applications. Therefore, a person who designing an upper-level application should consider processing latency when developing object vehicle estimation for driving safety.

The objective of this paper is to propose an object vehicle estimation scheme to improve radar accuracy. The scheme develops a data-driven object vehicle estimation scheme that can consider radar accuracy within a feasible set to solve the measurement uncertainty and latency problems. To resolve these problems, we first develop radar accuracy models by comparing the radar and ground truth data divided in each zone. Each zone's models are selected depending on where the object vehicle is located. We then solve the radar latency problem according to the relative velocity. The position error for the relative velocity data sets is stored in each vertex, and we find the solution in the feasible set for these data sets. By using the developed radar accuracy models with latency coordination, weighted interpolation is applied to estimate the object vehicle. This approach will allow the radar accuracy models to remove the measurement uncertainty and latency within the feasible set. We verify the utility of the proposed method through scenario-based experiments. The contribution of this paper is the developmetn of an accurate object vehicle estimation scheme that solves the radar measurement uncertainty and latency problems.

The remainder of this paper is organized as follows. Section 2 describes two problems to improve object vehicle estimation accuracy. The data-driven radar accuracy modeling with an occupancy zone is described in Section 3. In Section 4, weighted interpolation is applied to object estimation by considering error characteristics and latency. Section 5 describes the analysis and results by applying the proposed method to vehicle applications and mentions future work. Section 6 presents concluding remarks.

2. Problem Statement

The problem we are interested in is object vehicle estimation by considering a radar's measurement uncertainty and latency, as shown in Figure 1. There are two fundamental problems in accurate object vehicle estimation: measurement uncertainties in calculating an object's position with a virtual polygon box and latency due to the tracking algorithm of a commercial radar. To resolve these problems, we develop a data-driven object vehicle estimation scheme using a radar accuracy modeling method with weighted interpolation. The radar accuracy modeling is designed using an error model between the radar and the ground truth data, and taking into account the relative speed. We are also interested in demonstrating the utility of our method through experiments.

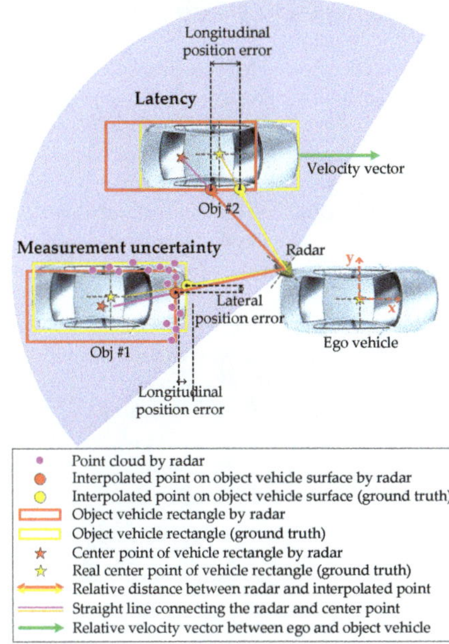

Figure 1. Example of object vehicle estimation by radar: the measurement uncertainty occurs due to insufficient point cloud and classification errors. This error occurs because the radar estimation algorithm (e.g., point cloud clustering, segmentation, single sensor tracking, multilateration, classification, and filtering) can only estimate an object vehicle's size with a virtual polygon box with partial information [7,8,14,35]. Furthermore, the latency that causes position errors occurs due to the tracking algorithm of a commercial radar. The error caused by the latency becomes larger depending on the relative velocity.

These problems are almost undetectable and unknown to those who develop high-level applications such as ADASs. Therefore, we propose a scheme modeling these undetectable and unknown error characteristics as noise characteristics based on each divided zone and design a data-driven object estimation scheme. In addition, we propose a method to reduce errors that occur in radar algorithms by developing data-driven latency coordination. We use the relative position and velocity, which are the only available data to the person designing an upper-level application.

3. Data-Driven Radar Accuracy Modeling

To improve radar accuracy, we first developed a model. Previous research analyzing radar accuracy [14,15] found that the error characteristics differ according to the mounting angle and detection area of the radar observing the object [16]. Since the radar error differs depending on the angle and the detection area, it is difficult to obtain an error characteristic solution for a radar's detection area. Therefore, we model these unknown error characteristics so that an error has the same value in each of the divided representative detection zones because the part of the object vehicle detected by the radar is similar to other object vehicles in the same detection area. In other words, each radar unit has a representative model for each zone. The measurement uncertainty of radar can be reduced by using an occupancy zone with the error characteristics. However, there is an error according to the object vehicle's velocity due to the radar's latency (the results of the analysis of the experimental data are shown in Section 4). This is caused by the object tracking algorithm of commercial radars [14,29]. This is a problem for anyone designing high-level applications for radar.

Therefore, we constructed an example of occupancy zones, as shown in Figure 2, taking into account the detectable area of the radar, where $\{X, Y\}$ is the global coordinate frame, $\{x, y\}$ is the ego vehicle coordinate frame, and $\dot{r}$ is the relative vehicle speed. The example of a divided occupancy zone configuration is divided by the x-axis (considering a multiple of the overall vehicle length), the y-axis (considering lane spacing), and the z-axis (considering experimental data analysis) based on the vehicle coordinate frame. Here, the z-axis is divided by data sets for each relative speed. The center point of each divided black quadrangle zone becomes each vertex of the red quadrangle (feasible set for error covariance). Then, the error characteristics analyzed in each zone are stored in each vertex. The radar accuracy in each divided occupancy zone detected by radar sensors is analyzed by comparing radar sensor data with ground truth (GT), as shown in Figure 1. An interpolated point on the object vehicle surface is calculated by a straight line connecting the ego vehicle's radar and the center point of the virtual polygon box of the object vehicle. Here, the real center point (ground truth) is calculated from the differential global positioning system (DGPS) mount point. Then, we can obtain the longitudinal and lateral position errors by comparing the interpolated point and the center point.

The model for object vehicle estimation can be expressed as a discrete-time state-space model assuming that the vehicle is moving with constant relative velocity in the longitudinal and lateral directions, respectively [36]. With the state $\mathbf{x}_k = \begin{bmatrix} r_x & r_y & \dot{r}_x & \dot{r}_y \end{bmatrix}^T$, the state-space model is defined as

$$\begin{aligned} \mathbf{x}_{k+1} &= \Phi \mathbf{x}_k + {}_{(m,n,s)}w_k, \\ \mathbf{y}_k &= C\mathbf{x}_k + {}_{(m,n,s)}v_k \end{aligned} \tag{1}$$

where

$$\Phi = \begin{bmatrix} 1 & T_c & 0 & 0 \\ 0 & 1 & 0 & 0 \\ 0 & 0 & 1 & T_c \\ 0 & 0 & 0 & 1 \end{bmatrix}, \tag{2}$$

$$\begin{aligned} {}_{(m,n,s)}w_k &\sim \mathcal{N}(0, {}_{(m,n,s)}Q_k), \\ {}_{(m,n,s)}v_k &\sim \mathcal{N}(0, {}_{(m,n,s)}R_k) \end{aligned}$$

with

$$m \in [1, 2, \cdots, M], \ n \in [1, 2, \cdots, N], \ s \in [1, 2, \cdots, S] \tag{3}$$

where $\mathbf{y}_k$ is the output variables at the measurement instant k, $_{(m,n,s)}w_k$ is the system noise, $_{(m,n,s)}v_k$ is the radar measurement accuracy, C is the identity matrix, r_x is the longitudinal relative distance, r_y is the lateral relative distance, $\dot{r}_x$ is the longitudinal relative velocity, $\dot{r}_y$ is the lateral relative velocity, m is the longitudinal relative positional zone index, n is the lateral relative positional zone index, s is the zone index for relative velocity, M is the zone number of the X-axis, N is the zone number of the Y-axis, and S is the zone number of the Z-axis.

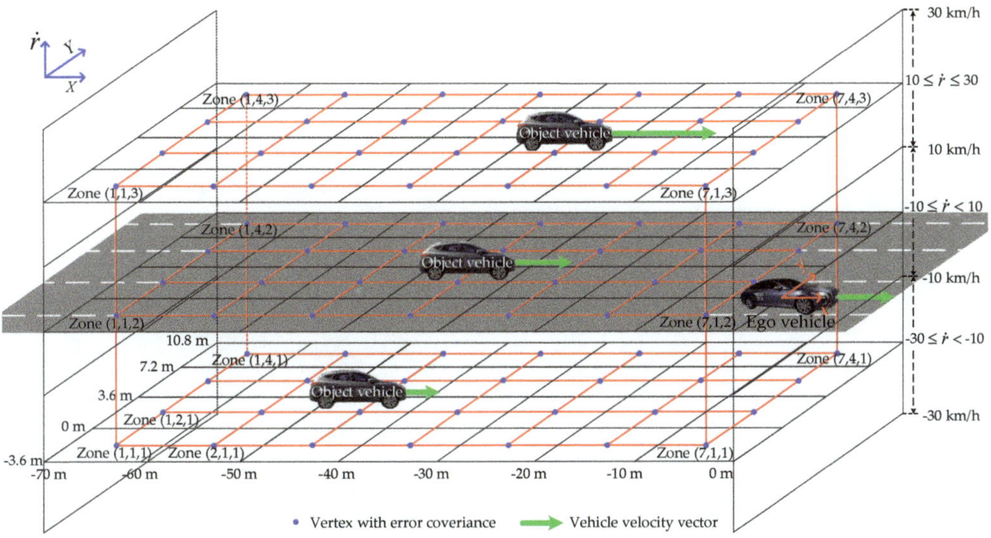

Figure 2. Example of a divided occupancy zone configuration: the occupancy zone is created by taking into account the detectable area of the radar, which is divided by the x-axis and y-axis based on the vehicle coordinate frame. The z-axis is divided by data sets for relative velocity. The center point of each divided black quadrangle zone becomes each vertex of the red quadrangle (feasible set for error covariance). Then, the error characteristics analyzed in each zone are stored in each vertex.

Assumption 1. *Radar measurement accuracy $_{(m,n,s)}v_k$ has a zero-mean white Gaussian distribution property in each zone* [37]. *The radar measurement accuracy covariance $_{(m,n,s)}R_k$ is a value determined by the characteristics of the sensor. The radar measurement accuracy covariance $_{(m,n,s)}R_k$ in each zone is set based on the error characteristics. The radar sensor is calibrated at each zone, such that the mean value of the position error becomes zero. Therefore, the zero-mean radar error becomes*

$$e = [r_x, \ r_y, \ \dot{r}_x, \ \dot{r}_y]^T_{RADAR} - [r_x, \ r_y, \ \dot{r}_x, \ \dot{r}_y]^T_{GT} \tag{4}$$

and its covariance is

$$\mathbb{E}[ee^T] \sim \mathcal{N}(0, {}_{(m,n,s)}R_k) \tag{5}$$

where subscript GT represents the ground truth data and subscript RADAR represents the calibrated radar data. Since it is not easy to obtain radar accuracy covariance values according to driving situations, we experimentally applied covariance values based on the method presented in [38]. *In this regard, the experimental analysis results with the calibrated radar accuracy are shown in Section 4.*

Remark 1. *By adjusting the system noise covariance $_{(m,n,s)}Q_k$ through the KF in which the previously set radar measurement accuracy covariance $_{(m,n,s)}R_k$ is used, estimation errors approaching* *the minimum value in each zone are obtained [39]. Then, we set the system noise covariance $_{(m,n,s)}Q_k$ for each zone.*

4. Object Tracking with Weighted Interpolation

4.1. Estimation with Error Characteristic

The weighted interpolation in the occupancy zone is applied to state estimation by considering the error characteristics. The weighted interpolation method is used to solve the ambiguity problem of moving from a zone to another zone caused by dividing the occupancy zone. To apply weighted interpolation, we create a feasible set f_c denoted by a red quadrangle relative to the center point of each black quadrangle zone in Figure 2. The center point of each zone is the vertex of the feasible set f_c for error covariance, and f_c takes into account the lane width. The data-driven covariance R^* calculated in the previous section is stored at each zone's vertex. This process is carried out offline using data analyzed in advance.

In online computation, the object vehicle positions x and y, and relative speed $\dot{r}$ are given by the radar sensor. Then, the data-driven covariance stored at the vertex is applied to state estimation. The three-dimensional parameter vector $\mathbf{P}_c = \begin{bmatrix} x & y & \dot{r} \end{bmatrix}^T \in \mathbb{R}^3$ can be represented in the polytopic form [15,40,41]:

$$\mathbf{P}_c = V_c \xi_c \tag{6}$$

where

$$\xi_c = \begin{bmatrix} \xi_{c,1} & \cdots & \xi_{c,8} \end{bmatrix}^T \in \mathbb{R}^8 \tag{7}$$

denotes a weighted interpolation parameter vector satisfying $\sum_{q=1}^{8} \xi_{c,q} = 1$, $\xi_{c,q} \geq 0$, and

$$V_c = \begin{bmatrix} \mathbf{P}_{c,1} & \cdots & \mathbf{P}_{c,8} \end{bmatrix} \in \mathbb{R}^{3 \times 8} \tag{8}$$

denotes each zone's vertices. When selecting the each zone's vertices V_c, we chose the eight vertices closest to the given x, y, and $\dot{r}$ measured by the radar in the feasible set f_c, as shown in Figure 2. Then, we can get

$$\xi_{c,q} = \frac{L_{c,sum}/L_{c,q}}{\sum_{i=1}^{8}(L_{c,sum}/L_{c,i})}, \quad q = 1, \cdots, 8 \tag{9}$$

where $L_{c,sum} = \sum_{i=1}^{8} L_{c,i}$ in which L_c is the Euclidean distance between each vertex and the given point $(x, y, \dot{r})$ measured by the radar. Using the interpolation parameters with the parameter vector at eight vertices, we can find an approximate data-driven covariance R^o from the precomputed data-driven covariance $R^*(V_c)$ calculated from Assumption 1 at each vertex. The approximate data-driven covariance R^o is expressed as follows:

$$R^o = \begin{bmatrix} R^*(\mathbf{P}_{c,1}) & \cdots & R^*(\mathbf{P}_{c,8}) \end{bmatrix} \xi_c. \tag{10}$$

From the KF using R^o, we can obtain the estimated object vehicle position $\hat{x}$ and $\hat{y}$ [39,42]. This approach satisfies the computational complexity because it does not consider all the zones' vertices. Here, we describe covariance related to the position for the object vehicle estimation; the covariance related to the velocity can be referred to [14] in a similar way.

4.2. Latency Coordination

To solve the aforementioned latency problem, weighted interpolation is applied to the state estimation similar to the previous subsection. As stated above, a radar's latency varies depending on the relative velocity. The velocity region is divided, and the average position error by latency that occurred in each velocity set is stored in each vertex. Detailed data analysis is provided in Section 4.

The object vehicle longitudinal relative velocity $\dot{r}_x$ and lateral relative velocity $\dot{r}_y$ are given by the radar sensor. Then, the position error average $\mathbf{E}_l^*$ by latency in each velocity region stored at the vertex is applied to the state estimation. Two-dimensional parameter vector $\mathbf{P}_l = \begin{bmatrix} \dot{r}_x & \dot{r}_y \end{bmatrix}^T \in \mathbb{R}^2$ can be represented in the polytopic form:

$$\mathbf{P}_l = V_l \zeta_l \tag{11}$$

where

$$\zeta_l = \begin{bmatrix} \zeta_{l,1} & \zeta_{l,2} \end{bmatrix}^T \in \mathbb{R}^2 \tag{12}$$

denotes the weighted interpolation parameter vector satisfying $\sum_{p=1}^{2} \zeta_{l,p} = 1$, $\zeta_{l,p} \geq 0$, and

$$V_l = \begin{bmatrix} \mathbf{P}_{l,1} & \mathbf{P}_{l,2} \end{bmatrix} \in \mathbb{R}^{2 \times 2} \tag{13}$$

denotes the vertices. When selecting two vertices from the given relative velocities $\dot{r}_x$ and $\dot{r}_y$ measured by the radar, we chose two vertices that matched the relative velocity data set from the viable set f_l. Then, we can get

$$\zeta_{l,p} = \frac{L_{l,sum}/L_{l,p}}{\sum_{j=1}^{2}(L_{l,sum}/L_{l,j})}, \ p = 1,2 \tag{14}$$

where $L_{l,sum} = \sum_{j=1}^{2} L_{l,j}$ in which L_l is the Euclidean distance between each vertex and the given relative velocities point $(\dot{r}_x, \dot{r}_y)$ measured by the radar. Using the interpolation parameters given the parameter vector for the relative velocity at two vertices, we can find an approximate position error from the interpolation between the precomputed average position error $\mathbf{E}_l^*(V_l) \in \mathbb{R}^{2 \times 2}$ at each vertex. The approximate position error $\mathbf{E}_l^o \in \mathbb{R}^2$ is expressed as follows

$$\mathbf{E}_l^o = \begin{bmatrix} \mathbf{E}_l^*(\mathbf{P}_{l,1}) & \mathbf{E}_l^*(\mathbf{P}_{l,2}) \end{bmatrix} \zeta_l. \tag{15}$$

Using the precomputed position error average $\mathbf{E}_l^*(V_l)$ at each vertex, we can calculate the approximate position error $\mathbf{E}_l^o = \begin{bmatrix} x^o & y^o \end{bmatrix}^T$ for a given relative velocity $\dot{r}_x$ and $\dot{r}_y$. The approximate position error $\mathbf{E}_l^o$, and $\hat{x}$ and $\hat{y}$ calculated by KF in the previous subsection are directly involved in the determination of estimated approximate position $\hat{x}^o$ and $\hat{y}^o$:

$$\hat{x}^o = \hat{x} - x^o, \ \hat{y}^o = \hat{y} - y^o. \tag{16}$$

Then, estimated approximate position $\hat{x}^o$ and $\hat{y}^o$ are applied to object vehicle tracking.

5. Application

We experimentally validated how useful the proposed data-driven weighted interpolation algorithm is when applied to object vehicle estimation of an autonomous vehicle.

5.1. Experimental Setup

For the experimental setup shown in Figure 3, the ego and object vehicles used were Genesis DH and Tucson IX from Hyundai, as shown in Figure 4, respectively. The rear left, and rear right view radars connected by a master and slave system with radar local control area network (CAN) were located on both sides of the rear of the ego vehicle and were rotated 23 degrees outward.

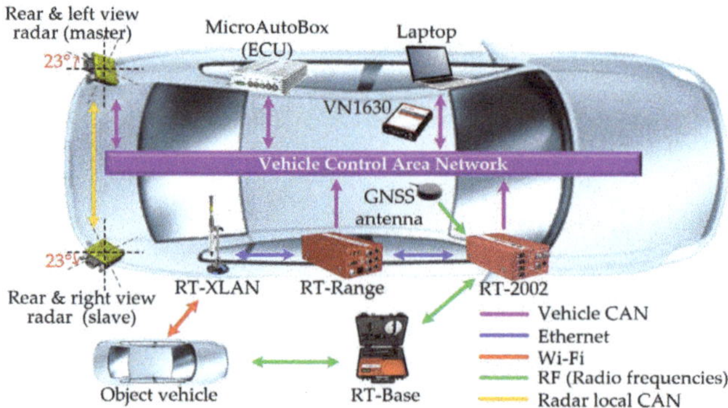

Figure 3. Hardware configuration of the experimental setup.

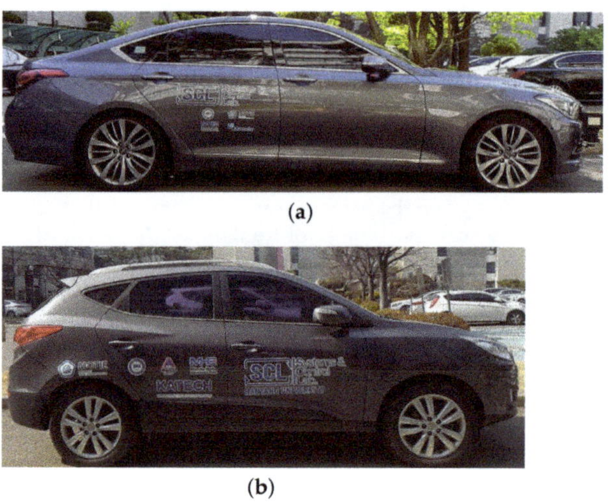

(a)

(b)

Figure 4. Vehicles used for experiment: (**a**) ego vehicle: Genesis DH from Hyundai and (**b**) object vehicle: Tucson IX from Hyundai.

The radar used 24 GHz BSD from Mando-Hella Electronics Corp., in Incheon, South Korea, the update sampling rate was 50 ms, and the distance detect range was up to 70 m. The ground truth data were collected at an update period of 10 ms using DGPS from OxTS (RT-2002, RT-Range, global navigation satellite system (GNSS) antenna, RT-XLAN, and RT-Base) with its real-time kinematic (RTK) positioning service ($1\sigma = 0.01$ m). We collected the object vehicle's ground truth data through the RT-Range and RT-XLAN Wi-Fi. Radar and DGPS data were collected through MicroAutoBox from dSPACE, analyzed with Vector's CANoe with VN1630, and evaluated using MATLAB/Simulink. These data were given by the ego and object vehicle driven manually on a high-speed circuit in the Korea Automobile Testing & Research Institute (KATRI) in South Korea, as shown in Figure 5.

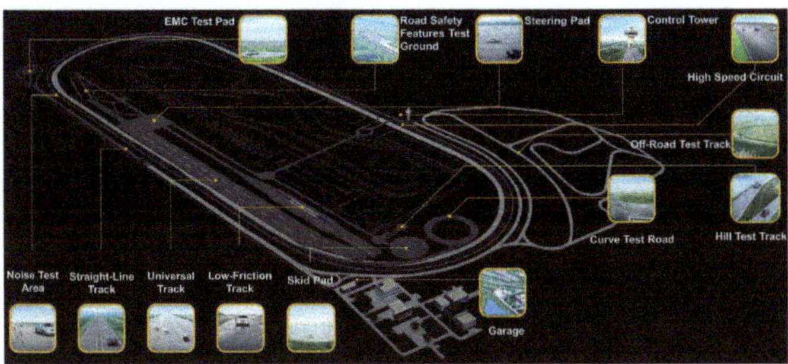

Figure 5. Test road: Korea Automobile Testing & Research Institute (KATRI).

5.2. Radar Accuracy Analysis

The radar accuracy was analyzed by the occupancy zone, as shown in Figure 6. The radar accuracy was analyzed by comparing the DGPS and radar data in each divided occupancy zone. Each zone shows the probability density function contour of the normal distribution based on DGPS, and blue ($10 \leq \dot{r} \leq 30$ km/h), black ($-10 \leq \dot{r} < 10$ km/h), and red ($-30 \leq \dot{r} < -10$ km/h) colors were plotted for each speed data set. It was found that the radar accuracy was different depending on the relative distance and speed. The error was increased as the relative distance and relative velocity between the ego vehicle and the object vehicle increased. Depending on the relative distance, measurement uncertainties by radar occurred [14,16]. We collected data through various real driving situations. The radar accuracy analysis was based on a total of 193,324 samples In this regard, the longitudinal relative velocity between the two vehicles was about -30 to 30 km/h.

Remark 2. *If the amount of sampled calibrated sensor data increases, the distribution of the measurement noise becomes the Gaussian distribution, as shown in Figure 6. Therefore, the system has better performance with more calibrated sensor data. Here is a reference if the measurement noise is not Gaussian [43].*

The longitudinal position error, which increases with relative velocity, was due to the latency of the radar, as shown in Figure 7. The average position error $\mathbf{E}_i^*$ of each velocity data set was analyzed as follows:

(i)   The average position error of the data set ($-30 \leq \dot{r}_x < -20$ km/h) is 1.968 m.

(ii)   The average position error of the data set ($-20 \leq \dot{r}_x < -10$ km/h) is 0.709 m.

(iii)   The average position error of the data set ($-10 \leq \dot{r}_x < 10$ km/h) is 0.018 m.

(iv)   The average position error of the data set ($10 \leq \dot{r}_x < 20$ km/h) is -0.511 m.

(v)   The average position error of the data set ($20 \leq \dot{r}_x \leq 30$ km/h) is -1.793 m.

The calculated position error average by latency was stored at each vertex of the velocity regions. Based on the analysis results, the position error and covariance values of each zone's error were obtained for the filter design.

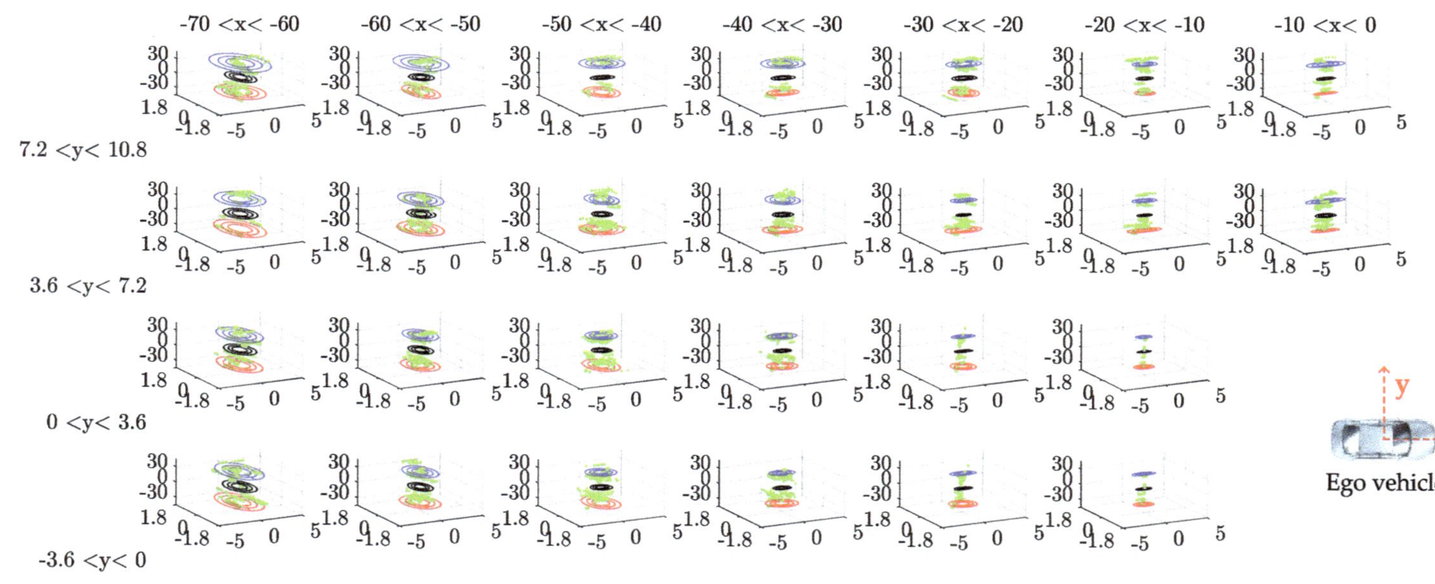

Figure 6. Data-driven radar accuracy analysis in the divided occupancy zone: each zone shows the probability density function contour ($1\,\sigma$, $2\,\sigma$, and $3\,\sigma$) of the normal distribution for radar's position error based on differential global positioning system (DGPS). Contour lines are shown in blue, black, and red for each speed data set.

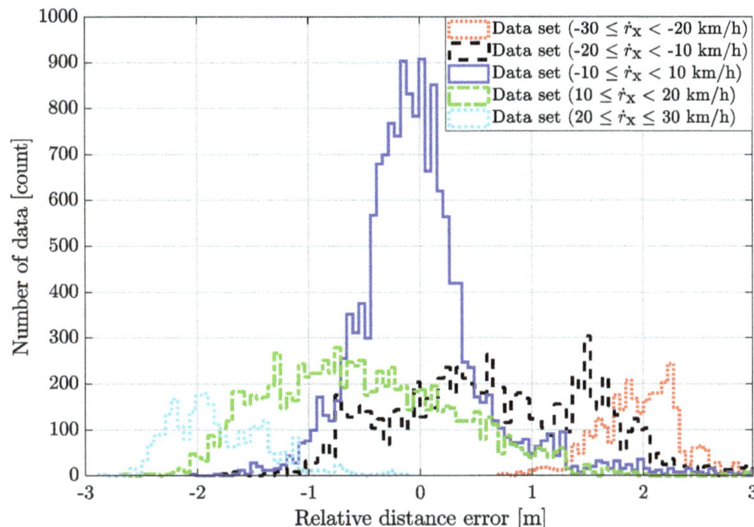

Figure 7. Histogram of the radar relative distance error depending on the vehicle speed.

The point to note here is that Figures 6 and 7 showed different results from what is generally understood, and therefore, careful attention is required. The radar's point cloud data is accurate in the longitudinal direction and inaccurate in the lateral direction. This is certain in radar's row data. However, providing a cloud data point to users makes it difficult for upper-level users to use radar data. Therefore, commercial radar represents an object as one tracking point data through an estimation algorithm (e.g., point cloud clustering, segmentation, single sensor tracking, multilateration, classification, and filtering) [7,8,14,35]. Therefore, there is latency (processing delay). The greater the difference in speed between the ego vehicle and the object vehicle, the greater the latency, and in a vehicle application with a velocity in the longitudinal direction, it causes a longitudinal error. As a result of this, unlike the general idea that the relative longitudinal distance of the radar is more accurate than the relative lateral distance, the experimental data with DGPS shows that the longitudinal direction is more inaccurate than the lateral direction. This means that, as the relative speed increases, the longitudinal error increases. Therefore, anyone designing an upper-level application needs to increase the radar accuracy. This is why we used object vehicle estimation with radar accuracy modeling.

Remark 3. *The latency of the relative lateral velocity is insignificant so it is not considered [44]. It can be calculated similarly to the method calculating the position error by latency.*

5.3. Scenario-Based Experimental Result

An object vehicle tracking scenario is constructed using data-driven object vehicle estimation with a radar sensor. For a comparative study of object vehicle tracking, we collected radar data while the object vehicle was driving in the detectable area of the rear left radar of the ego vehicle. As stated above, we determined the error characteristics in the occupancy zone by analyzing radar accuracy. Then, the approximate object estimation data were obtained by the data-driven weighted interpolation process using error characteristics data. When using weighted interpolation, the interpolation parameter vector was designed to satisfy $0 < \epsilon \le \zeta_{c,q}$ and $0 < \varepsilon \le \zeta_{l,p}$ for numerical stability, where ϵ and ε are small values. The proposed process improves the estimation performance of the commercial radar and the previously studied interpolation method [15].

The comparative results of the tracking performance are shown in Figure 8. The proposed weighted interpolation scheme improved the object vehicle tracking performance by reducing the estimation error. The scenario-based relative movement of object tracking was plotted in the top view. The proposed weighted interpolation scheme's performance was similar to that of DGPS. On the other hand, object tracking using a commercial radar had a larger tracking error than the proposed method due to measurement uncertainty and radar latency. Compared to DGPS, the root mean square error (RMSE) of the commercial radar and the proposed method are 3.04 m and 1.29 m in the longitudinal position and 0.57 m and 0.32 m in the lateral position, respectively. When estimated with a commercial radar, there is a longitudinal position error of about −5 m and a lateral position error of about −1.5 m between −43 m and −38 m. This is because the latency significantly affects the longitudinal position error. This error is also affected by the measurement uncertainty and relative acceleration. The influence of relative acceleration will be described in detail in the next paragraph. In addition, there is a lateral position error of about 1 m between −17 m and −15 m. This error is due to the influence of measurement uncertainty. In this regard, Figure 9 shows object tracking in the 3D view, including the relative velocity. The object vehicle changed lanes while increasing speed to overtake the ego vehicle. The proposed method outperforms the conventional radar estimation method and the previous interpolation method [15]. This is because the relative speed was not considered in the previous interpolation method. In this regard, the position error is covered in more detail in the next subsection, with Figures 10–14. The proposed weighted interpolation scheme reflects the average position error and covariance for the relative speed, even when there is speed variation. We observed that the proposed weighted interpolation scheme is robust against speed variation and that it outperforms the tracking performance of the commercial radar.

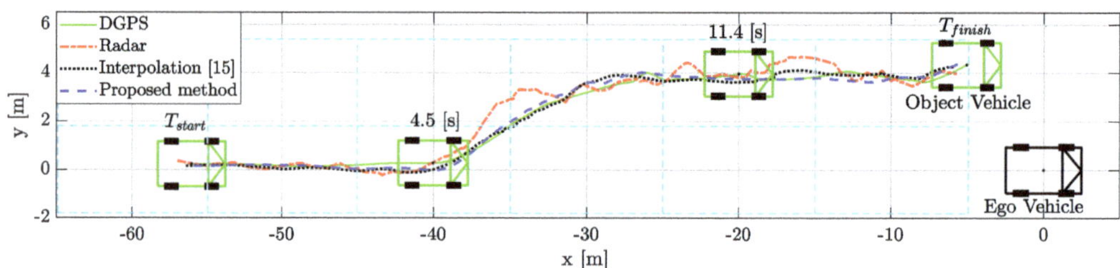

Figure 8. Scenario-based relative movement of object estimation from the top view: when estimated with a commercial radar, the longitudinal and lateral position errors (between −37 and −33 m) and lateral position error (between −17 and −15 m) occurred due to latency and measurement uncertainty.

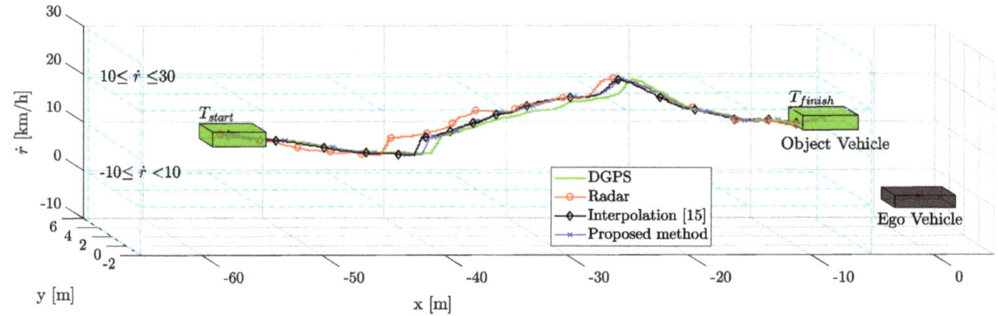

Figure 9. Scenario-based relative movement of object estimation with relative speed in a 3D view.

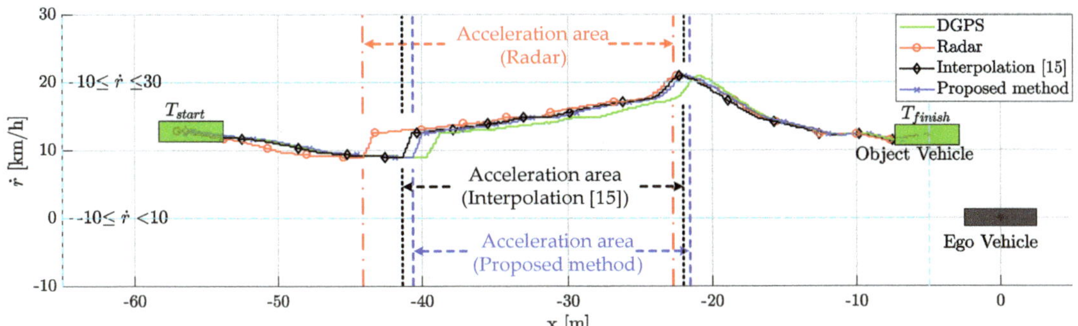

Figure 10. Scenario-based relative movement for relative longitudinal distance and relative speed replotted from Figure 9: there is an acceleration area because the object vehicle changes lanes with increasing speed to overtake the ego vehicle. After that, the relative speed decreases.

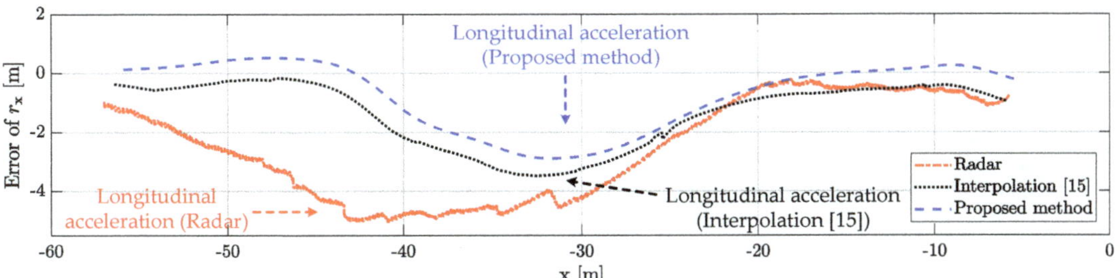

Figure 11. Longitudinal distance error for scenario-based object estimation: the proposed method outperforms the conventional radar estimation method and the previous interpolation method. However, there is a longitudinal error in all methods due to latency for longitudinal acceleration in the acceleration area.

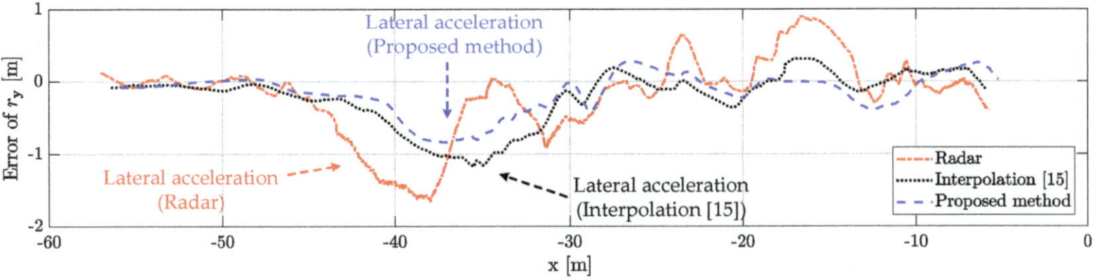

Figure 12. Lateral distance error for scenario-based object estimation: the proposed method outperforms the conventional radar estimation method and the previous interpolation method. When measured with radar, the lateral position error is heavily influenced by the measurement uncertainty. However, there is a lateral position error in all methods due to latency for lateral acceleration via the lane change motion of the object vehicle.

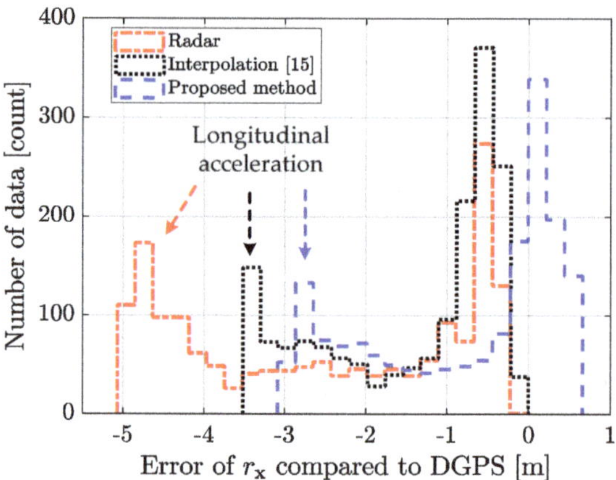

Figure 13. Histogram of the longitudinal distance error for scenario-based object estimation: using the weighted interpolation method improves the estimation performance statistically. However, there is a longitudinal error due to latency for longitudinal acceleration in the acceleration area.

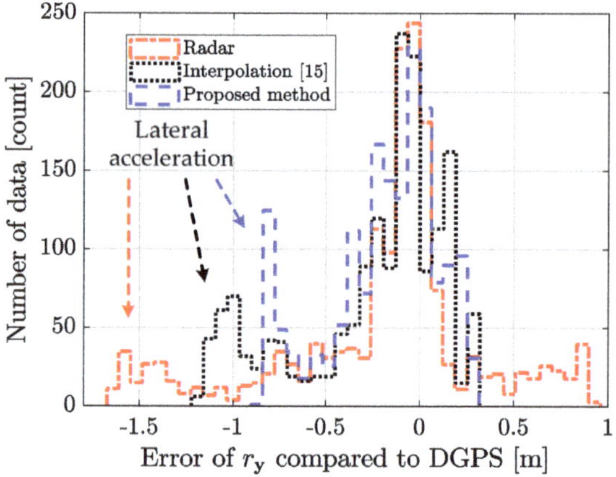

Figure 14. Histogram of the lateral distance error for scenario-based object estimation: using the weighted interpolation method improves the estimation performance statistically. However, there is a lateral position error due to latency for lateral acceleration via the lane change motion of the object vehicle.

5.4. Performance Analysis with Limitation

The proposed method outperforms the conventional radar estimation method and the previously researched interpolation method [15]. The performance for the scenario-based experimental result is shown in Figures 10–14. Figure 10 represents the relative longitudinal distance (x-axis) and relative velocity (y-axis) from Figure 9. There are acceleration areas (relative speed increase area) for radar, the previously researched interpolation method, and the proposed method. Figures 11 and 12 show the longitudinal and lateral position errors in terms of the x-axis position. The proposed method has a smaller position error than the conventional radar estimation method and the previous interpolation method compared to DGPS. Previously researched interpolation methods introduce measurement uncertainty and latency errors for speed. This is because speed is not considered. When measured with radar, the lateral position error is heavily influenced by the measurement uncertainty. Figures 13 and 14 show histograms of the relative position error for longitudinal and lateral, respectively. By using the proposed method, longitudinal errors due to latency and lateral errors due to measurement uncertainty are reduced.

However, there is a lateral position error in all methods due to latency for lateral acceleration via the lane change motion of the object vehicle. The estimation performance is improved by using the weighted interpolation method, but there is a limitation to the proposed method. The limitation arises because radar accuracy modeling is used only as the constant relative velocity model (2) and because relative acceleration is not considered. The longitudinal position error increases in the acceleration area, as shown in Figures 11 and 13. This is confirmed to be the effect of latency on relative acceleration. The position error was reduced outside of the acceleration area due to the proposed method. Since the object vehicle's lane change in the acceleration area is also performed, the lateral position error increases as the relative lateral acceleration increases, as shown in Figures 12 and 14. We have confirmed that the position error occurs in radar, the previously researched interpolation method, and the proposed method due to the influence of relative acceleration.

As future work, research should be conducted to reduce the effects of relative acceleration. The effect of relative acceleration can be reduced by using the relative acceleration model. In this regard, we will further consider the acceleration model using multiple models and expect to improve the collision risk performance using accurate radar estimation.

6. Conclusions

This paper proposed a data-driven object vehicle estimation scheme to solve the radar system accuracy problem. For object estimation considering the radar accuracy, we first developed an accuracy model that considers the different error characteristics depending on the zone. The accuracy model was used to solve the measurement uncertainty of radar. We also developed latency coordination for the radar system by analyzing the position error depending on the relative velocity. The developed accuracy modeling and latency coordination methods were applied to object vehicle estimation using weighted interpolation. The utility of the proposed method was validated through a scenario-based estimation experiment. The proposed data-driven object vehicle estimation outperformed the commercial radar algorithm and the previously researched interpolation method. The proposed method is expected to improve object vehicle estimation accuracy. Future work is expected to use an additional acceleration model as multiple models to reduce the effect of relative acceleration. This achievement is critical for autonomous driving technology for developing a high-level controller for functions such as collision risk decision, path planning with collision avoidance, and lane change system.

Author Contributions: Conceptualization, W.Y.C.; methodology, W.Y.C.; software, W.Y.C.; validation, J.H.Y. and C.C.C.; formal analysis, W.Y.C.; investigation, W.Y.C. and J.H.Y.; resources, C.C.C.; data curation, W.Y.C. and J.H.Y.; writing—original draft preparation, W.Y.C.; writing—review and editing, W.Y.C. and C.C.C.; visualization, W.Y.C.; supervision, C.C.C.; project administration, W.Y.C. and C.C.C.; funding acquisition, C.C.C. All authors have read and agreed to the published version of the manuscript.

Funding: This work was supported by the National Research Foundation of Korea (NRF) grant funded by the Korea government (MSIT) (No. 2021R1A2C2009908, Data-Driven Optimized Autonomous Driving Technology Using Open Set Classification Method) and by the Industrial Source Technology Development Programs (No. 20000293, Road Surface Condition Detection using Environmental and In-Vehicle Sensors) funded by the Ministry of Trade, Industry, and Energy (MOTIE, Korea).

Institutional Review Board Statement: Not applicable.

Informed Consent Statement: Not applicable.

Data Availability Statement: Not applicable.

Conflicts of Interest: The authors declare no conflict of interest.

References

1. Eskandarian, A. *Handbook of Intelligent Vehicles*; Springer: London, UK, 2012.
2. Rajamani, R. *Vehicle Dynamics and Control*; Springer Science & Business Media: New York, NY, USA, 2011.
3. Lin, P.; Choi, W.Y.; Chung, C.C. Local Path Planning Using Artificial Potential Field for Waypoint Tracking with Collision Avoidance. In Proceedings of the International Conference on Intelligent Transportation Systems (ITSC), Rhodes, Greece, 20–23 September 2020; pp. 1–7.
4. Choi, W.Y.; Lee, S.H.; Chung, C.C. Robust Vehicular Lane Tracking Control with Winding Road Disturbance Compensator. *IEEE Trans. Ind. Inform.* **2020**. [CrossRef]
5. Yang, J.H.; Choi, W.Y.; Chung, C.C. Driving environment assessment and decision making for cooperative lane change system of autonomous vehicles. *Asian J. Control* **2020**, 1–11. [CrossRef]
6. Lee, H.; Kang, C.M.; Kim, W.; Choi, W.Y.; Chung, C.C. Predictive risk assessment using cooperation concept for collision avoidance of side crash in autonomous lane change systems. In Proceedings of the International Conference on Control, Automation and Systems, Jeju, Korea, 18–21 October 2017; pp. 47–52.
7. Kellner, D.; Barjenbruch, M.; Klappstein, J.; Dickmann, J.; Dietmayer, K. Tracking of extended objects with high-resolution Doppler radar. *IEEE Trans. Intell. Transp. Syst.* **2015**, *17*, 1341–1353. [CrossRef]
8. Roos, F.; Kellner, D.; Klappstein, J.; Dickmann, J.; Dietmayer, K.; Muller-Glaser, K.D.; Waldschmidt, C. Estimation of the orientation of vehicles in high-resolution radar images. In Proceedings of the International Conference on Microwaves for Intelligent Mobility, Heidelberg, Germany, 27–29 April 2015; pp. 1–4.
9. Kim, B.; Yi, K.; Yoo, H.J.; Chong, H.J.; Ko, B. An IMM/EKF approach for enhanced multitarget state estimation for application to integrated risk management system. *IEEE Trans. Veh. Technol.* **2015**, *64*, 876–889. [CrossRef]
10. Yeddanapudi, M.; Bar-Shalom, Y.; Pattipati, K. IMM estimation for multitarget-multisensor air traffic surveillance. *Proc. IEEE* **1997**, *85*, 80–96. [CrossRef]
11. Gustafsson, F.; Gunnarsson, F.; Bergman, N.; Forssell, U.; Jansson, J.; Karlsson, R.; Nordlund, P.J. Particle filters for positioning, navigation, and tracking. *IEEE Trans. Signal Process.* **2002**, *50*, 425–437. [CrossRef]
12. Kulikov, G.Y.; Kulikova, M.V. Accurate continuous–discrete unscented Kalman filtering for estimation of nonlinear continuous-time stochastic models in radar tracking. *Signal Process.* **2017**, *139*, 25–35. [CrossRef]
13. Bar-Shalom, Y.; Li, X.R.; Kirubarajan, T. *Estimation with Applications to Tracking and Navigation: Theory Algorithms and Software*; John Wiley & Sons: Hoboken, NJ, USA, 2004.
14. Choi, W.Y.; Kang, C.M.; Lee, S.H.; Chung, C.C. Radar Accuracy Modeling and Its Application to Object Vehicle Tracking. *Int. J. Control Autom. Syst.* **2020**, *18*, 3146–3158. [CrossRef]
15. Choi, W.Y.; Yang, J.H.; Lee, S.H.; Chung, C.C. Object Vehicle Tracking by Convex Interpolation with Radar Accuracy. In Proceedings of the 2019 19th International Conference on Control, Automation and Systems (ICCAS), Jeju, Korea, 15–18 October 2019; pp. 1589–1593.
16. Alland, S.; Stark, W.; Ali, M.; Hegde, M. Interference in automotive radar systems: Characteristics, mitigation techniques, and current and future research. *Signal Process. Mag.* **2019**, *36*, 45–59. [CrossRef]
17. Averbuch, A.; Itzikowitz, S.; Kapon, T. Radar target tracking-Viterbi versus IMM. *IEEE Trans. Aerosp. Electron. Syst.* **1991**, *27*, 550–563. [CrossRef]
18. Kirubarajan, T.; Bar-Shalom, Y.; Pattipati, K.R.; Kadar, I. Ground target tracking with variable structure IMM estimator. *IEEE Trans. Aerosp. Electron. Syst.* **2000**, *36*, 26–46. [CrossRef]
19. Ebert, J.; Gumpp, T.; Münzner, S.; Matskevych, A.; Condurache, A.P.; Gläser, C. Deep Radar Sensor Models for Accurate and Robust Object Tracking. In Proceedings of the International Conference on Intelligent Transportation Systems (ITSC), Rhodes, Greece, 20–23 September 2020; pp. 1–6.
20. Alessandretti, G.; Broggi, A.; Cerri, P. Vehicle and guard rail detection using radar and vision data fusion. *IEEE Trans. Intell. Transp. Syst.* **2007**, *8*, 95–105. [CrossRef]
21. Cho, H.; Seo, Y.W.; Kumar, B.V.; Rajkumar, R.R. A multi-sensor fusion system for moving object detection and tracking in urban driving environments. In Proceedings of the IEEE International Conference on Robotics and Automation, Hong Kong, China, 31 May–7 June 2014; pp. 1836–1843.
22. Wu, S.; Decker, S.; Chang, P.; Camus, T.; Eledath, J. Collision sensing by stereo vision and radar sensor fusion. *IEEE Trans. Intell. Transp. Syst.* **2009**, *10*, 606–614.
23. Chavez-Garcia, R.O.; Aycard, O. Multiple Sensor Fusion and Classification for Moving Object Detection and Tracking. *IEEE Trans. Intell. Transp. Syst.* **2016**, *17*, 525–534. [CrossRef]
24. Kim, K.E.; Lee, C.J.; Pae, D.S.; Lim, M.T. Sensor fusion for vehicle tracking with camera and radar sensor. In Proceedings of the International Conference on Control, Automation and Systems, Jeju, Korea, 18–21 October 2017; pp. 1075–1077.
25. Kim, T.; Park, T.H. Extended Kalman filter (EKF) design for vehicle position tracking using reliability function of radar and lidar. *Sensors* **2020**, *20*, 4126. [CrossRef]
26. Nobis, F.; Geisslinger, M.; Weber, M.; Betz, J.; Lienkamp, M. A deep learning-based radar and camera sensor fusion architecture for object detection. In Proceedings of the 2019 Sensor Data Fusion: Trends, Solutions, Applications (SDF), Bonn, Germany, 15–17 October 2019; pp. 1–7.

27. Jha, H.; Lodhi, V.; Chakravarty, D. Object detection and identification using vision and radar data fusion system for ground-based navigation. In Proceedings of the International Conference on Signal Processing and Integrated Networks, Noida, India, 7–8 March 2019; pp. 590–593.
28. Muntzinger, M.M.; Aeberhard, M.; Zuther, S.; Maehlisch, M.; Schmid, M.; Dickmann, J.; Dietmayer, K. Reliable automotive pre-crash system with out-of-sequence measurement processing. In Proceedings of the 2010 IEEE Intelligent Vehicles Symposium, La Jolla, CA, USA, 21–24 June 2010; pp. 1022–1027.
29. Wielgo, M.; Misiurewicz, J.; Radecki, K. Processing latency effects on resource management in rotating AESA radar. In Proceedings of the 2017 18th International Radar Symposium (IRS), Prague, Czech Republic, 28–30 June 2017; pp. 1–10.
30. Supradeepa, V.; Long, C.M.; Wu, R.; Ferdous, F.; Hamidi, E.; Leaird, D.E.; Weiner, A.M. Comb-based radiofrequency photonic filters with rapid tunability and high selectivity. *Nat. Photonics* **2012**, *6*, 186–194. [CrossRef]
31. Klotz, M.; Rohling, H. 24 GHz radar sensors for automotive applications. *J. Telecommun. Inf. Technol.* **2001**, *4*, 11–14.
32. Parashar, K.N.; Oveneke, M.C.; Rykunov, M.; Sahli, H.; Bourdoux, A. Micro-Doppler feature extraction using convolutional auto-encoders for low latency target classification. In Proceedings of the 2017 IEEE Radar Conference (RadarConf), Seattle, WA, USA, 8–12 May 2017; pp. 1739–1744.
33. de Ponte Müller, F. Survey on ranging sensors and cooperative techniques for relative positioning of vehicles. *Sensors* **2017**, *17*, 271. [CrossRef]
34. Angelov, A.; Robertson, A.; Murray-Smith, R.; Fioranelli, F. Practical classification of different moving targets using automotive radar and deep neural networks. *IET Radar Sonar Navig.* **2018**, *12*, 1082–1089. [CrossRef]
35. Karunasekera, H.; Wang, H.; Zhang, H. Multiple object tracking with attention to appearance, structure, motion and size. *IEEE Access* **2019**, *7*, 104423–104434. [CrossRef]
36. Kang, C.M.; Lee, S.H.; Chung, C.C. Vehicle lateral motion estimation with its dynamic and kinematic models based interacting multiple model filter. In Proceedings of the Conference on Decision and Control (CDC), Las Vegas, NV, USA, 12–14 December 2016; pp. 2449–2454.
37. Li, X.R.; Jilkov, V.P. Survey of maneuvering target tracking. Part I. Dynamic models. *IEEE Trans. Aerosp. Electron. Syst.* **2003**, *39*, 1333–1364.
38. Myers, K.; Tapley, B. Adaptive sequential estimation with unknown noise statistics. *IEEE Trans. Autom. Control* **1976**, *21*, 520–523. [CrossRef]
39. Grewal, M.; Andrews, A. *Kalman Filtering: Theory and Practice with MATLAB*, 4th ed.; John Wiley & Sons: Hoboken, NJ, USA, 2015.
40. Choi, W.Y.; Kim, D.J.; Kang, C.M.; Lee, S.H.; Chung, C.C. Autonomous vehicle lateral maneuvering by approximate explicit predictive control. In Proceedings of the Annual American Control Conference (ACC), Milwaukee, WI, USA, 27–29 June 2018; pp. 4739–4744.
41. Rühaak, W. A Java application for quality weighted 3-d interpolation. *Comput. Geosci.* **2006**, *32*, 43–51. [CrossRef]
42. Stubberud, S.C.; Kramer, K.A. Monitoring the Kalman gain behavior for maneuver detection. In Proceedings of the 2017 25th International Conference on Systems Engineering (ICSEng), Las Vegas, NV, USA, 22–24 August 2017; pp. 39–44.
43. Zhang, D.; Xu, Z.; Karimi, H.R.; Wang, Q.G. Distributed filtering for switched linear systems with sensor networks in presence of packet dropouts and quantization. *IEEE Trans. Circuits Syst. I Regul. Pap.* **2017**, *64*, 2783–2796. [CrossRef]
44. Nishigaki, M.; Rebhan, S.; Einecke, N. Vision-based lateral position improvement of radar detections. In Proceedings of the 2012 15th International IEEE Conference on Intelligent Transportation Systems, Anchorage, AK, USA, 16–19 September 2012; pp. 90–97.

Article

An Evaluation of Executive Control Function and Its Relationship with Driving Performance

Lirong Yan [1,2,3,4], Tiantian Wen [1,2,3,4], Jiawen Zhang [1,2,3,4], Le Chang [2,3,4], Yi Wang [2,3,4], Mutian Liu [2,3,4], Changhao Ding [2,3,4] and Fuwu Yan [1,2,3,4,*]

[1] Foshan Xianhu Laboratory of the Advanced Energy Science and Technology Guangdong Laboratory, Foshan 528200, China; lirong.yan@whut.edu.cn (L.Y.); tiant_wen@163.com (T.W.); wen1160281134@163.com (J.Z.)

[2] Hubei Key Laboratory of Advanced Technology for Automotive Components, Wuhan University of Technology, Wuhan 430070, China; cl455508373@163.com (L.C.); wangyi_echo2020@163.com (Y.W.); liumutian0828@163.com (M.L.); teisyoukou@nsurg.med.osaka-u.ac.jp (C.D.)

[3] Hubei Collaborative Innovation Center for Automotive Components Technology, Wuhan 430070, China

[4] Hubei Research Center for New Energy & Intelligent Connected Vehicle, Wuhan University of Technology, Wuhan 430070, China

* Correspondence: yanfuwu@vip.sina.com; Tel.: +86-27-8785-7810

Citation: Yan, L.; Wen, T.; Zhang, J.; Chang, L.; Wang, Y.; Liu, M.; Ding, C.; Yan, F. An Evaluation of Executive Control Function and Its Relationship with Driving Performance. *Sensors* **2021**, *21*, 1763. https://doi.org/10.3390/s21051763

Academic Editor: Chao Huang

Received: 11 February 2021
Accepted: 25 February 2021
Published: 4 March 2021

Publisher's Note: MDPI stays neutral with regard to jurisdictional claims in published maps and institutional affiliations.

Copyright: © 2021 by the authors. Licensee MDPI, Basel, Switzerland. This article is an open access article distributed under the terms and conditions of the Creative Commons Attribution (CC BY) license (https://creativecommons.org/licenses/by/4.0/).

Abstract: The driver's attentional state is a significant human factor in traffic safety. The executive control process is a crucial sub-function of attention. To explore the relationship between the driver's driving performance and executive control function, a total of 35 healthy subjects were invited to take part in a simulated driving experiment and a task-cuing experiment. The subjects were divided into three groups according to their driving performance (aberrant driving behaviors, including lapses and errors) by the clustering method. Then the performance efficiency and electroencephalogram (EEG) data acquired in the task-cuing experiment were compared among the three groups. The effect of group, task transition types and cue-stimulus intervals (CSIs) were statistically analyzed by using the repeated measures analysis of variance (ANOVA) and the post hoc simple effect analysis. The subjects with lower driving error rates had better executive control efficiency as indicated by the reaction time (RT) and error rate in the task-cuing experiment, which was related with their better capability to allocate the available attentional resources, to express the external stimuli and to process the information in the nervous system, especially the fronto-parietal network. The activation degree of the frontal area fluctuated, and of the parietal area gradually increased along with the increase of CSI, which implied the role of the frontal area in task setting reconstruction and working memory maintaining, and of the parietal area in stimulus–Response (S–R) mapping expression. This research presented evidence of the close relationship between executive control functions and driving performance.

Keywords: attention; executive control; simulated driving; task-cuing experiment; electroencephalogram; fronto-parietal network

1. Introduction

Traffic safety has a great impact on the family and society. The World Health Organization (WHO) reported that approximately 1.25 million people died in road traffic accidents every year [1]. Among the traffic accidents, a very large proportion was caused by the drivers, which was nearly 90% according to the National Motor Vehicle Crash Causation Survey (NMVCCS) [2]. The driver, as the final service object, is the central node of sensation and control in the driver-vehicle-environment system and plays the most important role in traffic safety [3]. Drivers' physical and psychological state would greatly affect driving safety. The abnormal state of the driver, such as distraction and fatigue, would result in visual disturbances, which were related to most accidents [2]. Driving distraction and fatigues are the ubiquitous problems and the major cause of injury and death for the drivers throughout their life cycle [4]. Driving fatigue, usually resulted from lack of sleep or prolonged driving, would cause a decreased function of the sensory-motion system and a decline in driver's attention ability [5]. Driving distraction is defined as any activity that detracts the driver from the primary driving task, and is mainly reflected in three aspects, i.e., visual (taking one's eyes off the road), manual (taking one's hand off the wheel) and cognitive (taking one's mind away from the driving task) distraction [6]. Both driving fatigue and distraction are the manifestations of insufficient attention allocated for the driving tasks.

Several studies have investigated the monitoring method of the driving attentional state. Generally, several kinds of methods were developed, based on either the behaviors, the psychophysiological state of the driver, or the driving parameters of the vehicle. Some behaviors of the drivers, such as nodding, yawning and mouth movements were closely related to fatigue [7]. Usually, these behaviors were recorded and then analyzed to extract the fatigue-related features, such as Percentage of Eyelid Closure over the Pupil (PERCLOS) [8], the manipulation of the steering wheel [9], etc. Some studies demonstrated the correlation of the physiological parameters of the driver with driving attention, such as the high-frequency electrocardiogram (ECG) component [10] and the EEG (electroencephalogram) signals of the frontal areas [11]. The trajectory and the state of the vehicle, such as the speed, acceleration, and driving direction can also be utilized for distraction detection [12]. Studying the mechanism and the influential factors of attention can help to accurately evaluate the driver's alert state, replace the passive safety control strategy by active monitoring, improve the driving safety and effectively reduce the occurrence of traffic accidents.

The cognitive studies on attention included the behavioral [13], psychological [14,15], and neuroimaging schemas [16,17] both in subjects with attention-related disorders such as attention deficit hyperactivity disorder (ADHD) [18] and in normal people. Attention is characterized as the ability to effectively block outside distractions while focusing on a single object or task, which is a general function of the whole brain. The neuroimaging studies indicated that several neural networks were involved in attentional functions [19], among which three subsystems were specifically conceptualized, which were alerting, orienting and executive control [20]. Alerting is defined as reaching and maintaining a state that is highly sensitive to incoming stimuli, which would activate the anterior attention system, including the frontal cortex, posterior parietal cortex, and thalamus [20]. Alerting subsystem maintains the alert state and acts on the posterior attention system to support visual orienting. The orienting subsystem screens information from alert input to divert attention to the selected or focused stimulus, which is related with the activities of the frontal eye field, superior parietal cortex, temporal parietal junction, frontal eye fields, and superior colliculus [21]. The executive control subsystem monitors and resolves conflicts between thoughts, feelings, and responses, and plays a crucial role in attention, decision-making and complex conflict processing [20]. Currently, the most used paradigm to study the executive control function is the task-cuing paradigm. In this paradigm, the subjects would perform two or more types of tasks randomly under the instruction of a cue, which would be presented before or at the same time each target appears and prompt the type of task to be performed. The performance efficiency, such as RT and error rate, and the neuroimaging indexes, such as the EEG signal and the functional magnetic resonance imaging (fMRI) signal [19,22], would be recorded and compared between task switching and task repetition conditions. Results indicated that the response was slower, and the error rate was usually higher under the task switching condition, which was called the switch cost. Switch cost is an important indicator to quantify the function of executive control. Theoretical accounts of executive control assumed that multiple components were involved in activating a task-set, including paying attention to new cue-task connections, inhibiting the expression of previous task setting rules, shifting attention to relevant stimulus attributes, activating a goal representation, reconstructing the task's S–R (stimulus response) rules, setting response criteria and store task settings in working memory [23–27]. The switch cost was believed to occur during the active task setting reconstruction process. Better capability of the task setting reconstruction and complex cognitive processes optimization would result in the reduction of switching cost, which implied the higher efficiency of executive control function in cognitive processes coordination [25–27]. The switch cost, to some degree, is the behavioral manifestation of the executive control function. The spatiotemporal activities of the brain, on the other hand, laid the psychophysiological foundation of the executive control function. Several brain areas, including the prefrontal cortex, temporal cortex and anterior cingulate gyrus [18,22,28], were involved. Their activities varied among people with different attentional states, such as stronger activation of the dorsal anterior cingulate cortex, middle temporal gyrus, precuneus, lingual gyrus, precentral gyrus and insula in ADHD patients compared with the healthy adults under the task switching condition [18]. Besides, the psychological experiments demonstrated that the attentional state and CSIs were closely related. For example, the switch cost would increase if the CSI was too short [29,30]. The dynamic relationship between switch cost and brain activities is important to evaluate the executive control function, and is worthy of further research.

The executive control functions should be closely related to the driving performance. To test this hypothesis, quantitatively analyze the behavioral manifestations of the executive control functions, and explore the underlying cognitive mechanism, a total of 35 subjects were recruited to participate in a simulated driving experiment and a task-cuing experiment. The dataset including their driving behavior and EEG signals were acquired. The subjects were divided into three groups according to their driving performance (aberrant driving behaviors, namely lapses and errors). The performance efficiency and brain activation characteristics under different task transition types and CSI levels in different groups were analyzed. The results demonstrated the close relationship between driving performance and executive control efficiency. The fronto-parietal network participated in the executive control process and had a specific function in task setting construction and working memory maintenance.

2. Materials and Methods

2.1. Method Overview

The main research work was organized as follows: (i) simulated driving experiment and task-cuing experiment; (ii) systematic clustering (SPSS20.0, United States) to divide the subjects into different groups based on the driving performance; (iii) three-way repeated measures ANOVA for behavioral and EEG data among different groups of subjects; (iv) one-way repeated measures ANOVA and paired T-test to analyze differences between CSI and task transition types under different groups; (v) one-way ANOVA and two independent sample T-test to test the differences among different groups.

2.2. Subjects and Experiment Design

A total of 35 right-handed healthy adults (26 males and 9 females; 4 undergraduates, 28 postgraduates, 2 PhD candidates and 1 PhD) with no history of neurological disease were recruited, ranging in age from 21 to 46 (24.9 ± 5.7) years. Their visions were normal or corrected normal. All subjects had a Chinese C1 type (small car) driver's license with 1 to 17 (3.7 ± 3.1) driving years. They signed the written informed consent. The research was granted by the ethical review committee of Wuhan University of Technology. All subjects participated in two experiments: the simulated driving and the task-cuing experiment.

The simulated driving platform was built by Unity3D (Unity Technologies, Austin, TX, USA) and the Logitech G29 driving simulator (Logitech, Zurich, Switzerland), as shown in Figure 1a. The simulated driving scenario was an approximately 7 km circular orbital road, including slopes, turns, bridge holes, and other elements. Subjects were instructed to sit comfortably wearing the 64-channel Ag/AgCl electrode EEG cap (actiCHamp, Brain Products GmbH, Gilching, Germany), focus on driving along the road, and perform the operation of twisting the steering wheel or braking. The electrodeposition of the EEG electrode cap is shown in Figure 1b. Before the experiment, all subjects had enough time

(15 min or so) to familiarize themselves with the driving scene, brake pedal, acceleration torque, and steering wheel sensitivity to prepare for the experiment. During the driving process, each subject was required to complete three driving tasks at a speed limit of 70 km per hour, and each driving task included four laps. After each task, the participants took a short break of five minutes to avoid driving fatigue. The Logitech G29 provided similar force feedback of the steering wheel and brake as real driving. No subjects reported discomfort or driving sickness.

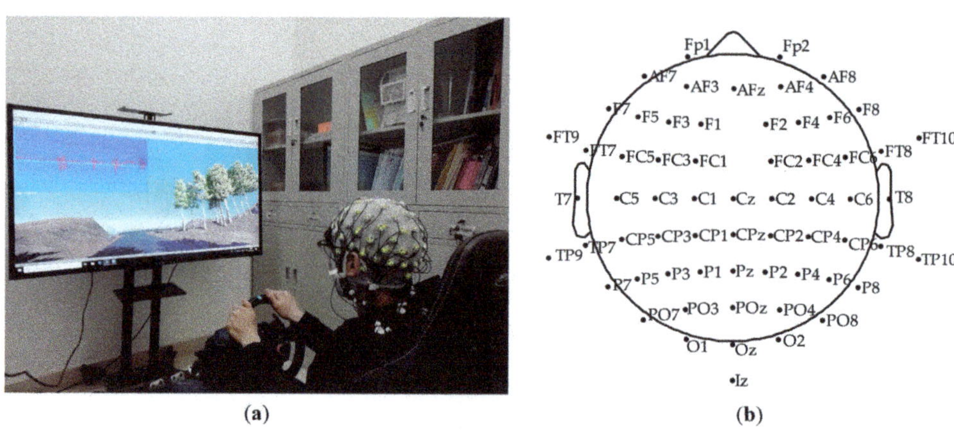

(a)   (b)

Figure 1. (**a**) Simulated driving platform; (**b**) EEG (electroencephalogram) cap electrode location map.

The task-cuing experiment was designed by E-Prime3.0 (Psychology Software Tools Inc., Sharpsburg, PA, USA) and presented on a 19-inch liquid crystal display (LCD) monitor with a screen resolution of 1600*900 (Figure 2). The task-cue was a white picture of a circle or triangle (6 cm × 6 cm) in a black background. The stimulus was a random number from 1 to 9 (except for 5) in red or green. The subjects sat in front of the screen with their sightline on the screen center, wore the EEG cap (actiCHamp, Brain Products GmbH, Gilching, Germany), and were instructed to respond to two types of tasks according to the task-cue. Task A: If the task-cue was a triangle, the subjects needed to judge the color of the number, and press "1" for red or "2" for green. Task B: If the task-cue was a circle, the subjects needed to judge the size of the number, and press "1" for numbers smaller than 5 or "2" for bigger than 5. There is also a task transition type that needed to be reminded about the trials. The task was either repeated or switched relative to the previous trial. According to the execution instructions of the task-cue, the participants were required to distinguish the color or size of the number. If the current task was different from the previous one, the current trial was classified as a switching trial; if the current task was the same as the previous one, the current trial was classified as a repeat trial. This factor was checked to see whether or not the switching trial has an impact on executive control over the repeat trial.

All subjects conducted seven sessions of the task-cuing experiment. Each session contained 42 trials, in which two kinds of tasks appeared randomly and evenly. The occurrence of different events in the same task was different, which was 2:1 of red to green ratio, and 2:1 of bigger-than-5-number to smaller-than-5-number ratio. In each trial, a "+" was shown for 100 ms, then an empty screen for 250 ms, followed by the cue for 100 ms and then the stimulus. The CSI between the cue and the stimulus was set at seven levels, i.e., 200 ms, 400 ms, 600 ms, 800 ms, 1000 ms, 1200 ms, and 1400 ms, which distributed randomly and evenly in each session. The stimulus would not disappear until the subjects pushed a button. After the reaction of the subjects, an empty screen would be shown for 500 ms.

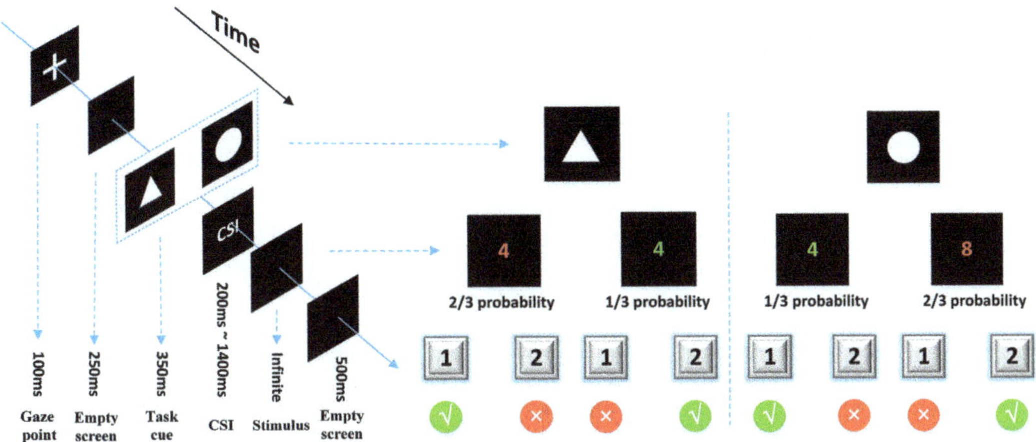

Figure 2. A single trial presentation process and task operation rules of the task-cuing experiment.

All subjects practiced before the formal experiment to get familiar with the task protocols. During the experiment, they could take a short break between two sessions.

2.3. Data Acquisition

In the simulated driving experiment, the driving data and EEG data were recorded simultaneously. The driving data, including the vehicle position and the steering wheel rotation angle, were acquired by the C# scripts based on Unity3D. The EEG data was collected at 1000 Hz by the Biopac actiCHamp Amplifier and BrainVision PyCorder (Brain Products GmbH, Gilching, Germany). The cap worn by the subjects was referenced to the FCz electrode according to the international 10–20 system protocol. The whole driving process of the vehicle on the screen was recorded by Apowersoft (Apowesoft, Hong Kong, China). For the task-cuing experiment, the behavioral data, including the RTs and error rates of the subjects were recorded by the E-DataAid module of E-Prime. The task transition type of each trial except the first one was defined as either repeated or switched relative to the previous trial, i.e., task repetition or switching.

2.4. Analysis of Behavioral Data

The driving performance of the subjects was evaluated according to the recorded screen video in the driving process. Specifically, the errors (severe accidents of driving out of the road or car collisions in which situation the vehicle was out of control and needed to be reset to the normal state by the experimenter) and lapses (moderate accidents resulted in off-road but under-control vehicle) made during the simulated driving experiment were counted. Systematic clustering was applied to these two types of errors to divide the subjects into different groups.

The behavioral data in the task-cuing experiment, including the RTs and the error rates under different conditions (group category, task transition type, and CSI) were analyzed. The differences in task activation among the subjects were tested using a 3 (group category: group 1, group 2, group 3) × 2 (task transition type: task repetition, task switch) × 7 (CSI: 200 ms, 400 ms, 600 ms, 800 ms, 1000 ms, 1200 ms, 1400 ms) repeated measures ANOVA (SPSS20.0, United States).

2.5. Analysis of EEG Data

The preprocessing of the EEG data was carried out using the EEGLAB toolbox (Swartz Center for Computational Neuroscience, San Diego, CA, USA) in MATLAB (R2013b, MathWorks, Natick, MA, USA). The signal in Fp1 and Fp2 channels were removed from the subsequent statistical analysis due to the disturbance of the electrooculogram (EOG). TP9 and TP10 were selected as the re-reference electrodes. Bandpass filtering (0.1–35 Hz) was applied to remove the noise. By extracting data epochs (200 milliseconds before stimulation to 1500 ms after stimulation) from the continuous EEG signal and data averaging, event information was obtained and event-related potential (ERP) images were created. Finally, independent component analysis (ICA) was applied to remove eye artifacts (including the signal artifacts due to the movement of the eyeball, ocular muscles, and eyelid), ECG artifacts, electromyography (EMG) artifacts, and other noises.

By behavioral data analysis, the EEG data were also analyzed by $3 \times 2 \times 7$ repeated measures ANOVA. The F values in the analysis result of variance were extracted to draw the topographic maps, the interactions and single-factor effects were analyzed. Paired T-test (testing for activation differences between different task transition types), one-way repeated measures ANOVA (testing for activation differences under different CSI conditions), one-way ANOVA (test the differences among three groups) and two independent sample T-test (testing for activation differences between any two groups) were used to explore the effects of various factors on the implementation of executive control mechanisms.

3. Results

3.1. Behavioral and EEG Characteristics of Different Groups of Subjects

3.1.1. Grouping Results

The 35 subjects were divided into different groups according to their aberrant behaviors (errors and lapses) using the systematic clustering ("bottom-up" aggregation, Euclidean distances, shortest distance algorithm). Initially each subject belonged to the different categories. Then, the pair of subjects with the shortest distance were merged into one category. The distance between this category and the other categories were calculated and merged the two nearest categories. Continue this procedure until all the categories were merged into one. Three categories were set in advance and the subjects were classified according to the pedigree cluster diagram. The clustering results based on the driving data are shown in Figure 3. Subject 7, 17, 21, 23, 27 and 28 were classified as group 1, subject 1, 8, 10, 14, 15, 16, 19, 24, 25, 26, 29, 30, 32, 33, and 34 were classified as group 2, and the rest were classified as group 3. There was no significant difference in genders, ages, driving years and education levels among the three groups ($\chi^2 = 4.836$, $P = 0.089$; $F = 0.149$, $P = 0.862$; $F = 0.102$, $P = 0.903$; $\chi^2 = 2.978$, $P = 0.561$ respectively). The average numbers of errors, lapses and all aberrant driving behaviors (summed numbers of lapses and errors) in group 1, group 2 and group 3 were (7.67 ± 5.75, 20.5 ± 1.61, 28.17 ± 6.85), (9.73 ± 3.83, 10.87 ± 2.29, 20.6 ± 4.32) and (2.93 ± 2.58, 2.86 ± 1.75, 5.79 ± 3.51) respectively. The mean occurrence of the errors, lapses and total occurrence in the three groups were significantly different ($F = 12.152$, 170.065 and 64.951 respectively). The post hoc pair-wise comparison indicated significant difference of lapses ($T = 9.202$, 20.644, and 10.515), and all the aberrant driving behaviors ($T = 3.065$ of group 1 vs. group 2, 9.789 of group 1 vs. group 3, and 10.084 of group 2 vs. group 3, all $P < 0.01$). The occurrence of errors was significantly different between group 2 and group 3 ($T = 5.709$, $P < 0.01$). The difference of the errors in group 1 vs. group 2 and group 3 was not significant ($T = -0.969$, $P = 0.345$; $T = 1.941$, $P = 0.102$ respectively).

3.1.2. Effect of Task Transition Types, CSIs and Group on the Behavioral Data

The three-way repeated measures ANOVA on RTs (reaction times) analysis indicated that the main effects of task transition type and CSI on RTs were significant ($F (1, 32) = 35.531$, $P = 0.000$, and $F (6, 192) = 7.769$, $P = 0.000$ respectively). The main effect of group ($F (2, 32) = 2.986$, $P = 0.065$), the threesome interaction effect ($F (12, 192) = 1.532$, $P = 0.115$) and pair-wise interaction effects ($F (6, 192) = 1.541$, $P = 0.167$; $F (2, 32) = 1.233$, $P = 0.305$; $F (12, 192) = 0.772$, $P = 0.679$) were not significant. Generally, in the three groups, the RTs were basically smaller under the task repetition condition than those under the task switching condition (Figure 4). The mean RT of group 1 (Figure 4a) was 799 ms and 909 ms for the task repetition and task switching condition respectively, of group 2 (Figure 4c) was 866 ms and 980 ms respectively, and of group 3 (Figure 4e) was 744 ms and 809 ms respectively. The RT in group 1 was shorter than that of group 2 but longer than that of group 3 (not significant). In group 1, when the CSI was lower than 800 ms, the switch cost fluctuated around 150 ms to 200 ms, as the CSI continued to increase, the switch cost first decreased and then increased, reaching the minimum when the CSI was 1200 ms. In group 2, the switch cost first increased along with the increasing of CSI and then fluctuated around 100 ms. In group 3, the switch cost fluctuated around 70 ms and reached minimum when the CSI was 400 ms. The mean switch cost of RT in group 1, group 2 and group 3 was 110 ms, 114 ms and 65 ms respectively.

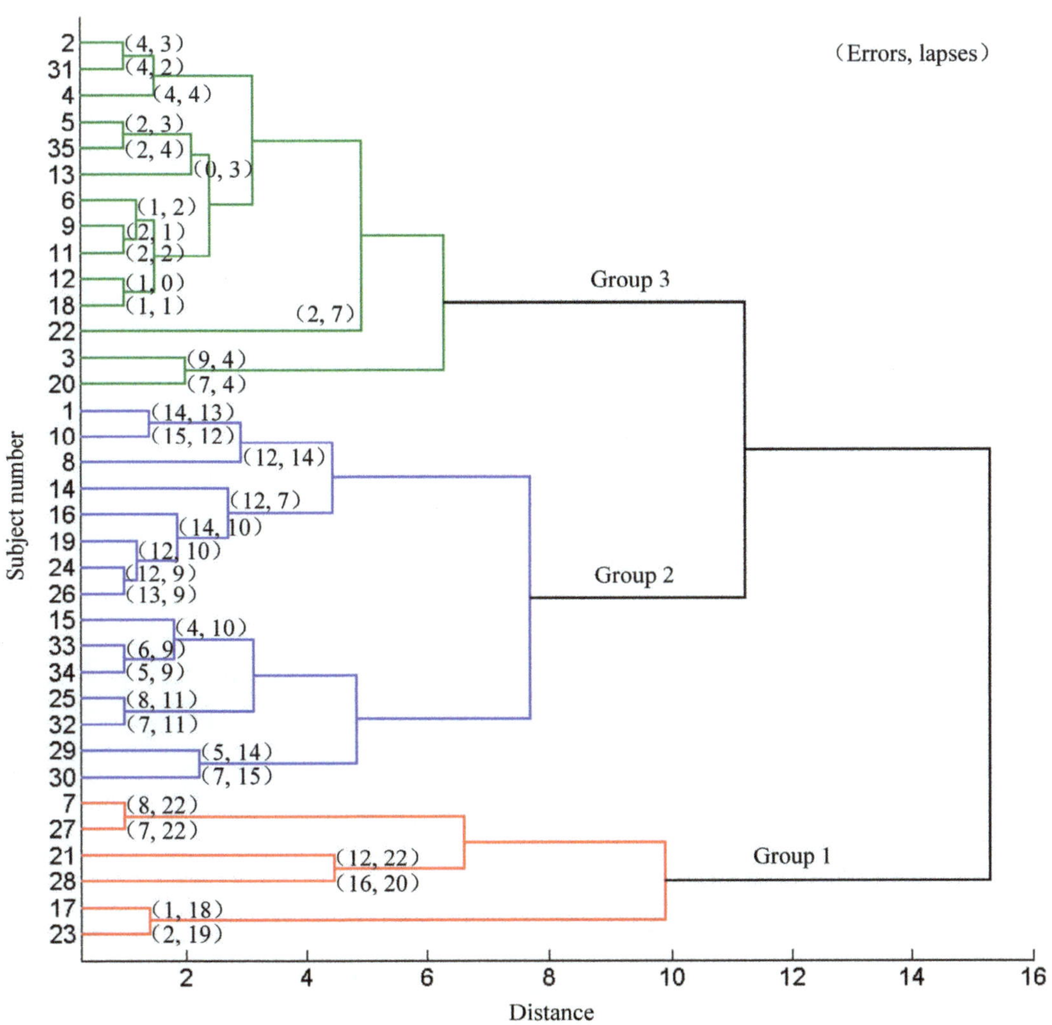

Figure 3. Results of systematic clustering based on the driving data.

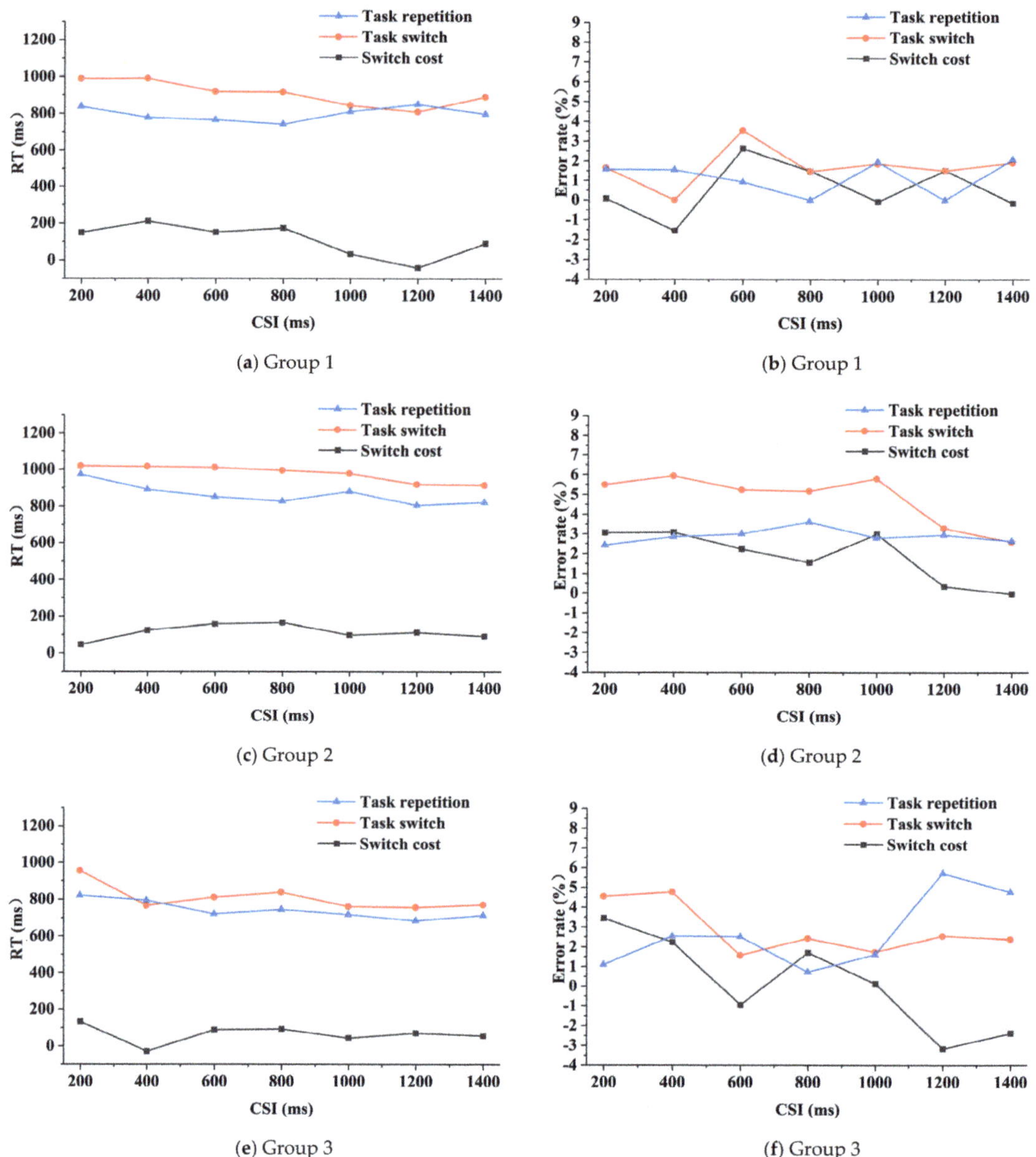

Figure 4. RT (reaction time) and error rate of the three groups as functions of task transition type and CSI (cue-stimulus interval).

The main effect of task transition type on the error rates was significant (F (1, 32) = 6.154, P = 0.019). The main effects of CSI and group (F (6, 192) = 0.546, P = 0.773; F (2, 32) = 2.673, P = 0.084), the threesome interaction effect (F (12, 192) = 0.661, P = 0.787) and the pair-wise interaction effects (F (6, 192) = 0.563, P = 0.759; F (2, 32) = 0.979, P = 0.387; F (12, 192) = 0.933, P = 0.515) were not significant. The mean error rate of group 1 was 1.2% and 1.7% under the task repetition and switching condition respectively; of group 2 was 2.9% and 4.8%

respectively, and of group 3 was 2.7% and 2.9% respectively. Generally, for the three groups, the error rates were smaller under the task repetition condition than that under the task switching condition, and the error rate of group 2 was higher than the other two groups. In group 1, the switch cost fluctuated around 0.5%, and the smallest absolute value appeared when the CSI was 200 ms (Figure 4b). The switch cost of group 2 had a downward trend as the CSI increased, except for an abnormal increase when the CSI was 1000 ms (Figure 4d). In group 3, the switch cost generally decreased along with the increasing of CSI, but increased when the CSI was 800 ms and 1400 ms, respectively (Figure 4f). The mean switch cost of error rate in group 1, 2 and 3 was 0.6%, 1.9% and 0.1% respectively.

3.1.3. Effect of Task Transition Types, CSIs and Group on the EEG Data

The main effect of CSI on brain activity was significant in the most frontal and parietal (Figure 5b, all electrodes except TP7, P7, PO7, O1, Oz and Iz), of group in the prefrontal (AFz, AF3, AF4 and AF8), the frontal (F1, F2, F3, F4, F6, F7 and F8), the frontal-central (FC1, FC2, FC4 and FC6), the central (Cz and C1) and the right fronto-temporal regions (FT8, Figure 5c). The main effect of task transition type (Figure 5a), the threesome interaction effect (Figure 5d) among group, task transition type and CSI was not significant. The pairwise interaction effects were significant at several limited electrodes (AFz, F1, F2, T8 and PO4 in Figure 5e, FC2, Cz and C2 in Figure 5f).

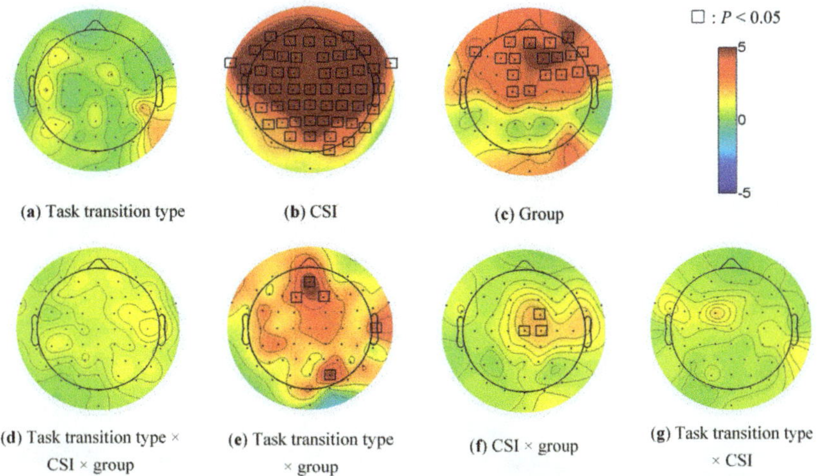

Figure 5. The main effects and interaction of group, task transition type and CSI in EEG data.

Considering the existence of the interaction effects, the post hoc comparison was performed to test the simple effects of the task transition type, CSI and group.

In group 1 (Figure 6a), the repeat trials caused stronger activation in the left prefrontal region (AF3) when the CSI was 800 ms, and the switching trials caused significantly stronger activation in the right parietal cortex (P8) as the CSI increased to 1400 ms. The difference in CSI was mainly concentrated in the prefrontal (AFz, AF7 and AF8), frontal (F1, F3 and FC1), fronto-temporal (FT7 and FT8) central (C2, C4 and C6) and central-parietal (CP2 and CP4) regions under the task repetition condition, and in the frontal (centered at the F1 electrode), right fronto-temporal (FT8) and right central parietal (centered at the C2 electrode) regions under the task switching condition.

In group 2 (Figure 6b), the repeat trials caused stronger activation in the left central parietal (CP1) and parietal regions (P1) when the CSI was 200 ms, the switching trials caused significantly stronger activation in the right prefrontal region (AF8) when the CSI was 1400 ms. The difference in CSI was mainly concentrated in the frontal (F2, FC2 and

FC4), parietal (P6) and parietal-occipital (PO8) regions under the task switching condition, while not significant under the task repetition condition.

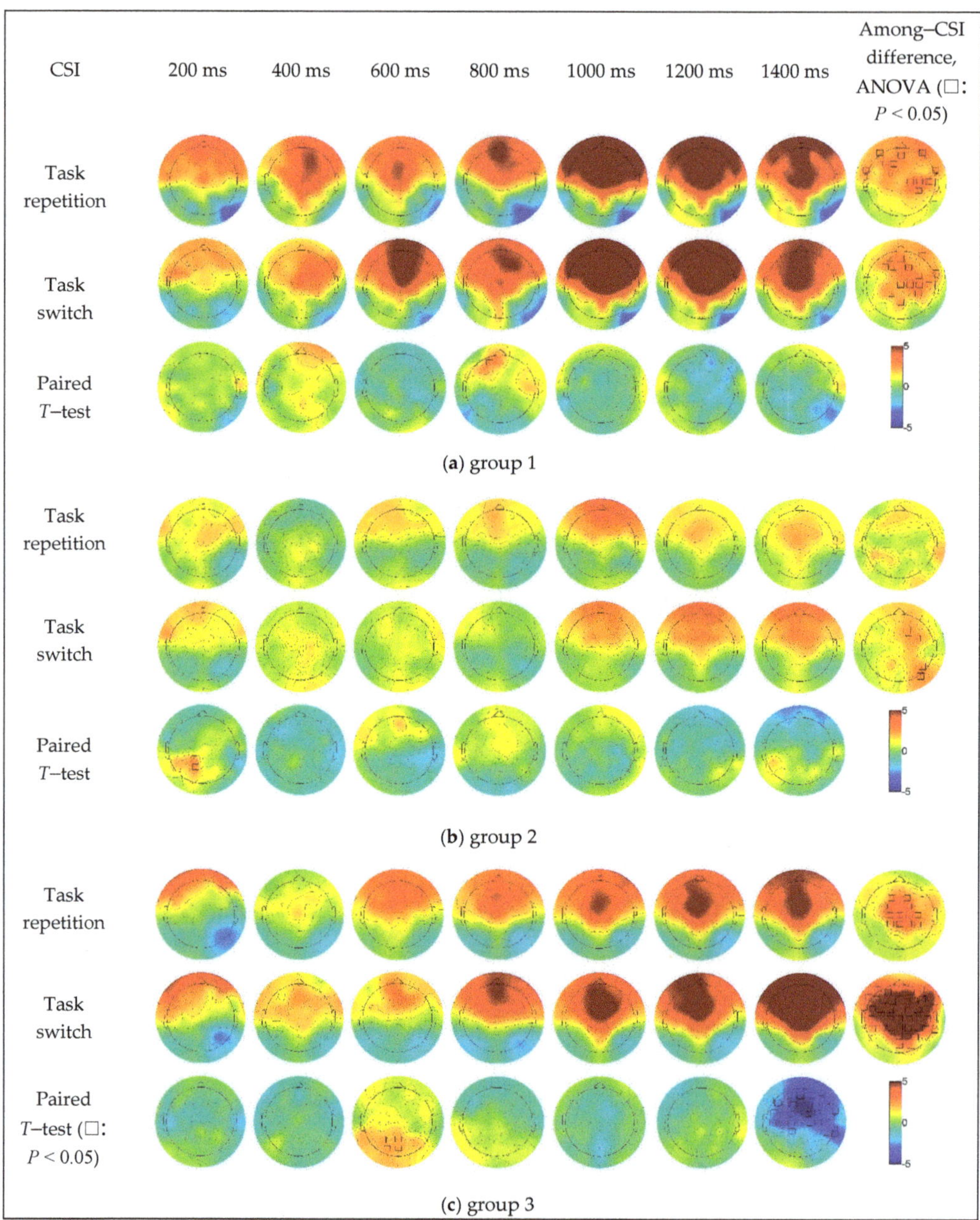

Figure 6. The simple effect of CSI and task transition types on EEG data in different groups.

In group 3 (Figure 6c), compared to the switch trials, the repeat trials caused significantly stronger activation in the parietal (Pz and P3) and parietal-occipital (POz and PO3) regions when the CSI was 600 ms, and the switching trials caused significantly stronger in the right fronto-temporal (FT8), temporal (T8), temporal parietal (TP8), most central frontal (a large area centered in the FCz electrode) regions when the CSI was 1400 ms. The difference in CSI was mainly concentrated in the frontal (F1, F2, FC1 and FC2), central (Cz, C1, C2 and C4) and central parietal (CPz, CP1, CP2 and CP4) regions under the task repetition condition, and under the task switching condition, in most of the prefrontal, frontal, parietal, temporal and parietal-occipital regions (up to AFz, down to POz, left to FT7, and right to FT8 electrodes), which was the fronto-parietal network.

For the difference among the three groups (Figure 7a), under the task repetition condition, the brain activation differences were in the right frontal (F2, F4, F6, F8 and FC6) and right fronto-temporal (FT8) regions when the CSI was 400 ms, in the right frontal-central (FC6), left fronto-temporal (FT7 and FT9), central (Cz, C1, C2, C3, C4 and C6) and central-parietal (CP1) regions when the CSI was 1000 ms, in the frontal-central (FC2), right fronto-temporal (FT8) and central (Cz) regions when the CSI was 1200 ms, and in most of the prefrontal and central regions (centered at the F1 electrode) when the CSI was 1400 ms. Under the task switching condition, the brain activation differences were in the right parietal-occipital (PO8) region when the CSI was 400 ms, in the prefrontal (AFz and AF4), frontal (F4), right parietal-occipital (PO8) and occipital (O2 and Iz) regions when the CSI was 600 ms, in the prefrontal (AFz), right frontal (F2, F4, F6 and FC4), parietal (CP1, P1 and P5), left parietal-occipital (PO7) and occipital (O1) regions when the CSI was 800 ms, in the prefrontal (AFz), frontal (F1, F2, F3, F5, FC2 and FC5), left fronto-temporal (FT7) and central (Cz, C1 and C6) regions when the CSI was 1000 ms, in the frontal and central parietal regions (part of the area centered at the Cz electrode) when the CSI was 1200 ms. When the CSI increased to 1400 ms, the brain activation differences occurred in almost all areas of the prefrontal, frontal, bilateral temporal and central parietal regions.

For the difference between group 1 and group 2 (Figure 7b), the brain activations in group 1 were more intense, in the frontal (centered at the F2 electrode, CSI = 400 ms), central parietal (CPz and CP1, CSI = 600 ms) and frontal-parietal (most areas of the frontal and parietal regions, CSI > 800 ms) regions under task repetition condition, in the frontal (centered at the F2 electrode, CSI = 600 ms), parietal (centered at the CP1 electrode, CSI = 800 ms and 1400 ms) and frontal-parietal (most areas of the frontal and parietal regions, CSI = 1000 ms and 1200 ms) regions under task switching condition. The brain activations in group 2 were stronger in the right parietal-occipital (PO8, CSI = 400 ms, 600 ms, 800 ms and 1000 ms) and occipital (Iz, CSI = 600 ms) regions.

For the difference between group 1 and group 3 (Figure 7c), the brain activations in group 1 were stronger, in the right central (C2, C4, FC4 and FC6, CSI = 400 ms) and bilateral fronto-temporal (FT7 and FT8, CSI = 400 ms and 1000 ms) regions under task repetition condition, in the parietal-occipital (centered at the PO3 electrode, CSI = 800 ms) and right fronto-temporal (FT8, CSI = 1200 ms) regions under task switching condition. The brain activation in group 3 was stronger in the frontal central region (FC2, CSI = 1400 ms) under task switching condition.

As for the difference between group 2 and group 3 (Figure 7d), the brain activation in group 3 was stronger, in the central (FC3, Cz and CP1, CSI = 1000 ms and 1200 ms) and fronto-parietal (most areas of the frontal and parietal regions, CSI = 1400 ms) regions under task repetition condition. Under the task switching condition, the brain activations in group 2 were stronger in the left parietal (P7, CSI = 800 ms), parietal-occipital (PO7 and PO8, CSI = 400 ms, 600 ms and 800 ms) and occipital (Oz, O1, O2 and Iz, CSI < 1000 ms) regions. With the increase of CSI, the brain activation intensity and activation range in group 3 gradually increased, and when the CSI was 1400 ms, the brain activation area almost covered the entire frontal-parietal network.

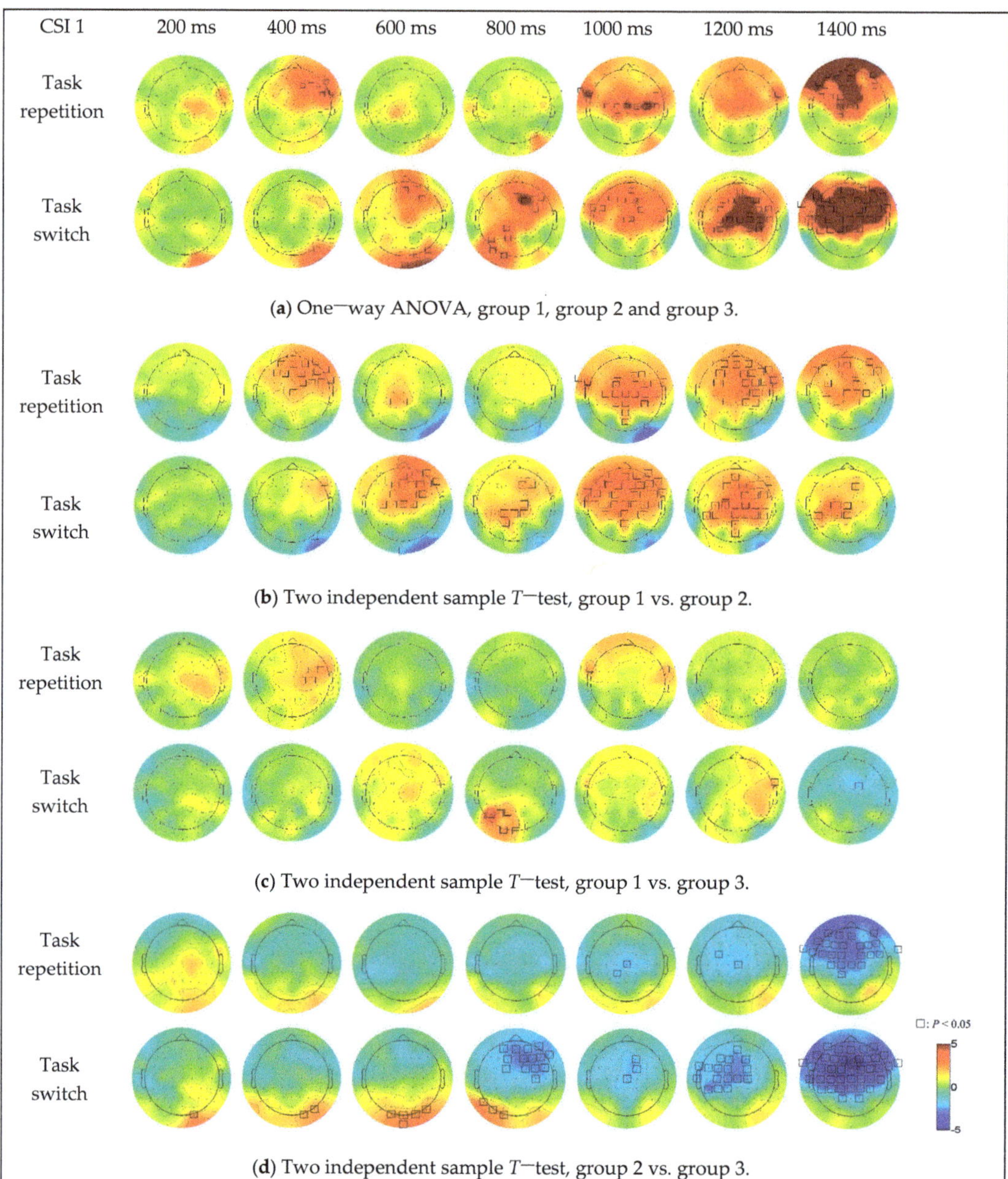

(**a**) One—way ANOVA, group 1, group 2 and group 3.

(**b**) Two independent sample *T*—test, group 1 vs. group 2.

(**c**) Two independent sample *T*—test, group 1 vs. group 3.

(**d**) Two independent sample *T*—test, group 2 vs. group 3.

Figure 7. The simple effect of group on EEG data under different CSI conditions and task transition types.

4. Discussion

In this work, a total of 35 healthy subjects were recruited to participate in a simulated driving experiment and a task-cuing experiment. The subjects were divided into three groups according to their driving performance. Then the performance efficiency and EEG data acquired in the task-cuing experiment were compared among the three groups, and the effect of task transition types and CSIs was statistically analyzed. The performance efficiency and the underlying cognitive mechanism of the executive control function, and its relationship with the driving performance was investigated.

4.1. Relationship between Driving Performance and Executive Control Efficiency

Driving is a very complicated procedure, which is composed of a series of behavioral operations, and resulted from the dependable perception-decision-execution cycle of the brain. The driving performance can be studied using the number of crashes, the number of incorrect use of turn signals, overtaking distance, vehicle trajectory and speed, etc. [12,31]. Reason et al. [32] presented a useful theoretical model by using the risky driving behaviors for the driving performance evaluation. Particularly three categories of aberrant behaviors related to different cognitive and decisional processes were defined, i.e., errors, lapses and violations. Errors were defined as failures to achieve the intended consequences of planned actions (e.g., braking too quickly on a road with low friction), largely representing information-processing deficits. Lapses were defined as failures of attention or memory (e.g., attempt to drive away from traffic light in third gear), largely representing information-sensory deficits. Violations were defined as deliberate violation of rules or failure to follow safe driving practices (e.g., decide to continue driving at the red light). Enlighted by this definition, we defined the errors and lapses in our work according to the severity of the accidents and the controllability of the vehicle. The severe accidents were caused by a series of mistakes made during the information processing procedure, while the moderate accidents were usually resulted from negligence of the external information. Errors and lapses constructed two dimensions to depict the aberrant driving behaviors in our driving scene. Accordingly, the enrolled 35 subjects were divided into three groups. The driving performance was the best in group 3 with the fewest errors and lapses. Group 1 had the highest occurrence of lapses and medium occurrence of errors, and group 2 had the highest occurrence of errors and medium occurrence of lapses.

Executive control refers to the coordination of multiple tasks to complete complex cognitive control processes. task-cuing experiment is a common paradigm to study the underlying mechanism of executive control function. The subjects needed to perform the same task as the former one (task repetition) or quickly switch to another kind of task (task switching). During the experiment, the subjects would maintain a specific cognitive state and construct a task setting process involving perception, attention, memory, and response [33]. Under the task repetition condition, the subjects only needed to implement previously configured task settings. While under the task switching condition, the subjects needed more effort to complete the configuration of a new kind of task. The executive control demands were greater, due to the working memory requirements to maintain multiple tasks in memory, the inhibition of the previous task, and the activation of the current task [34]. Consequently the subjects' response was usually slower and the accuracy lower, which was considered as the switch cost phenomenon [30]. The switch cost could be utilized as a quantitative indicator and was positively correlated with the subjects' executive control efficiency [23,27]. In our work, the task transition type had the significant independent impact on RT and error rate, which was significantly larger under the task switching condition for all the groups. Though group effect was not significant, the average switch costs of RTs and error rates in group 1, 2, and 3 were 110 ms and 0.6%, 114 ms and 1.9%, and 65 ms and 0.1%, respectively (Figure 4), which indicated the best executive control performance of group 3, and the worst of group 2. The behavioral performance of group 3 in the task-cuing experiment revealed that group 3 obviously had better capability to allocate the available attentional resources when the demands for the working memory maintaining former information inhibition, and reconfiguration of the current task was greater. This capability also resulted in better driving performance of group 3, which was highly correlated with their attentional and cognitive states. Although the total number of the abnormal driving behaviors of group 2 was lower than that of group 1, group 2 had the most errors, the largest switch cost, and the worst executive control function.

This implied that in the two aberrant driving behaviors, error clearly better reflected the executive control function. This would be further ascertained by the EEG results. CSI had a significantly independent impact on RT (Figure 4). As CSI increased, the performance efficiency (including RT and error rate) was significantly improved. The switch cost of the error rate decreased as the CSI increased, especially in group 3. The impact of CSI on behavioral performance and switch cost proved that the task setting reconstruction process, i.e., the preparation effect for new trials [35], was an important source of switch cost. During the experiment, once the task-cue appeared, the brain began to complete the control conversion process from the initial abstract rule representation to the actual representation [22]. When the CSI was short, this process cannot be well executed due to the pressure and insufficiency of the preparation time, which would result in an unstable characterization of the actual stimulus. Whereas when the CSI was longer, the conversion process would be much smoother, and better performance efficiency could be achieved.

4.2. The Underlying Cerebral Network for the Executive Control Function

The brain activities under different task transition conditions reflected how the brain was organized to fulfill the executive control function. As can be seen in Figure 5, generally the main effect of task transition type on the EEG activities was not significant (Figure 5a). However, it had the interaction effect with group factor in some electrodes (Figure 5e). The following post hoc simple effect analysis indicated that group 1 (Figure 6a) had stronger activation in AF3 (CSI = 800 ms), group 2 (Figure 6b) in CP1 and P1 (CSI = 200 ms), and group 3 (Figure 6c) in Pz, P3, POz and PO3 (CSI = 600 ms) under the task repetition condition; while group 1 had stronger activation in P8 (CSI = 1400 ms), group 2 in AF8 (CSI = 1400 ms) and group 3 in most electrodes of the fronto-parietal network (CSI = 1400 ms) under the task switching condition. These results suggested the different activation patterns during the executive control procedure in three groups of subjects. The ANOVA results did reveal the significant main effect of group on the EEG data, specifically in the prefrontal (AFz, AF3, AF4 and AF8), the frontal (F1, F2, F3, F4, F6, F7 and F8), the frontal-central (FC1, FC2, FC4 and FC6), the central (Cz and C1) and the right fronto-temporal regions (FT8, Figure 5c), which constituted the fronto-parietal network [17,29]. In general, the activation levels were stronger and activation ranges in the fronto-parietal network were wider in group 3 under the task switching condition. This was responsible for their better capability to reallocate the attentional resources, which also proved that most of the brain regions of the fronto-parietal network were required to complete the task setting process [33–35].

The activation comparison among three groups indicated that their difference was stronger under the task switching condition, when the regions extended from a small area in the frontal and central regions (centered at F2) to most areas of the fronto-parietal network with increased intensity as well (Figure 7a). Under both task switch and task repetition conditions, the activation degree of group 1 and group 3 was significantly stronger than group 2 (Figure 7b,d). The weakest activation intensity of group 2 was responsible for their worst performance of the executive control function and their highest occurrence of the driving errors. The underlying regions of interest for the executive control functions and their activity changes, along with the CSI were further analyzed. As for the comparison between group 1 and 2 (Figure 7b), when the CSI was 200–800 ms, the activation range and intensity varied and the difference was mainly concentrated in the prefrontal and frontal regions (around F2). When the CSI was 1000–1200 ms, the difference was stable in most brain regions including the prefrontal, frontal, central, temporal and superior parietal regions. As for the comparison between group 2 and 3 (Figure 7d), the activation difference was mainly concentrated in the frontal (centered at F2, CSI = 800 ms) and frontal-central regions (centered at the FC1, CSI = 1200ms) under the task switching condition, and occupied most of the fronto-parietal network under both conditions when the CSI was 1400 ms. In general, our results indicated the significant effect of the group factor in the frontal-parietal network. Additionally, the instability of frontal region activation revealed its specific role in executive control. It has been suggested that a superordinate fronto-cingulo-parietal network supporting cognitive control may also underlie a series of distinct executive functions, including attention, signal recognition, behavior strategy formulation, motion control, impulse control, and information feedback recognition [22,28,36]. In the task-cuing experiment, during the period from the end of the previous trial to the end of the next task-cue, there was a process of working memory maintenance of multiple task settings, evaluation and reconstruction of the current task setting [22,37]. When the task to perform was switched, the task settings needed to be updated and the extra work was required to suppress the previous task setting, which resulted in stronger cerebral activities. Consistent with the behavioral results, the better performance efficacy and stronger activation of group 3 indicated their better capability to allocate the available attentional resources, to express the external stimuli, and to process the information in the nervous system, especially the fronto-parietal executive control network.

4.3. Effect of CSI Level on the Brain Activities

The main effect of CSI on EEG data was significant. Besides, CSI had the interaction effect with groups in some channels (FC2, Cz and C2 in Figure 5f). The influence of CSI levels on the executive control process can be also observed in the following post hoc simple effect analysis. The among-CSI difference of the brain activation existed in mostly the left prefrontal, frontal, right frontal-parietal and bilateral fronto-temporal regions under task repetition condition and in fronto-parietal network (centered at FC2) under task switching condition in group 1 (Figure 6a), in F2, FC2, FC4, P6 and PO8 under task switching condition in group 2 (Figure 6b), in the frontal and central regions (a small area centered at Cz) under the task repetition condition and in the fronto-parietal network (most areas of frontal, central, bilateral temporal and parietal and parietal-occipital regions) under task switching condition in group 3 (Figure 6c). The simple effect analysis indicated that the activation degree and range of brain regions increased along with the increase of CSI. The brain activation differences among different CSIs were mainly concentrated in the fronto-parietal network, which was most strong in group 3, secondly strong in group 1, and the weakest in group 2 (Figure 6). The results indicated that subjects with better attention status and better executive control efficiency were more sensitive to CSI.

In general, the increase of the CSI is helpful for the activation of the task settings and the more effective conversion among different tasks. It is noted that when CSI was 1000 ms, the intensity and range of the brain activation were significantly increased in all three groups and then remained at a high level (Figure 6). Additionally, 1000 ms seemed to be also a key downtrend point of the switch cost, especially for the RT of group 2 and 3 (Figure 4). Both the performance efficiency and the underlying cognitive process reached an optimal level at this CSI. An appropriate CSI might be helpful for the subjects to maintain the balance of task setting reconstruction and S–R mapping expression. Both the task setting reconstruction and S–R mapping expression relied on the working memory, which was an iterative process including encoding, storage, recognition and recall. When the CSI was relatively short, the task-cue processing and task setting reconstruction was needed to be performed synchronously, there was no time for enough iterations, and resultantly, the accuracy of the executive control function could not be guaranteed. On the other hand, when the CSI was long, a series of iterations could be fulfilled. Besides, according to the experience of the subjects, they might even have time for the rehearsal of the expected task. Under this circumstance, the pre-task setting reconstruction process might have already started before the stimulus, and the working memory load would remain at a high level.

Though there existed differences among the three groups, both frontal and parietal regions were involved. The activities of the frontal cortex fluctuated along with the CSI. It was activated when the CSI was 200 ms, and the activation seemed to be weakened when the CSI increased to 400 ms in group 2 and group 3 (Figure 6b,c). As the CSI continued to increase, the frontal activation increased again. The frontal cortex was responsible for maintaining task settings, regulating and controlling task-related behaviors [16]. When the CSI was short, the subjects did not have enough time to classify the stimulus and complete the task setting reconstruction process [38]. The time pressure and the conflict between task setting and S–R mapping required the high degree of participation of the frontal region [29]. When the preparation time is sufficient, the time pressure was reduced, the conflict between task setting and S–R mapping expression was reduced, and the requirement of the cognitive control was also decreased. As a result, the activity of the frontal cortex was weakened. When the CSI increased further, the activation of the frontal cortex was restrengthened, which was related to the increased load of the working memory due to the prolonged time of S–R mapping expression [29,34]. This can also explain the phenomenon that the activity of the parietal cortex, which was sensitive to the conflict of S–R mapping [39], was stronger when the CSI was larger. These results indicated the possible role of the frontal area in task setting reconstruction and working memory maintaining, and of the parietal area in S–R mapping expression.

4.4. Novelty and Limitations

In this research, a unified experimental and analytical schema for multimodal data including the behavior, EEG activity and psychological performance was presented to explore the underlying cognitive mechanism of attention in driving. Through the comparative analysis of different groups of participants, the quantitative correlation between executive control function and driving performance was established, and the spatiotemporal activity of the brain during this procedure was revealed. The relationship between dangerous driving behavior and attention, along with CSIs and other parameters, was disclosed. Based on the presented methods and the acquired results, the attentional state of the driver could be monitored through EEG signals to avoid the distraction, and the dangerous driving behaviors including errors and elapses could be prevented. Besides, the attentional and behavioral characteristics of the drivers can be analyzed in advance and the subject-specific driving style can be evaluated. Accordingly, different real-time online human-computer interaction schemes can be provided for different kinds of drivers. Furthermore, the individual's driving performance and their EEG performance could be mutually corroborated, which would supply new reference for driving training and administration. In general, the presented schema supplies a new kind of intelligent human-computer interaction method, and this active safety control would significantly improve the driving safety. Except for driving, the research findings would be applied to other life risk activities.

The present study is limited principally by the unbalanced gender proportion, uneven ages and driving ages of the subjects. Age, gender and educational background are all the crucial factors affecting the executive control functions, performance efficiency, and brain activities of the human [40,41]. A total of 35 subjects were studied and there are only 6 subjects in group 1. Though the meaningful results were found and no significant difference was detected for ages, genders, driving ages and education backgrounds among the three groups, these results need to be replicated in much larger sample size and the general population. Besides these factors, the other demographic factors of different groups, such as the driving experience including the driving frequency and the load, might all have significant effect on the cognition and behavior of the drivers. The definition of all the related parameters and a larger sample size would be crucial to help consolidate a bigger picture of our work. The individual's driving performance and their EEG performance could be mutually corroborated, which would supply new reference for driving training and administration. However, in our work, their interaction dynamics cannot be analyzed because of the insufficient repeated measures of the subjects from both the cognitive and the behavioral sides. We would like to conduct a longitudinal cohort study and we believe that very interesting and more robust results would be obtained. Second, the spatiotemporal characteristics of the underlying brain function need to be further studied. In our work, we analyzed the brain electrical activity mapping. As EEG is a kind of scalp electrical signal and has a limited spatial resolution, the location of the anatomical areas might not be very accurate. We observed the participation of the frontal and parietal regions in the executive control process. However, deeper areas in the brain, such as the cingulate gyrus, which has been demonstrated to be a critical part of the fronto-cingulo-parietal network for executive functions [28], cannot be located. The EEG source localization technique might be helpful. The causal relationship among the regions and their dynamic activity can be further analyzed by using the time series analysis methods, such as dynamic causal modeling [42]. Finally, the executive control function is one of the three sub-functions (altering, orienting and executive control) of attention. According to our understanding, alerting and orienting subsystems acted mainly in the information perception level. Compared with them, the executive control subsystem acted mainly in the higher decision and control level. The executive control function would have a direct relationship with the behaviors, such as the switch cost of reaction time and error rate in the psychological experiment and driving performance in the simulated driving experiment. Hence in this work, we focused on the executive control function and found its positive correlation with these behavioral performances. However, the other two sub-functions are also important for the whole process, and their implication in the driving performance is worthy of study. Till now, the relationship among these sub-functions and the underlying mechanism of attention is not yet clear [43]. This warrants further synthetic research of the sub-functions of attention. The driving performance of the drivers was evaluated based on two types of aberrant driving behaviors, i.e., errors and lapses. Although this definition method was relatively common in the research of human factors in engineering and driving behavior [31,32], it was still a subjective judgment method. The objective data such as steering wheel angle and driving route was planned to define abnormal driving behavior in the future research.

5. Conclusions

In this work, the simulated driving and task-cuing experiments were conducted, and the correlation between driving performance and executive control function was analyzed. The subjects with lower driving error rates had better performance efficiency as indicated by the RT and error rate in the task-cuing experiment, which was related with their better capability to allocate the available attentional resources, to express the external stimuli and to process the information in the nervous system, especially the fronto-parietal executive control network. The activation degree of the frontal area fluctuated, and of the parietal area gradually increased along with the increase of CSI, which implied the possible role of the frontal area in task setting reconstruction and working memory maintaining, and of the parietal area in S–R mapping expression. This research provided evidence of a close relationship between executive control functions and driving performance, which supplies new reference for intelligent human-computer interaction and active safety control in driving.

Author Contributions: Experimental design, L.Y., T.W., J.Z., L.C. and C.D.; data processing, T.W., Y.W. and M.L.; statistical analysis, T.W. and L.Y.; writing—original draft preparation, T.W.; writing—review and editing, L.Y. and T.W.; supervision, F.Y.; project administration, L.Y.; funding acquisition, L.Y. All authors have read and agreed to the published version of the manuscript.

Funding: This research was funded by the Natural Science Foundation of China (grant number 61876137), 111 Project (B17034), Innovative Research Team Development Program of the Ministry of Education of China (IRT_17R83).

Institutional Review Board Statement: The study was conducted according to the guidelines of the Declaration of Helsinki, and approved by the Ethics Committee of Wuhan University of Technology (protocol code 202003001, 6 March 2020).

Informed Consent Statement: Informed consent was obtained from all subjects involved in the study.

Data Availability Statement: The data presented in this study are available on request from the corresponding author.

Acknowledgments: The authors are grateful to the subjects in the experiment and appreciate the reviewers for their helpful comments and suggestions in this study.

Conflicts of Interest: The authors declare no conflict of interest.

References

1. Sta, G. Global status report on road safety. *Inj. Prev.* **2015**, *15*, 286. Available online: https://www.afro.who.int/publications/global-status-report-road-safety-2015 (accessed on 1 February 2021).
2. Singh, S. *Critical Reasons for Crashes Investigated in the National Motor Vehicle Crash Causation Survey*; No. DOT HS 812 115; 2015. Available online: http://www-nrd.nhtsa.dot.gov/Pubs/812115.pdf (accessed on 1 February 2021).
3. Lerner, N.; Robinson, E.; Singer, J.; Jenness, J.; Huey, R.; Baldwin, C.; Fitch, G. Human Factors for Connected Vehicles: Effective Warning Interface Research Findings. Available online: https://www.nhtsa.gov/sites/nhtsa.dot.gov/files/812068-humanfactorsconnectedvehicles.pdf (accessed on 20 February 2021).
4. Smith, A.P. A UK survey of driving behaviour, fatigue, risk taking and road traffic accidents. *BMJ Open* **2016**, *6*, e011461. [CrossRef] [PubMed]
5. Walker, H.E.; Trick, L.M. Mind-wandering while driving: The impact of fatigue, task length, and sustained attention abilities. *Transp. Res. Part F Traffic Psychol. Behav.* **2018**, *59*, 81–97. [CrossRef]
6. Garner, A.; Fine, P.R.; A Franklin, C.; Sattin, R.W.; Stavrinos, D. Distracted driving among adolescents: Challenges and opportunities. *Inj. Prev.* **2011**, *17*, 285. [CrossRef]
7. Jimenez-Moreno, R.; Orjuela, S.A.; Van Hese, P.; Prieto, F.A.; Grisales, V.H.; Philips, W. Video surveillance for monitoring driver's fatigue and distraction. *Photonics Eur.* **2012**, *8436*, 84360. [CrossRef]
8. Sigari, M.-H.; Pourshahabi, M.-R.; Soryani, M.; Fathy, M. A Review on Driver Face Monitoring Systems for Fatigue and Distraction Detection. *Int. J. Adv. Sci. Technol.* **2014**, *64*, 73–100. [CrossRef]
9. Li, Z.; Li, S.E.; Li, R.; Cheng, B.; Shi, J. Online Detection of Driver Fatigue Using Steering Wheel Angles for Real Driving Conditions. *Sensors* **2017**, *17*, 495. [CrossRef]
10. Deshmukh, S.V.; Dehzangi, O. ECG-Based Driver Distraction Identification Using Wavelet Packet Transform and Dis-criminative Kernel-Based Features. In Proceedings of the IEEE International Conference on Smart Computing, Hong Kong, China, 29–31 May 2007.
11. Yan, L.; Wang, Y.; Ding, C.; Liu, M.; Yan, F.; Guo, K. Correlation Among Behavior, Personality, and Electroencephalography Revealed by a Simulated Driving Experiment. *Front. Psychol.* **2019**, *10*, 1524. [CrossRef] [PubMed]
12. Jin, L.; Niu, Q.; Hou, H.; Xian, H.; Wang, Y.; Shi, D. Driver Cognitive Distraction Detection Using Driving Performance Measures. *Discret. Dyn. Nat. Soc.* **2012**, *2012*, 1–12. [CrossRef]
13. Alavi, S.S.; Mohammadi, M.R.; Soori, H.; Ghanizadeh, M. The Cognitive and Psychological Factors (Personality, Driving Behavior, and Mental Illnesses) As Predictors in Traffic Violations. *Iran. J. Psychiatry* **2017**, *12*, 78–86.
14. Go, R.; Wu, J.; Liu, Y.; Kuroda, Y.; Wu, Q. Cognitive Psychological Study on The Occurrence of Microsaccades in Visual Spatial Attention. In Proceedings of the 2018 IEEE International Conference on Mechatronics and Automation (ICMA), Changchun, China, 5–8 August 2018.
15. Liu, M.; Zhang, J.; Jia, W.; Chang, Q.; Shan, S.; Hu, Y.; Wang, D. Enhanced executive attention efficiency after adaptive force control training: Behavioural and physiological results. *Behav. Brain Res.* **2019**, *376*, 111859. [CrossRef] [PubMed]
16. Funahashi, S. Neuronal mechanisms of executive control by the prefrontal cortex. *Neurosci. Res.* **2001**, *39*, 147–165. [CrossRef]
17. Shi, L.; Sun, J.; Xia, Y.; Ren, Z.; Chen, Q.; Wei, D.; Yang, W.; Qiu, J. Large-scale brain network connectivity underlying creativity in resting-state and task fMRI: Cooperation between default network and frontal-parietal network. *Biol. Psychol.* **2018**, *135*, 102–111. [CrossRef] [PubMed]
18. Dibbets, P.; Evers, E.A.T.; Hurks, P.P.M.; Bakker, K.; Jolles, J. Differential brain activation patterns in adult attention-deficit hyperactivity disorder (ADHD) associated with task switching. *Neuropsychology* **2010**, *24*, 413–423. [CrossRef] [PubMed]
19. Dove, A.; Pollmann, S.; Schubert, T.; Wiggins, C.J.; von Cramon, D.Y. Prefrontal cortex activation in task switching: An event-related fMRI study. *Cogn. Brain Res.* **2000**, *9*, 103–109. [CrossRef]
20. Posner, M.I.; Rothbart, M.K. Research on Attention Networks as a Model for the Integration of Psychological Science. *Annu. Rev. Psychol.* **2007**, *58*, 1–23. [CrossRef] [PubMed]
21. Corbetta, M.; Shulman, G.L. Control of goal-directed and stimulus-driven attention in the brain. *Nat. Rev. Neurosci.* **2002**, *3*, 201–215. [CrossRef]
22. Braver, T.S.; Reynolds, J.R.; I Donaldson, D. Neural Mechanisms of Transient and Sustained Cognitive Control during Task Switching. *Neuron* **2003**, *39*, 713–726. [CrossRef]
23. Brown, J.W.; Reynolds, J.R.; Braver, T.S. A computational model of fractionated conflict-control mechanisms in task-switching. *Cogn. Psychol.* **2007**, *55*, 37–85. [CrossRef] [PubMed]
24. Logan, G.D.; Gordon, R.D. Executive Control of Visual Attention in Dual-Task Situations. *Psychol. Rev.* **2001**, *108*, 393–434. [CrossRef]
25. Meiran, N. Modeling cognitive control in task-switching. *Psychol. Res.* **2000**, *63*, 234–249. [CrossRef] [PubMed]
26. Monsell, S. Task Switching. *Trends Cogn. Sci.* **2003**, *7*, 134–140. [CrossRef]
27. Rogers, R.D.; Monsell, S.D. Costs of a predictable switch between simple cognitive tasks. *J. Exp. Psychol.* **1995**, *124*, 207. [CrossRef]
28. Cole, M.W.; Schneider, W. The cognitive control network: Integrated cortical regions with dissociable functions. *NeuroImage* **2007**, *37*, 343–360. [CrossRef] [PubMed]
29. Jamadar, S.; Hughes, M.; Fulham, W.; Michie, P.; Karayanidis, F. The spatial and temporal dynamics of anticipatory preparation and response inhibition in task-switching. *NeuroImage* **2010**, *51*, 432–449. [CrossRef] [PubMed]

30. Weiler, J.; Hassall, C.D.; Krigolson, O.E.; Heath, M. The unidirectional prosaccade switch-cost: Electroencephalographic evidence of task-set inertia in oculomotor control. *Behav. Brain Res.* **2015**, *278*, 323–329. [CrossRef] [PubMed]

31. Menno, N.; Borst, J.P.; Hedderik, V.R.; Taatgen, N.A. Driving and Multitasking: The Good, the Bad, and the Dangerous. *Front. Psychol.* **2016**, *7*, 1–16. [CrossRef]

32. Parker, D.; Reason, J.T.; Manstead, A.S.R.; Stradling, S.G. Driving errors, driving violations and accident involvement. *Ergonomics* **1995**, *38*, 1036–1048. [CrossRef]

33. Monsell, S.; Mizon, G.A. Can the task-cuing paradigm measure an endogenous task-set reconfiguration process? *J. Exp. Psychol. Hum. Percept. Perform.* **2006**, *32*, 493–516. [CrossRef]

34. Sakai, K. Task Set and Prefrontal Cortex. *Annu. Rev. Neurosci.* **2008**, *31*, 219–245. [CrossRef] [PubMed]

35. Altmann, E.M. Advance Preparation in Task Switching. *Psychol. Sci.* **2004**, *15*, 616–622. [CrossRef]

36. Scott, J.G.; Schoenberg, M.R. *The Little Black Book of Neuropsychology*; Springer: Boston, MA, USA, 2011.

37. Altmann, E.M.; Gray, W.D. An integrated model of cognitive control in task switching. *Psychol. Rev.* **2008**, *115*, 602–639. [CrossRef]

38. Stuss, D.T.; Murphy, K.J.; Binns, M.A.; Alexander, M.P. Staying on the job: The frontal lobes control individual performance variability. *Brain* **2003**, *126*, 2363–2380. [CrossRef] [PubMed]

39. Brass, M.; Ullsperger, M.; Knoesche, T.R.; Von Cramon, D.Y.; Phillips, N.A. Who Comes First? The Role of the Prefrontal and Parietal Cortex in Cognitive Control. *J. Cogn. Neurosci.* **2005**, *17*, 1367–1375. [CrossRef]

40. Guo, F.; Klauer, S.G.; Fang, Y.; Hankey, J.M.; Antin, J.F.; A Perez, M.; E Lee, S.; A Dingus, T. The effects of age on crash risk associated with driver distraction. *Int. J. Epidemiol.* **2017**, *46*, 258–265. [CrossRef] [PubMed]

41. Ftouni, S.; Patterson, J.; Dubaj, M.; Michael, N. Driving while distracted: The effects of experience, age and gender. In Proceedings of the AUTOCRC CONFERENCE, Melbourne, Australia, 30 November–2 December 2009.

42. Friston, K.J.; Harrison, L.; Penny, W. Dynamic causal modelling. *NeuroImage* **2003**, *19*, 1273–1302. [CrossRef]

43. Rosenberg, M.D.; Finn, E.S.; Scheinost, D.; Papademetris, X.; Shen, X.; Constable, R.T.; Chun, M.M. A neuromarker of sustained attention from whole-brain functional connectivity. *Nat. Neurosci.* **2016**, *19*, 165–171. [CrossRef] [PubMed]

sensors

Article

Multi-Task End-to-End Self-Driving Architecture for CAV Platoons

Sebastian Huch [1,2,*], Aybike Ongel [1], Johannes Betz [3] and Markus Lienkamp [1,2]

1   TUMCREATE, 1 CREATE Way, #10-02 CREATE Tower, Singapore 138602, Singapore; aybike.ongel@tum-create.edu.sg (A.O.); lienkamp@ftm.mw.tum.de (M.L.)
2   Institute of Automotive Technology, Technical University of Munich, Boltzmannstr. 15, 85748 Munich, Germany
3   Department of Electrical and Systems Engineering, School of Engineering and Applied Science, University of Pennsylvania, 220 South 33rd Street Philadelphia, Philadelphia, PA 19104, USA; joebetz@seas.upenn.edu
*   Correspondence: sebastian.huch@tum-create.edu.sg

Abstract: Connected and autonomous vehicles (CAVs) could reduce emissions, increase road safety, and enhance ride comfort. Multiple CAVs can form a CAV platoon with a close inter-vehicle distance, which can further improve energy efficiency, save space, and reduce travel time. To date, there have been few detailed studies of self-driving algorithms for CAV platoons in urban areas. In this paper, we therefore propose a self-driving architecture combining the sensing, planning, and control for CAV platoons in an end-to-end fashion. Our multi-task model can switch between two tasks to drive either the leading or following vehicle in the platoon. The architecture is based on an end-to-end deep learning approach and predicts the control commands, i.e., steering and throttle/brake, with a single neural network. The inputs for this network are images from a front-facing camera, enhanced by information transmitted via vehicle-to-vehicle (V2V) communication. The model is trained with data captured in a simulated urban environment with dynamic traffic. We compare our approach with different concepts used in the state-of-the-art end-to-end self-driving research, such as the implementation of recurrent neural networks or transfer learning. Experiments in the simulation were conducted to test the model in different urban environments. A CAV platoon consisting of two vehicles, each controlled by an instance of the network, completed on average 67% of the predefined point-to-point routes in the training environment and 40% in a never-seen-before environment. Using V2V communication, our approach eliminates casual confusion for the following vehicle, which is a known limitation of end-to-end self-driving.

Keywords: connected and autonomous vehicles; artificial neural networks; end-to-end learning; multi-task learning; urban vehicle platooning; simulation

Citation: Huch, S.; Ongel, A.; Betz, J.; Lienkamp, M. Multi-Task End-to-End Self-Driving Architecture for CAV Platoons. *Sensors* **2021**, *21*, 1039. https://doi.org/10.3390/s21041039

Academic Editor: Chao Huang
Received: 21 December 2020
Accepted: 31 January 2021
Published: 3 February 2021

Publisher's Note: MDPI stays neutral with regard to jurisdictional claims in published maps and institutional affiliations.

Copyright: © 2021 by the authors. Licensee MDPI, Basel, Switzerland. This article is an open access article distributed under the terms and conditions of the Creative Commons Attribution (CC BY) license (https://creativecommons.org/licenses/by/4.0/).

1. Introduction

One major trend in intelligent transportation systems is the development of connected and autonomous vehicles (CAVs), which has seen vast progress over the past few years. However, research on autonomous vehicles has a long history, starting in the 1980s with the PROMETHEUS project and Autonomous Land Vehicle in a Neural Network (ALVINN), an experimental vehicle that made use of neural networks for the driving task. Over 30 years later, vehicles with lower levels of driving automation—up to Level 2 as specified by SAE J3016 [1]—have become commercially available. Researchers and the industry now focus on the development of vehicles with higher levels of driving automation, such as Levels 4 and 5. These enable autonomous driving without the need for human interaction. Furthermore, highly automated vehicles can increase road safety and reduce emissions [2]. Higher levels of driving automation also allow vehicles to drive in a platoon and reduce the inter-vehicle distance. A platoon generally consists of a leading vehicle with one or more followers. The leading vehicle is driven autonomously or manually, while the following

207

vehicle usually acts autonomously, especially at close inter-vehicle distances. Driving with a close inter-vehicle distance not only has the potential to reduce emissions on highways because of the reduction in drag and hence energy consumption, but also saves space in an already limited urban area [3,4]. For example, CAV platoons on a dedicated lane can reduce the reverse-accordion effect when jointly accelerating at a green traffic light. In this work, we focus on CAV platoons deployable in an urban area.

Key to the success of CAVs are the recent advances in machine learning (ML). Different ML techniques, e.g., deep learning, facilitate the extraction of information from a large amount of sensor data to understand the CAV's environment. Although there exist different approaches to achieve the ultimate goal of developing self-driving cars, most of them rely on the usage of ML methods to a high degree. These approaches can be classified into the following three categories: modular pipelines, direct perception, and end-to-end deep learning.

The approach primarily used for self-driving cars employs modular pipelines and is based on the mediated-perception principle [5]. This approach makes use of the decomposition of the driving task to split the complex task into successive modules (perception, prediction, planning, and control) [6]. Each module consists of a range of specialized submodules. Using onboard sensors like the camera, LiDAR, radar, GNSS, and the inertial measurement unit, the perception module creates a comprehensive internal representation of the self-driving car's environment. Given the internal environment's representation, the planning module is responsible for the route, path, and motion planning [7]. The subsequent control module converts the trajectories generated by the planning system into commands for the self-driving car's actuators of the steering wheel, engine, and brakes.

Although the modular pipeline approach has good interpretability, the information computed at every time step adds unnecessary complexity to the system. Additionally, a large amount of labeled data (e.g., 3D bounding boxes [8] or pixel-wise semantic segmentation [9]) are needed for supervised training of the submodules of the perception module. These data are costly and hard to obtain. Additionally, some localization algorithms require accurate high-definition road maps [10], which have to be generated offline. Another problem is the error propagation that can occur in multiple sections of the modular pipeline [11].

Based on the psychological theory about perception [12], the direct-perception approach was proposed by [13] to solve the self-driving challenge. Rather than splitting the entire driving task into smaller submodules, the authors used a single neural network in the perception module to predict meaningful affordances. These affordances are a low-dimensional compact and intermediate representation of the CAV's surroundings, as opposed to the high-dimensional output of the modular pipeline approach. Examples of affordances used by the authors are the car's angle relative to the lane, distances to lane markings, and cars in the current and adjacent lanes. Based on the affordances, a simple car controller outputs the commands to drive the vehicle. The authors proved the generalization ability of their system by testing the model and trained with data captured in the TORCS simulator, on real-driving videos. Reference [14] generalized the direct-perception approach to address the task of driving in an urban environment, which demands additional affordance indicators. A general problem posed by the direct-perception approach is that the affordances are manually chosen and may not be suitable for the complex driving task, i.e., not all situations can be covered by a few low-dimensional affordances.

The third paradigm for autonomous driving is end-to-end learning [15]. In general, the idea of end-to-end learning is to train a neural network that directly maps the raw sensor data to driving commands, i.e., it outputs the steering angle and throttle or brake values. The input data may come from different sensors, such as mono cameras, stereo cameras, LiDAR, or any combination thereof. However, the most widely used sensor is the mono camera. In recent years, convolution neural networks (CNN) [16] have been the most widely used method for feature extraction of image data, since CNNs achieved the best results for image recognition at the ImageNet Large Scale Visual Recognition

Challenge (ILSVRC) [17]. The training is based on supervised learning. During the training phase, the networks process the images and the corresponding labels. During inference, the networks predict these labels, with images as the only input. In direct comparison, the driving performance of the end-to-end approach is comparable to the driving performance of the modular pipeline, but it is less fragile in new environments [18]. A benefit of this approach is that there is no need for hand-engineered heuristics, and hence, the system is more robust for unpredictable situations. Furthermore, collecting labeled training data on an image-level for end-to-end learning is easier than on a pixel-level for the modular pipeline approach.

In this work, we propose a multi-task deep learning architecture for CAV platoons that is trained in an end-to-end fashion. This architecture is based on vision only and can be used in both the leading vehicle and following vehicles to maneuver them. The vehicles communicate via V2V communication to improve the performance. The neural network was tested in a never-seen-before urban environment in a simulation under different weather conditions and with dynamic traffic. A two-vehicle platoon followed the road while obeying traffic rules, such as traffic lights at intersections. We benchmarked the performance of our approach with state-of-the-art metrics and demonstrated the capability our model taking over control of a two-vehicle platoon to complete predefined routes.

2. Related Work

Research on the platooning of vehicles has a long history, although it is mainly focused on the energy savings or communication topology of platooning vehicles. Preliminary work on autonomous truck platooning was undertaken by [19] in 1995. The authors tested a two-vehicle platoon with a manually-driven leading truck followed by an autonomous truck. The sensor setup for the following truck consisted of a single front-facing camera together with the vehicle states (e.g., acceleration and velocity) of the leading truck, which were obtained via V2V communication. This vision-based approach estimated the truck's heading based on the lane markings and the distance to the leading truck by visual detection of active infrared lights on the rear of the leading truck. Major large-scale pilot studies were also conducted to determine the feasibility and benefits of truck platooning on highways. These works include the European research programs CHAUFFEUR and CHAUFFEUR2 in the early 2000s [20]. Other similar projects were the California PATH truck-platooning program and KONVOI [21]. A more comprehensive review of these projects can be found in [22].

End-to-end self-driving: The origin of this approach dates back to the late 1980s. In 1989, ALVINN [23] used a small (by today's standards) fully connected network to predict the turn curvature in front of the car. The inputs were images captured by a camera and scans from a laser range finder. The training data were collected using a simulated road generator, which created video images, as well as laser range finder images. As a result, ALVINN was capable of following a 400 m section of the road at a speed of 0.5 m/s. Instead of a fully connected network, Reference [24] implemented a six layer CNN in the remote-controlled mobile robot DAVE. DAVE was intended to operate in unknown off-road terrain and avoid smaller obstacles such as trees and ponds. The input of this network consisted of images from a front-facing stereo camera, and the output was the steering angle. The network was trained with images and control commands recorded while a human was maneuvering the robot in different terrains and light conditions. After the training, DAVE managed to drive approximately 20 m on average until it hit an obstacle [25]. Reference [26] used the TORCS simulator to train an agent to drive on a race track by employing reinforcement learning. This approach is solely vision-based, with a recurrent neural network receiving front-facing images and predicting the steering angle, brake, and throttle. The performance is comparable to that of hard-coded controllers, which have access to all vehicle states. A seminal work on end-to-end learning is DAVE-2 [25]. In 2016, the authors designed and trained a CNN named PilotNet to steer a full-sized vehicle. Unlike the mobile robot DAVE, which was tested a decade earlier, DAVE-2 was trained with about 72 h of real-world driving data, captured by three front-facing cameras while driving on public roads under different light and weather conditions. The network training benefited from two off-center cameras, which implied a shift from the center of the lane and thereby enabled the network to learn to recover from mistakes. Only the center camera was used for testing.

Similar systems have been tested in recent years, with training data either taken from real-world datasets [27–32] or generated in simulations [33–36]. Some of this research explored the benefits of including temporal dependencies between consecutive images by adding recurrent layers to the network. For example, References [27,30,31] added long short-term memory (LSTM) layers on top of the CNN. The CNN was responsible for feature extraction of the images, while the LSTM incorporated features of the last time steps for the driving decision of the current time step. Reference [36] followed the same approach and additionally implemented the end-to-end approach in a simulated urban environment. To obey the traffic rules, the authors added additional inputs to the network, such as the relevant traffic light state and speed limit. Application of end-to-end self-driving to vehicle platooning: The above works did not address end-to-end deep learning algorithms that are specifically designed for the application in vehicle platoons. This topic was first addressed in [32,34].

Reference [34] used the end-to-end learning of CNNs to maneuver a truck behind a leading truck in a simulation based on vision only. The output of the network was branched and predicted the steering angle and throttle/brake. The simulation environment was a flat terrain in an off-road area. The leading truck made random driving decisions, while the following truck was operated manually during data collection. The labels associated with each image consisted of discrete steer and throttle/brake values, making the training a classification problem. After supervised training, several test runs were conducted. Although this was the first work to address end-to-end self-driving for CAV platoons, it did not consider an urban environment and focused on the following vehicle only. Since the leading vehicle was manually driven, the platoon could not operate fully autonomously. Furthermore, the system setup produced discrete output values that would not allow smooth driving behavior. Additionally, because the network was tested in the same simulation environment that it was trained in, its ability to generalize and extrapolate data to unseen environments was not assessed.

In 2019, Reference [32] was the first to mention the additional implementation of V2V communication to enhance the performance of end-to-end self-driving. The authors proposed a network architecture that processed images recorded by two cooperative self-driving vehicles simultaneously. The goal was to predict the steering angle of the following vehicle for lane-keeping on highways. Via V2V communication, the image data from a leading vehicle were transmitted to the following vehicle, where they were merged with the follower's images. The authors argued that adding images from the leading vehicle improved the performance of the system because there was more available information. Using information from the leading vehicle can improve driving behavior, but the image stream between the cars as proposed by the authors transmits information at a high level. Simple low-level vehicle states, e.g., acceleration and velocity, are easier to transmit because of the reduced data size. Moreover, the images of the leading vehicle might already be processed by the network of the leading vehicle itself, making a second processing in the following car redundant.

With our work, we want to address the limitations of [32,34]. Our contributions are the implementation of an end-to-end self-driving model that:

- is not limited for deployment in following vehicles, but can also be used in leading vehicles of a CAV platoon, hence performing multiple tasks;
- is predicting continuous instead of discrete control commands for a smooth driving behavior;
- is not limited to operating in off-road or highway environments, but can be used in a challenging urban environment;

- and is using V2V communication, but transmitting low-level vehicle states instead of a high-level image stream.

3. Models

In this section, we introduce our proposed architecture for CAV platoons and modified architectures of this baseline for comparison. Before going into the details of the architectures in the following subsections, some features that all models have in common are discussed. The last part of this section addresses the specifics of training that come along with multi-task learning.

The general vision-based end-to-end approach for CAVs maps observations s_t, e.g., images, at each time step t to an action a_t. This implies that there exists a driving policy that satisfies a function F, which describes the relationship between s_t and a_t, that is $a_t = F(s_t)$. Given a dataset $D = \{(s_i, a_i)\}_{i=1}^{N}$ with expert driving demonstrations, the goal of the network training process is to optimize the parameters θ of a function $G(s, \theta)$ to approximate $F(s_t)$ by minimizing the loss $\mathcal{L}$:

$$\min_{\theta} \sum_{i} \mathcal{L}(G(s_i, \theta), a_i). \tag{1}$$

While the observations s_t usually contain images, they can be extended by including vehicle states or other measurements.

To operate a CAV platoon, at least two vehicles are controlled simultaneously with slightly different tasks. The leading vehicle follows a route while staying in the lane, obeying traffic rules and avoiding collisions with static and dynamic objects. It makes driving decisions based on the perceived environment and the high-level plan (e.g., navigation to a predefined destination). Following vehicles in a platoon, however, have limited front visibility because of their close proximity to the preceding car. Furthermore, as long as they are part of the platoon, they follow their immediate predecessor while staying in the lane. Following a preceding vehicle can be further simplified for a homogeneous CAV platoon, in which all vehicles are identical. In this case, any following vehicle of the platoon follows a preceding vehicle with similar visual features to all other vehicles, independent of its position.

The different tasks of leading and following vehicles have different requirements for the inputs and outputs of the network. The basic input for the multi-task network in both vehicles is images from a single front-facing camera. Other sensor modalities are not taken into account to avoid sensor fusion. We chose a single camera over stereo cameras to overcome the calibration problem. When used in the leading vehicle, two additional inputs are added to the network, that is the current vehicle speed and the high-level command. The current vehicle speed is added as an input since it influences the longitudinal driving decisions [37]. The high-level command provides information about the driving direction. Possible discrete states are ⟨*Left, Straight, Right, No Intersection*⟩. The state *No Intersection* is active when the leading vehicle is not in close proximity to an intersection, whereas the other three states indicate the driving direction at an upcoming intersection. These states can be provided by a high-level planner, e.g., a navigation system. As stated before, the following vehicle always follows the leading vehicle, and therefore, the velocity and high-level command inputs are not required.

The outputs for the leading vehicle are the control commands, including steering angle and throttle/brake. For the following vehicle, the outputs are the steering angle and gap between the vehicles. We did not choose the throttle/brake as the output for the following vehicle because of the ambiguity of the images, known as casual confusion [38]. The images showing an acceleration and deceleration phase may look similar, but have different labels, and hence, the network cannot distinguish between these situations. To overcome the casual confusion for the following vehicle, we chose to predict the gap between the cars, which is unique for each image. This can further improve the steering-angle prediction, since the network learns to focus on the leading vehicle during training.

We combined the driving task for the leader and follower into a single multi-task network, which is based on the following constraints. First, we used identical vehicles as the leader and follower, so that the camera position in both vehicles was the same. Second, the shared layers perceived a similar visual input, namely the image of the environment in front of the vehicle. We argue that each task can benefit from the other: the leading vehicle trains the network to focus on the road and lanes in general, while the following vehicle strengthens the network to detect dynamic objects. Additionally, the joint training with different datasets can improve the generalization ability of the network.

The information flow topology of the V2V communication follows the predecessor-following principle as described in [39], in which following vehicles in the platoon receive information from their immediate predecessor. The control of following vehicles is a combination of a neural network for the lateral control and a conventional car-following model for the longitudinal control. The car-following model follows a simple constant spacing control strategy, which is sufficient for our experiments (ensures weak string stability), but it could be replaced with different control strategies.

3.1. Model A: Multi-Task Network Baseline

Our proposed multi-task network consists of a shared CNN acting as a feature extractor and two prediction heads (fully connected (FC) networks) for the leader or follower, respectively. As illustrated in Figure 1, the same trained network can be used for the leading or following vehicle. Different inputs and outputs are activated or deactivated, depending on in which vehicle the network is used. In the leading vehicle, the network takes images as the input for the CNN. Together with the high-level command and the velocity, the output of the CNN is processed in an FC network to predict the steering angle and throttle/brake, which are directly applied as the vehicle controls. In the following vehicle, the network also uses images as the input for the CNN. The subsequent FC network predicts the steering angle and the gap between the vehicles. As in the leading vehicle, the steering angle is directly used for vehicle control. The predicted gap, together with the acceleration and velocity of the leading vehicle (transmitted via V2V communication), serves as the input for a simple car-following model, which is similar to [40]. The desired acceleration $\ddot{x}_{\mathrm{Follower}}(t)$ of the vehicle is calculated based on the acceleration of the leading vehicle $\ddot{x}_{\mathrm{Leader}}(t)$, the velocity difference between both vehicles $\Delta\dot{x}(t)$, the predicted gap $s(t)$, and the desired gap s_{des}. The influence of the leader's acceleration, the velocity difference, and the gap is tuned with α, β, and γ, respectively ($\alpha = 1.0$, $\beta = 0.75$, and $\gamma = 0.2$ in our experiments, similar to [40]). As with [41], $\alpha = 1.0$ guarantees weak string stability for this constant spacing control strategy.

$$\ddot{x}_{\mathrm{Follower}}(t) = \alpha\ddot{x}_{\mathrm{Leader}}(t) + \beta\Delta\dot{x}(t) + \gamma(s(t) - s_{\mathrm{des}}) \tag{2}$$

The detailed multi-task neural-network architecture is shown in Figure 2. The feature extractor is similar to the CNN of PilotNet [25]; however, two additional max pooling layers are inserted after the last two convolutional layers to reduce the dimensions. The input of the feature extractor is RGB images with 400×132 pixels normalized to $[-1, 1]$. The feature extractor output is shared by the prediction heads of the leader and the follower.

The prediction head for the leader is split into two branches predicting the steering angle and the throttle/brake. These two branches include the additional inputs, namely the high-level command and the ego velocity. The inputs are concatenated with the feature extractor output and jointly serve as the input for the subsequent shared fully connected layer with 1000 neurons. An auxiliary classification task is added within the leader head to classify the status of a traffic light within sight. This auxiliary task is activated only during training and helps the network to focus on traffic lights to improve the throttle/brake prediction. We define three different output states ⟨*no traffic light, red, green*⟩. The state *no traffic light* is active if there is no traffic light within 30 m along the path. Below 30 m, the auxiliary task classifies the state of the traffic light as either *red* or *green*. The auxiliary task is directly connected to the feature extractor output without the influence of the additional inputs. This way, the CNN learns to predict the traffic light state using visual clues only. The prediction head of the follower is also split into two branches for the prediction of steering angle and vehicle gap.

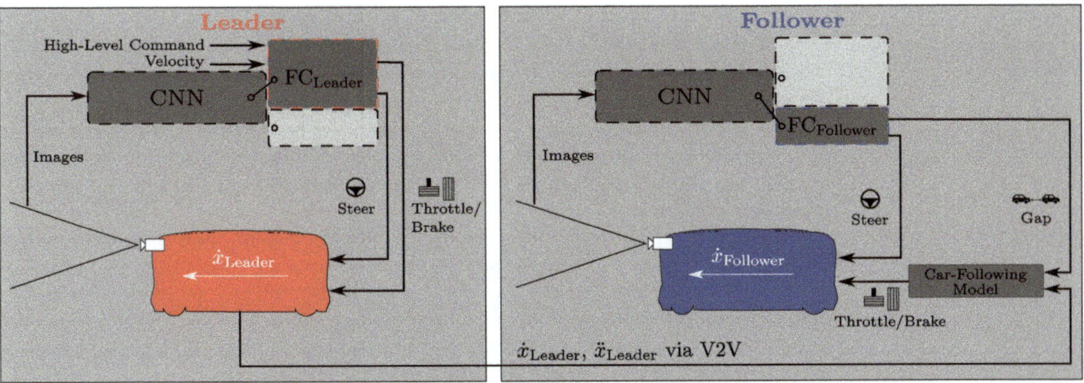

Figure 1. Overview of the proposed system. The leading and following vehicle use the same multi-task network architecture (for details, see Figure 2), but activate different prediction heads. The acceleration and velocity of the leading vehicle are transmitted to the following vehicle via V2V. Details on the car-following model used in the following vehicle can be found in (2).

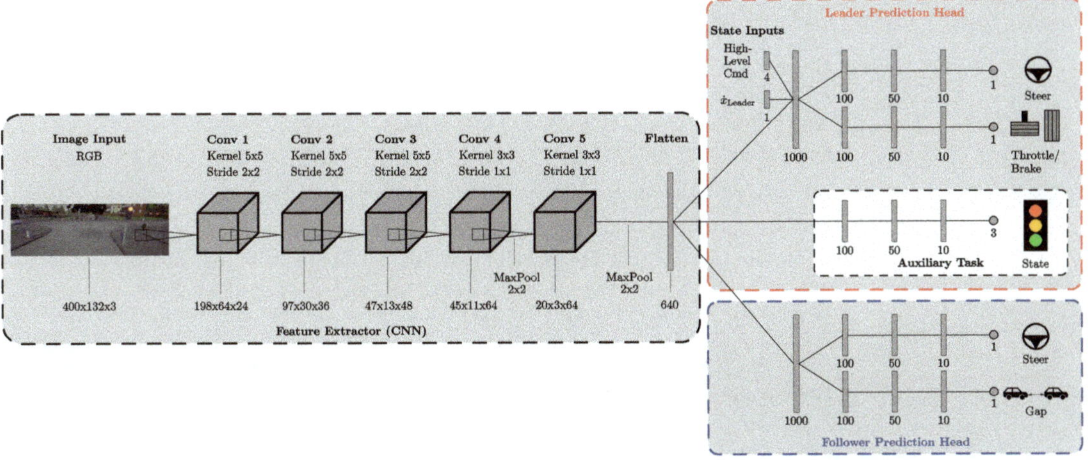

Figure 2. Detailed network architecture. The proposed multi-task deep learning network for connected and autonomous vehicle (CAV) platoons (Model A) with the details of the architecture. The feature extractor is a CNN based on PilotNet [25]. Following the feature extractor, the network consists of two prediction heads to serve its tasks in the leading and following vehicle, respectively. The prediction heads consist of FC networks with multiple branches for the individual outputs.

The structure of all branches was inspired by the FC network of [25]. We use fully connected layers with a decreasing number of neurons at each subsequent layer. All layers use dropout to improve the generalization of the network, with a dropout probability of 0.2 for convolutional layers and 0.4 for fully connected layers (at each batch, we ignore 20% and 40% of the neurons per layer, respectively). The activation function for all layers except the output layers is the rectified linear unit (ReLU). The steering and throttle/brake output layers use tanh (limits outputs to $(-1, 1)$); the traffic light output (auxiliary task) uses softmax for classification (outputs class probabilities); and the gap prediction uses a linear activation function for regression. Contrary to [34], the outputs for the throttle/brake and steering are continuous values instead of discrete values, which results in more realistic and smoother driving. In total, the baseline model has 1,910,747 trainable parameters. This relatively small number of parameters may help to reduce overfitting and decreases training time.

3.2. Model B: Multi-Task Network with LSTM Extension

Model B seizes the idea of [32] to incorporate time-distributed images using a recurrent neural network and is implemented to compare this approach with Model A. Different from [32], V2V communication is not used to send an image stream, but to transmit low-level vehicle states as in Model A. Furthermore, as [32] was limited to lateral vehicle control, the outputs of Model B are identical to the outputs of Model A for lateral and longitudinal control.

We add three LSTM layers between the feature extractor and the prediction heads. Each LSTM layer has 128 neurons to keep the overall parameter count comparable to that of Model A (multi-task network baseline). In total, the multi-task network with the LSTM extension has 1,492,443 parameters. The activation function of the LSTM layers is tanh, and the weights are initialized with the Xavier initialization (most widely used for the tanh activation function). We do not use dropout for the LSTM layers.

The LSTM network has two variables, namely the sequence length and the sampling interval. The sequence length ℓ defines the number of images from the past, including the image at the current time step, which are used to make the prediction at the current time step. The sampling interval τ describes the number of steps between the consecutive images used in the LSTM network. For example, a sequence length of $\ell = 5$ combined with a sampling interval of $\tau = 2$ takes every other image of the training data five times in order to fill one sequence. Therefore, this exemplary sequence covers a period of nine consecutive images. We use a sliding window to generate the sequences during training, following [30]. Instead of using fixed consecutive sequences, the sliding window ensures that, within a training epoch, every image is used once at all positions of the sequence. The principle of the sliding window with a sequence length of three is depicted in Figure 3.

3.3. Model C: Multi-Task Network with Pre-Trained Feature Extractor

Instead of training a network from scratch, transfer learning can be applied to use pre-trained networks. The idea behind transfer learning is that a network trained on a large dataset, such as a dataset for object detection, can be applied in a different domain. We replace the feature extractor (PilotNet CNN) in our architecture with a CNN with more convolutional layers; in particular, we use ResNet-50 [42] pre-trained on the ImageNet (ILSVRC) dataset. To adapt this network to our existing architecture, we remove the last classification layer of ResNet-50 and replace it with a max pooling layer. This last layer serves as the input for the prediction heads. The prediction heads are identical to the prediction heads of Model A. The weights of the layers in the first seven of 16 residual blocks are not trainable and are therefore blocked from updating to retain the lower-level features. Embedded in our architecture, the complete network has 28,300,999 parameters.

3.4. Model D: Single-Task Networks

To compare the multi-task network baseline with single-task networks, we split our multi-task network into two single-task networks for the leading and following vehicle. Each network is trained separately with the corresponding part of the dataset. These models should provide information on whether multi-task learning is beneficial for the driving performance. The single-task network for the follower is similar to the model of [34] and serves as a comparison with their approach. Different from [34], Model D uses continuous control commands and V2V communication identical to Model A.

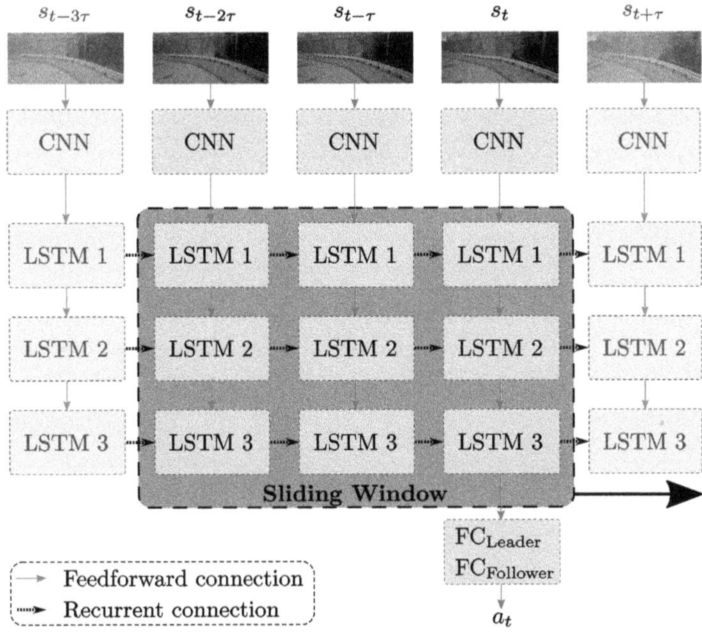

Figure 3. Model B: multi-task network with LSTM extension. Three LSTM layers follow the feature extractor output. The sliding window method includes the last ℓ images with a time step of τ between the images.

3.5. Model E: Multi-Task Network without Auxiliary Task

The auxiliary task is added to the multi-task network to raise the network's awareness for traffic lights during training. In Model E, the auxiliary task is removed, while the rest of Model E is identical to the multi-task network baseline (Model A). This model is used in a separate experiment to provide information on whether the auxiliary task leads to a performance increase of the multi-task network.

3.6. Multi-Task Network Training

The multi-task networks are trained in single supervised training runs, where the training data contain images from both the leading and following vehicle. This has advantages over successive training, as in the latter, the network is prone to developing a bias towards the data with which it was initially trained. To jointly train the network to achieve good performance for both tasks (i.e., in the leading and following vehicle), we use a dynamically weighted loss function,

$$\mathcal{L} = (1 - \eta)\mathcal{L}_{\text{Leader}} + \eta\mathcal{L}_{\text{Follower}} \tag{3}$$

where η weights the loss functions of the leader and follower prediction heads. For each batch, we train with either $\mathcal{L}_{\text{Leader}}$ or $\mathcal{L}_{\text{Follower}}$, i.e., a batch never consists of samples from both a leading and a following vehicle. If a batch consists of samples only showing images captured by the leading vehicle, we set $\eta = 0$, otherwise $\eta = 1$. This way, we can control in which prediction head the weights are updated during gradient backpropagation.

The leader and follower loss functions are summations of the individual loss functions of each branch within the prediction heads. For the leader loss function, that is:

$$\mathcal{L}_{\text{Leader}} = \sum_{i=1}^{3} \lambda_{\text{Leader}}^{i}\mathcal{L}_{\text{Leader}}^{i} \tag{4}$$

with the two individual loss functions for the two branches of steering and throttle/brake and the auxiliary loss for the traffic light classification. The factor $\lambda^i_{\text{Leader}}$ weights the individual and auxiliary loss functions. Similarly, the follower loss function is defined as:

$$\mathcal{L}_{\text{Follower}} = \sum_{i=1}^{2} \lambda^i_{\text{Follower}} \mathcal{L}^i_{\text{Follower}} \tag{5}$$

with the two individual loss functions for steering and gap. We use the mean squared error for all individual loss functions except for the auxiliary loss $\mathcal{L}^{\text{Traffic Light}}_{\text{Leader}}$, where we use the categorical cross-entropy since it is a classification output.

4. Experimental Setup

This section describes the simulation environment, the training dataset, the training process, and the experiments conducted for evaluation.

4.1. Simulation Environment

For training and validation data, we used data generated in a simulator rather than real-world datasets. This was because we needed a special dataset containing vehicle platooning scenarios, which is not available in common datasets. The simulator also allowed us to test the performance of the networks in different environments, compared with only using a subset of the dataset as test data. We recorded the dataset in the CARLA simulator [18], which provides different environments with rural or urban landscapes and allows the configuration of environmental conditions such as weather or light.

Our experimental vehicles in the simulation were minibuses with a length of 6 m; see Figure 4a. The specifications of these minibuses are similar to 2getthere's GRT [43] or Navya's Autonom Shuttle Evo [44]. Each vehicle was equipped with three front-facing cameras: a center camera at a height of 2.7 m and two cameras located at the left and right at a distance of 0.5 m from the center camera. Only the center camera was used during testing. The off-center cameras simulated a shift from the center of the lane and improved the ability of the network to recover from disturbances, as with [25]. The image resolution of each camera was 400×132 pixels with a field of view of $100°$ and a pitch of $-15°$. We recorded at 10 fps, as a higher sample rate would produce more similar images.

During the dataset recording, both vehicles of the platoon drove autonomously and randomly in the simulated urban environment, capturing images and their corresponding labels. Both vehicles used a Stanley controller for the lateral control (minimizing heading error and cross-track error). The gap between both vehicles was varied to capture a broad range of gaps. The leading vehicle obeyed traffic rules such as traffic lights and speed limits and paid attention to other road users. The following vehicle followed the leader all the time. We injected noise into both vehicles by occasionally shifting them laterally. This happened every 10 to 15 s with a probability of ⅔. The training data were captured in CARLA *Town01*, and testing was conducted in *Town01* and *Town02*. *Town01* (see the map in Figure 4b) has 2.9 km of drivable roads in total, and *Town02* has 1.4 km of drivable roads [18]. We used several weather and light settings during training.

(a)

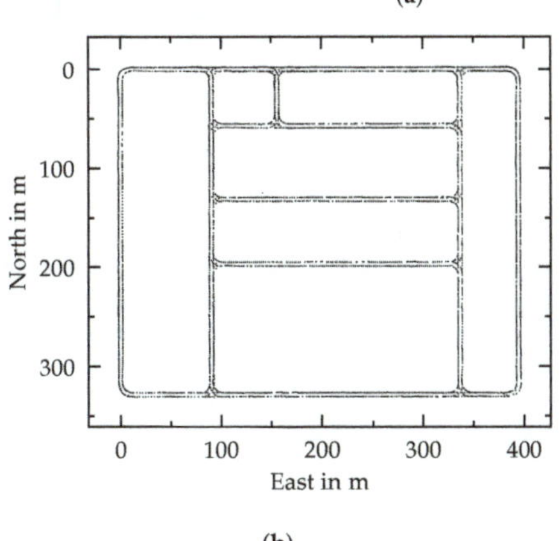

(b)

Figure 4. Simulation environment. (**a**) Two-vehicle platoon with minibuses used for dataset recording in the CARLA simulator [18]. (**b**) Map of *Town01*, which serves as the urban environment for the training and validation dataset recording. Each point represents a position visited by the vehicles during dataset recording. Smaller deviations from the center line are due to injected noise (lateral shifts).

4.2. Training Dataset

In total, we collected 160,000 images from both the leading and following vehicles, which corresponds to about 4.5 h or 75 km of driving. We split our dataset into a training part and validation part. The validation part was used to monitor the validation loss during training and consisted of 10,000 samples, and the remaining 150,000 samples were used for training. The 10,000 samples of the validation part were the last 6.25% of the complete dataset, a similar ratio to [14]. As with [45], we did not randomly select samples for the validation dataset from the complete dataset, as this would have led to the validation dataset being similar to the training dataset. The testing was directly conducted in the simulation, so we did not need a testing dataset.

The majority of the training data consisted of images showing the vehicle driving straight, as seen in the histogram in Figure 5a. The higher the steering angle was, the fewer training samples were in the dataset. Note the asymmetry with more samples showing higher right steering angles (positive) than left steering angles (negative). This is because of the right-hand traffic in the simulation, which usually leads to higher steering angles for right turns than for left turns. Training with a dataset containing mainly steering angles around zero degrees could have led to a network that was biased towards driving straight. To compensate for this, we upsampled images showing higher steering angles. We selected a steering angle threshold of $\pm 20°$, and images satisfying this threshold were upsampled with a factor of three. Many samples of the training dataset showed the vehicle waiting at a red traffic light with a velocity of 0 m/s (Figure 5b). While staying static at a red traffic light, the images were nearly identical. For this reason, we downsampled these images by reducing their occurrence by 50%.

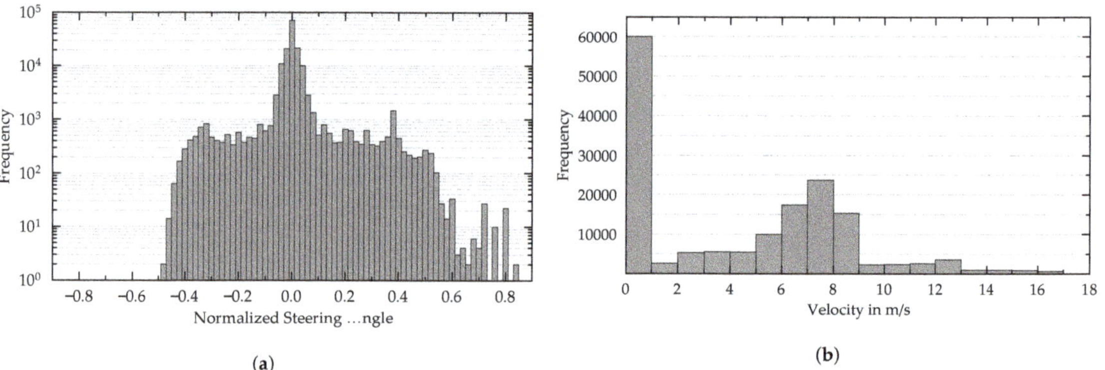

Figure 5. Distribution of the captured dataset for (**a**) the steering angle and (**b**) the velocity. The dataset contains 160,000 samples in total.

In the dataset, the actions for every state are the steering and throttle/brake values present at the moment the state, i.e., the image, was captured. However, during testing, the network should predict the control commands for the next discrete time step, since the current time step was already executed. For this reason, we shifted all actions to the previous state, i.e., an image captured at time t is labeled with the control commands that were applied at the next time step $t + 1$.

4.3. Network Training

The goal of the training was to minimize the dynamic weighted loss function described in (3). We used the AdaMax optimizer with an initial learning rate of 10^{-4}. However, the learning rate was reduced when the training stagnated. We tried different initial learning rates from 10^{-3} to 10^{-6}, but achieved the best results with 10^{-4}. Network weights were initialized with the He or Xavier initialization for layers with ReLU or tanh activation functions, respectively. The performance decreased with other weight initialization techniques (e.g., He initialization with a normal distribution instead of a uniform distribution). Training data normalization was performed for all input and output data. The images, steering angle, and throttle/brake were normalized to $[-1, 1]$, and the velocity and the gap were normalized to $[0, 1]$. The high-level command was encoded in a one-hot vector, and we used a batch size of 128 samples. Larger batch sizes not only require more memory, but also can lead to a lower generalization ability [46].

Prior to feeding the data into the network, we augmented the images with random rotations, shifts, zooming, and brightness changes in the following ranges, each with a continuous uniform distribution:

- Rotation in the range $[-5° \ldots 5°]$
- Horizontal shift in the range $[-5\% \ldots 5\%]$
- Vertical shift in the range $[-5\% \ldots 5\%]$
- Zoom in the range $[-10\% \ldots 0\%]$
- Brightness change in the range $[-10\% \ldots 10\%]$

During training, we observed the training and validation loss of the joint and individual loss functions. We saved the model after each training epoch and stopped the training as the loss converged after about 100 epochs (~27 h of training time), as shown in Figure 6 for Model A. We used the model with the lowest validation loss for testing, which was not necessarily the last model.

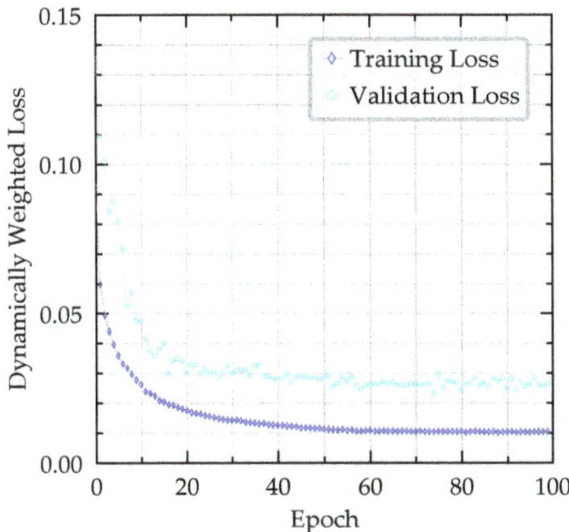

Figure 6. Dynamically weighted loss curve for Model A. The validation loss is calculated at the end of every epoch. Note that for this model, the lowest validation loss was achieved at Epoch 74.

4.4. Experiments and Evaluation Metrics

We conducted two experiments to evaluate the performance of the multi-task network: predefined point-to-point routes and free driving. In both experiments, each trained Model A–D was tested in three different scenarios separately. First, we assessed the performance of the model taking over control of the leading vehicle, without considering any following vehicles. In the second scenario, the model took over control of only the following vehicle, while the leading vehicle was controlled by the controller of the simulation. Finally, we evaluated the performance of a two-vehicle platoon in which two instances of the model took over the vehicle control of both vehicles. In this scenario, an episode ends if any vehicle could not complete the objective due to a crash.

The models were tested in two different environments in the simulation, a similar arrangement to that for the tests conducted by [14,35,36]. The first environment was the same as the one in which the training data were collected (*Town01*). Thus, the network had already processed similar images during training. The second environment was a never-seen-before map with an urban landscape (*Town02*). This environment was used to analyze the generalization ability of the network for unseen places.

Point-to-point routes: As with the test scheme of [18], we used point-to-point routes where the vehicle had to make several turns and drive through intersections to reach its destination. Models A–D were tested on 25 randomly generated routes, with each route repeated twice, which gave a total of 50 episodes per tested model. The length of each route was 500 m, with zero to eight signalized intersections per route. An episode was finished as soon as the vehicle reached the destination or hit an obstacle. The average completion rate (ACR) of the episodes was reported for every model.

Free driving: In addition to the average route completion, we recorded the mean time to failure (MTTF) for all models, as with [34]. A failure was a crash with an object, a stop longer than 60 s, or a high-level command not being obeyed. Traffic light violations did not count as failures, but were recorded for a separate evaluation. For the calculation of this metric, we placed the vehicles in the simulated environment and let them drive randomly. At intersections, the high-level command was chosen randomly based on the intersection layout. If a crash occurred, the episode was stopped and the time to failure was logged. The MTTF was calculated by taking the average of the individual TTFs of 10 episodes per model and scenario. The maximum TTF per episode was limited to 1440 s, and exceeding this limit triggered a new episode. This setup was also used to assess the impact of the auxiliary task, via calculation and comparison of the average violation rate of red traffic lights (AVR) for Model A and Model E. The AVR is the sum of red traffic light violations divided by the total number of traffic lights passed.

5. Results

Point-to-point routes: The results of the quantitative evaluation of the point-to-point routes with a static environment are summarized in Figure 7. The graph shows the average completion rate of the point-to-point routes for Models A–D. The results for the platoon refer to a two-vehicle platoon consisting of one leading and one following vehicle. In general, all models performed better in the training environment than the never-seen-before environment, which was to be expected.

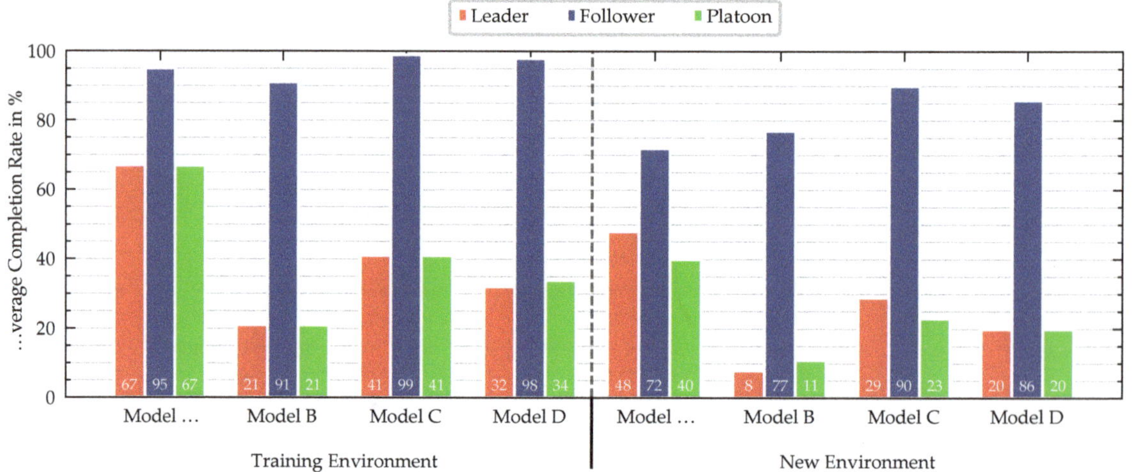

Figure 7. Average completion rate of the routes in % for four different models in two environments. The results for the platoon refer to a two-vehicle platoon consisting of one leading and one following vehicle (**A**: multi-task network baseline, **B**: multi-task network with LSTM extension, **C**: multi-task network with pre-trained feature extractor, **D**: single-task networks).

The multi-task network baseline (Model A) achieved the highest ACRs in the vehicle platoon and leader scenario in both simulation environments. The performance of the two-vehicle platoon is limited by the leading vehicle's performance, which is reflected by similar results for leader and platoon. Whenever the following vehicle could not complete an episode, it was in all cases the result of a crash into static objects. However, we observed that, in 35% of the uncompleted episodes for the leader, the vehicle did not accelerate after a full stop or directly at the beginning of an episode. This is due to the casual confusion, as shown in [47]. We could solve the casual confusion for the following vehicle by implementing V2V communication, but it still persisted for the leading vehicle.

The LSTM extension (Model B), similar to [32], achieved a performance similar to that of Model A for the following vehicle. We observed low-frequency lateral oscillations of both vehicles within the lane. These oscillations led to the vehicles leaving the lane and resulted in crashes in some cases. The casual confusion was more severe than with Model A and in many cases caused zero acceleration at the beginning of an episode.

Replacing the feature extractor with the pre-trained ResNet-50 (Model C) led to superior performance of the following vehicle and driving that was nearly without failures. Although the driving performance of the leading vehicle was good, we observed that this model did not always follow the high-level command.

Even though the single-task network (Model D) of the leader did not share the feature extractor with the follower and therefore could specialize the feature extractor to its sole needs, the performance was worse compared with the multi-task network baseline.

We tested the driving behavior of the multi-task network baseline (Model A) with dynamic traffic. The performance of both the leader and follower decreased mainly as a result of crashes with other vehicles, with ACRs of 57 and 72%, respectively, in the training environment. Casual confusion after stopping behind other vehicles was another reason for the reduction in the leader's performance.

Free driving: The MTTF results are summarized in Figure 8, separated into model (A–D), scenario (leader, follower, platoon), and environment. In case the failure was caused by casual confusion, the TTF was taken at the moment the vehicle stopped. The MTTF results confirm the ACR results of the point-to-point routes, with minor exceptions. We limited the maximum TTF per episode to 1440 s; however, the longest episode without failure for the following vehicle lasted close to one hour and covered nearly 14 km of autonomous driving before it was stopped manually.

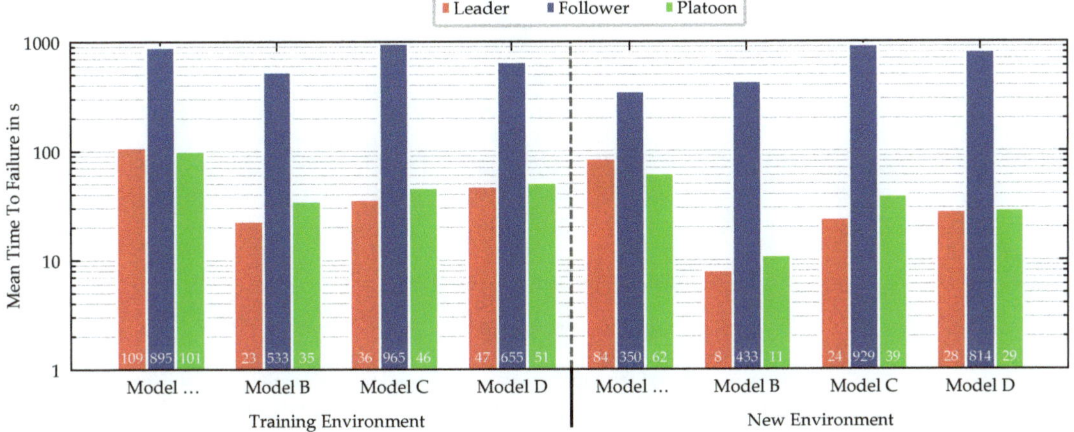

Figure 8. Mean time to failure (MTTF) in seconds for four different models in two environments. The results for the platoon refer to a two-vehicle platoon consisting of one leading and one following vehicle (**A**: multi-task network baseline, **B**: multi-task network with LSTM extension, **C**: multi-task network with pre-trained feature extractor, **D**: single-task networks).

Auxiliary task: The auxiliary task was used to set the focus of the feature extractor on traffic lights during training, but this output was not used during testing. Table 1 shows the AVR results of the comparison of the multi-task model with (Model A) and without (Model E) the auxiliary task; a lower AVR is better. The model with the auxiliary task violated one out of three red traffic lights. Without the prediction of the traffic light state, the vehicle violated nearly two out of three traffic lights, an increase of almost 100%. The overall driving performances of both models were comparable, which is reflected by the similar MTTF scores (Model A had 109 s and 84 s, while Model E had 313 s and 78 s for training and the new environment, respectively). The lower MTTF of Model A in the training environment was induced by casual confusion, since this model performed more full stops. These results show that the network learned to focus on traffic lights without significantly losing driving performance.

Gap estimation: For the following vehicle, we predicted the gap to the leader instead of throttle/brake commands to avoid casual confusion. Figure 9 shows the accuracy of the gap prediction. Each predicted gap was compared with the ground-truth gap at the time of prediction. All points were captured on a single route containing 14,400 predictions in total, with the desired gap set to 7.5 m. This route included multiple poses of the leading vehicle and situations in which only parts of the leading vehicle were visible to the following vehicle due to the curvature of the road. The RMSE of the prediction on this route was 0.44 m, and the R^2 was 0.532. In general, a higher ground-truth gap led to a higher prediction, with a few exceptions.

Table 1. Average violation rate (AVR) of red traffic lights in % for the multi-task network with (Model A) and without (Model E) the auxiliary task.

	Training Environment		New Environment	
	Model A	**Model E**	**Model A**	**Model E**
AVR	**33**	64	**34**	65

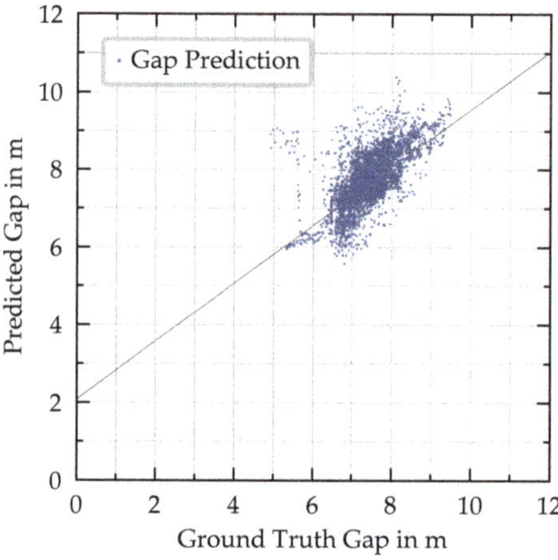

Figure 9. Accuracy of the gap prediction. Every point is a gap prediction captured on a single route, with 14,400 predictions in total. The solid black line indicates the linear regression of all points.

6. Discussion

The results show that networks trained in an end-to-end fashion can be used to drive a CAV platoon with two vehicles. In our experiments, the performance of the leading vehicle, in particular, improved in comparison with single-task networks, which was demonstrated by the fact that the ACR more than doubled in both environments. If the following vehicle lost track of the leader, e.g., during sharp turns, the vehicle could benefit from the knowledge it had gained through the joint training. If the following vehicle separates from the platoon, it could simply switch the prediction head and therefore change its state to act as a leader.

For the following vehicle, we achieved an MTTF of 895 s with the multi-task network baseline (Model A) in the training environment. Reference [34] reported an MTTF of 58 s for a following vehicle in a simulation with their best model. Although the results are not directly comparable, since we used a different simulation, our higher MTTF is still a significant improvement and may be attributed to the joint training of our multi-task network and to our model's use of V2V communication.

In contrast to [32], the overall performance of the multi-task network with LSTM extension (Model B) was the lowest among the evaluated models. We used a single set of sequence lengths and sampling intervals (three each). However, an extensive study of these parameters could improve the performance. Tuning of other hyperparameters such as the number of LSTM layers, the number of neurons per layer, and the position of the LSTM layers within the network could also lead to a performance increase. In addition, we observed that Model B tended to oscillate within the lane, which may lower the riding comfort for passengers. It is not known if the LSTM model of [32] is also prone to oscillations, since their offline evaluation was limited to a dataset in which the control command of the current time step has no influence on the vehicle's position in the next time step.

The results of the multi-task network with a pre-trained feature extractor (Model C) varied for the leading and following vehicle. This model achieved the best performance when used in the following vehicle, but used in the leading vehicle, it was outperformed by the multi-task network baseline using PilotNet as a feature extractor with fewer convolutional layers. The leading vehicle drove in a stable manner, but did not always obey the given high-level command, which led to lower performance in both the ACR and MTTF.

The performance of the single-task networks (Model D) is significantly worse than the performance of the multi-task network (Model A) in both conducted experiments. Since other parameters of the models are identical, this performance difference can be attributed to multi-task learning. This proves the initial assumption that multi-task learning can improve the platoon's performance.

The auxiliary task for the traffic light prediction proved able to raise the network's awareness for traffic lights while maintaining the driving performance. Without the auxiliary task (Model E), the average violation rate of red traffic lights increased by about 100% compared to the model with the auxiliary task (Model A). However, Model A still violated one-third of all red traffic lights. Further investigations have be done to improve driving behavior with respect to traffic lights, since this is not only limited to platooning, but a general challenge for all end-to-end self-driving approaches.

The longitudinal control of the following vehicle was based on information transmitted via V2V communication and the predicted gap. This structure was not fully end-to-end, but it solved the casual confusion for the following vehicle, whereas the casual confusion still existed for the leading vehicle and lowered its performance significantly. Reference [47] suggested including an auxiliary task that predicts the ego velocity to solve the problem of casual confusion. However, even though the authors noticed a performance increase, the casual-confusion problem was still not solved.

Although the latency of V2V communication in our simulation is nonexistent, real-world applications of V2V communication following the IEEE 802.11p standard with an update rate of 10 Hz show a distance-dependent latency. Since we are using data from the previous time step to predict the next control commands as described in Section 4.2, the transmission of the data via V2V communication must be completed within one interval of the update rate, which is 100 ms for our model. Therefore, any latency below 100 ms has no impact on the computed control commands of our model. For a close range (<35 m between the vehicles), Reference [48] measured an average latency of less than 50 ms. Based on these measurements, we assume that the latency, including all delays, is always below 100 ms. In case a message is not received because of wireless dropout, the following vehicle could switch to the leader mode at any time.

In this work, we focused on using the camera as the only sensor. Since the camera is not able to measure the gap to the preceding vehicle and its velocity directly, we predict the gap within the neural network and send the velocity information via V2V communication. The accuracy of the distance prediction (RMSE 0.44 m) is sufficient for the downstream car-following model to follow the leading vehicle at a safe distance (7.5 m in the experiments). Using an additional sensor such as LiDAR or radar could yield the gap directly and substitute or support the prediction.

As this work introduces the multi-task network for CAV platoons and is targeted to prove the model's general ability to drive a CAV platoon, we limit our experiments to a platoon consisting of two vehicles, i.e., one leading and one following vehicle. Another direction for future work could be the extension of this model and the experiments to cover a vehicle platoon with more than two vehicles.

7. Conclusions

This paper presents a new multi-task end-to-end self-driving architecture for the special use case of CAV platoons. The architecture consists of a CNN and two fully connected subsequent prediction heads. The CNN serves as a feature extractor to process images captured by a front-facing camera. Only one prediction head is activated at a time, depending on where the network is used—in the leading or following vehicle of the platoon.

Our experiments show that neural networks trained in an end-to-end fashion are capable of driving two vehicles in a platoon autonomously. In particular, we proved that joint training of a network with two similar tasks can increase the overall performance. We modified our architecture to include different concepts proposed in the literature, such as the incorporation of temporal dependencies by using LSTM layers and transfer learning with a pre-trained network. In our application, LSTM layers were not shown to improve the performance. However, transfer learning had a positive impact on the performance of the following vehicle, although the performance of the leading vehicle suffered as it sometimes ignored the high-level navigational command. Adding an auxiliary task, in our case the prediction of the traffic light state, can assist the network to focus on certain features of the images.

In the experiments performed in the simulation, the following vehicle completed most of the routes or achieved high completion rates. The problem of the casual confusion could be solved for the following vehicle by incorporating V2V communication and predicting the gap instead of throttle/brake values. The main limitation of the platoon is the casual confusion of the leading vehicle, which resulted in no acceleration after full stops. Further research could therefore be undertaken to investigate possible solutions to this limitation.

Author Contributions: S.H., as the first author, conceived of the idea for this paper and was responsible for the presented concept, the implementation of the software, and the results. A.O. and J.B. contributed to the overall concept and system design and supported the critical discussion of the results. M.L. contributed to the conception of the research project and revised the paper critically for important intellectual content. He also gave final approval of the version to be published and agrees with all aspects of the work. As guarantor, he accepts responsibility for the overall integrity of the paper. All authors read and agreed to the published version of the manuscript.

Funding: This work was financially supported by the National Research Foundation (NRF) Singapore under its Campus for Research Excellence And Technological Enterprise (CREATE) program.

Conflicts of Interest: The authors declare no conflict of interest. The funders had no role in the design of the study; in the collection, analyses, or interpretation of data; in the writing of the manuscript; nor in the decision to publish the results.

References

1. SAE. *J3016: Taxonomy and Definitions for Terms Related to Driving Automation Systems for On-Road Motor Vehicles*; SAE International: Warrendale, Germany, 2018.
2. Fagnant, D.J.; Kockelman, K. Preparing a nation for autonomous vehicles: Opportunities, barriers and policy recommendations. *Transp. Res. Part A Policy Pract.* **2015**, *77*, 167–181. [CrossRef]
3. Sethuraman, G.; Liu, X.; Bachmann, F.R.; Xie, M.; Ongel, A.; Busch, F. Effects of Bus Platooning in an Urban Environment. In Proceedings of the 2019 IEEE Intelligent Transportation Systems Conference—ITSC, Auckland, New Zealand, 27–30 October 2019; IEEE: Piscataway, NJ, USA, 2019; pp. 974–980. [CrossRef]
4. Sethuraman, G.; Reddy Ragavareddy, S.S.; Ongel, A.; Lienkamp, M.; Raksincharoensak, P. Impact Assessment of Autonomous Electric Vehicles in Public Transportation System. In Proceedings of the 2019 IEEE Intelligent Transportation Systems Conference–ITSC, Auckland, New Zealand, 27–30 October 2019; IEEE: Piscataway, NJ, USA, 2019; pp. 213–219. [CrossRef]
5. Ullman, S. Against direct perception. *Behav. Brain Sci.* **1980**, *3*, 373–381. [CrossRef]
6. Buehler, M.; Iagnemma, K.; Singh, S. (Eds.) *The DARPA Urban Challenge: Autonomous Vehicles in City Traffic*; Springer Tracts in Advanced Robotics, 1610–7438; Springer: Berlin, Germany, 2009; Volume 56.
7. Badue, C.; Guidolini, R.; Carneiro, R.V.; Azevedo, P.; Cardoso, V.B.; Forechi, A.; Jesus, L.; Berriel, R.; Paixão, T.; Mutz, F.; et al. Self-Driving Cars: A Survey. *Expert Syst. Appl.* **2021**, *165*. [CrossRef]

8. Geiger, A.; Lenz, P.; Stiller, C.; Urtasun, R. Vision meets robotics: The KITTI dataset. *Int. J. Robot. Res.* **2013**, *32*, 1231–1237. [CrossRef]

9. Long, J.; Shelhamer, E.; Darrell, T. Fully convolutional networks for semantic segmentation. In Proceedings of the 2015 IEEE Conference on Computer Vision and Pattern Recognition (CVPR), Boston, MA, USA, 7–12 June 2015; IEEE: Piscataway, NJ, USA, 2015; pp. 3431–3440. [CrossRef]

10. Ziegler, J.; Bender, P.; Schreiber, M.; Lategahn, H.; Strauss, T.; Stiller, C.; Dang, T.; Franke, U.; Appenrodt, N.; Keller, C.G.; et al. Making Bertha Drive—An Autonomous Journey on a Historic Route. *IEEE Intell. Transp. Syst. Mag.* **2014**, *6*, 8–20. [CrossRef]

11. McAllister, R.; Gal, Y.; Kendall, A.; van der Wilk, M.; Shah, A.; Cipolla, R.; Weller, A. Concrete Problems for Autonomous Vehicle Safety: Advantages of Bayesian Deep Learning. In Proceedings of the International Joint Conferences on Artificial Intelligence, Melbourne, Australia, 19–25 August 2017; pp. 4745–4753. [CrossRef]

12. Gibson, J.J. The Ecological Approach to the Visual Perception of Pictures. *Leonardo* **1978**, *11*, 227. [CrossRef]

13. Chen, C.; Seff, A.; Kornhauser, A.; Xiao, J. DeepDriving: Learning Affordance for Direct Perception in Autonomous Driving. In Proceedings of the 2015 IEEE International Conference on Computer Vision, Santiago, Chile, 7–13 December 2015; IEEE: Piscataway, NJ, USA, 2015; pp. 2722–2730. [CrossRef]

14. Sauer, A.; Savinov, N.; Geiger, A. Conditional Affordance Learning for Driving in Urban Environments. In Proceedings of the 2nd Conference on Robot Learning, Zurich, Switzerland, 29–31 October 2018; pp. 237–252.

15. Gaussier, P.; Moga, S.; Quoy, M.; Banquet, J.P. From Perception-action Loops to Imitation Processes: A Bottom-up Approach of Learning by Imitation. *Appl. Artif. Intell.* **1998**, *12*, 701–727. [CrossRef]

16. LeCun, Y.; Boser, B.; Denker, J.S.; Henderson, D.; Howard, R.E.; Hubbard, W.; Jackel, L.D. Backpropagation Applied to Handwritten Zip Code Recognition. *Neural Comput.* **1989**, *1*, 541–551. [CrossRef]

17. Russakovsky, O.; Deng, J.; Su, H.; Krause, J.; Satheesh, S.; Ma, S.; Huang, Z.; Karpathy, A.; Khosla, A.; Bernstein, M.; et al. ImageNet Large Scale Visual Recognition Challenge. *Int. J. Comput. Vis.* **2015**, *115*, 211–252. [CrossRef]

18. Dosovitskiy, A.; Ros, G.; Codevilla, F.; Lopez, A.; Koltun, V. CARLA: An Open Urban Driving Simulator. In Proceedings of the 1st Annual Conference on Robot Learning, Mountain View, CA, USA, 13–15 November 2017, Volume 78; pp. 1–16.

19. Franke, U.; Bottiger, F.; Zomotor, Z.; Seeberger, D. Truck platooning in mixed traffic. In Proceedings of the IEEE Intelligent Vehicles '95, Detroit, MI, USA, 25–26 September 1995; pp. 1–6. [CrossRef]

20. Gehring, O.; Fritz, H. Practical results of a longitudinal control concept for truck platooning with vehicle to vehicle communication. In Proceedings of the IEEE Conference on Intelligent Transportation Systems, Boston, MA, USA, 9–12 November 1997; pp. 117–122. [CrossRef]

21. Kunze, R.; Haberstroh, M.; Ramakers, R.; Henning, K.; Jeschke, S. Automated Truck Platoons on Motorways – A Contribution to the Safety on Roads. In *Automation, Communication and Cybernetics in Science and Engineering 2009/2010*; Jeschke, S., Isenhardt, I., Henning, K., Eds.; Springer: New York, NY, USA, 2010; pp. 415–426. [CrossRef]

22. Tsugawa, S.; Jeschke, S.; Shladover, S.E. A Review of Truck Platooning Projects for Energy Savings. *IEEE Trans. Intell. Veh.* **2016**, *1*, 68–77. [CrossRef]

23. Pomerleau, D.A. ALVINN: An Autonomous Land Vehicle in a Neural Network. In *Advances in Neural Information Processing Systems 1*; Touretzky, D.S., Ed.; Morgan Kaufmann Publishers Inc.: San Francisco, CA, USA, 1989; pp. 305–313.

24. Muller, U.; Ben, J.; Cosatto, E.; Flepp, B.; Cun, Y.L. Off-Road Obstacle Avoidance through End-to-End Learning. In *Advances in Neural Information Processing Systems 18*; Weiss, Y., Schölkopf, B., Platt, J.C., Eds.; MIT Press: Vancouver, BC, Canada, 2006; pp. 739–746.

25. Bojarski, M.; Testa, D.D.; Dworakowski, D.; Firner, B.; Flepp, B.; Goyal, P.; Jackel, L.D.; Monfort, M.; Muller, U.; Zhang, J.; et al. End to End Learning for Self-Driving Cars. *arXiv* **2016**, arXiv:1604.07316.

26. Koutník, J.; Cuccu, G.; Schmidhuber, J.; Gomez, F. Evolving large-scale neural networks for vision-based reinforcement learning. In *GECCO '13, Proceedings of the 15th Annual Conference on Genetic and Evolutionary Computation*; Association for Computing Machinery: New York, NY, USA, 2013; p. 1061. [CrossRef]

27. Xu, H.; Gao, Y.; Yu, F.; Darrell, T. End-to-End Learning of Driving Models from Large-Scale Video Datasets. In Proceedings of the 30th IEEE Conference on Computer Vision and Pattern Recognition, Honolulu, HI, USA, 21–26 July 2016; IEEE Computer Society, Conference Publishing Services: Los Alamitos, CA, USA, 2017; pp. 3530–3538. [CrossRef]

28. Chen, Z.; Huang, X. End-to-end learning for lane keeping of self-driving cars. In Proceedings of the 2017 IEEE Intelligent Vehicles Symposium (IV 2017), Los Angeles, CA, USA, 11–14 June 2017; IEEE: Piscataway, NJ, USA, 2017; pp. 1856–1860. [CrossRef]

29. Chi, L.; Mu, Y. Deep Steering: Learning End-to-End Driving Model from Spatial and Temporal Visual Cues. *arXiv* **2017**, arXiv:1708.03798.

30. Eraqi, H.M.; Moustafa, M.N.; Honer, J. End-to-End Deep Learning for Steering Autonomous Vehicles Considering Temporal Dependencies. In Proceedings of the 31st Conference on Neural Information Processing Systems, NIPS 2017, Long Beach, CA, USA, 4–9 December 2017.

31. Hecker, S.; Dai, D.; van Gool, L. End-to-End Learning of Driving Models with Surround-View Cameras and Route Planners. In *Computer Vision—ECCV 2018*; Springer: Cham, Germany, 2018; Volume 11211, pp. 449–468.

32. Valiente, R.; Zaman, M.; Ozer, S.; Fallah, Y.P. Controlling Steering Angle for Cooperative Self-driving Vehicles utilizing CNN and LSTM-based Deep Networks. In Proceedings of the IV19, Paris, France, 9–12 June 2019; IEEE: Piscataway, NJ, USA, 2019; pp. 2423–2428. [CrossRef]

33. Rausch, V.; Hansen, A.; Solowjow, E.; Liu, C.; Kreuzer, E.; Hedrick, J.K. Learning a deep neural net policy for end-to-end control of autonomous vehicles. In Proceedings of the 2017 American Control Conference (ACC), Seattle, WA, USA, 24–26 May 2017; IEEE: Piscataway, NJ, USA, 2017; pp. 4914–4919. [CrossRef]
34. Pierre, J.M. End-to-End Deep Learning for Robotic Following. In Proceedings of the ICMSCE 2018, Bandung, Indonesia, 5 May 2018; The Association for Computing Machinery: New York, NY, USA, 2018; pp. 77–85. [CrossRef]
35. Codevilla, F.; Muller, M.; Lopez, A.; Koltun, V.; Dosovitskiy, A. End-to-End Driving Via Conditional Imitation Learning. In Proceedings of the 2018 IEEE International Conference on Robotics and Automation (ICRA), Brisbane, Australia, 21–25 May 2018; Lynch, K., Ed.; IEEE: Piscataway, NJ, USA, 2018; pp. 4693–4700. [CrossRef]
36. Haavaldsen, H.; Aasboe, M.; Lindseth, F. In Proceedings of the Autonomous Vehicle Control: End-to-end Learning in Simulated Urban Environments. *arXiv* **2019**, arXiv:1905.06712.
37. Yang, Z.; Zhang, Y.; Yu, J.; Cai, J.; Luo, J. End-to-end Multi-Modal Multi-Task Vehicle Control for Self-Driving Cars with Visual Perception. *arXiv* **2018**, arXiv:1801.06734.
38. de Haan, P.; Jayaraman, D.; Levine, S. Causal Confusion in Imitation Learning. *arXiv* **2019**, arXiv:1905.11979.
39. Wang, Z.; Bian, Y.; Shladover, S.E.; Wu, G.; Li, S.E.; Barth, M.J. A Survey on Cooperative Longitudinal Motion Control of Multiple Connected and Automated Vehicles. *IEEE Intell. Transp. Syst. Mag.* **2020**, *12*, 4–24. [CrossRef]
40. Van Arem, B.; van Driel, C.J.G.; Visser, R. The Impact of Cooperative Adaptive Cruise Control on Traffic-Flow Characteristics. *IEEE Trans. Intell. Transp. Syst.* **2006**, *7*, 429–436. [CrossRef]
41. Swaroop, D.; Hedrick, J.K. Constant Spacing Strategies for Platooning in Automated Highway Systems. *J. Dyn. Syst. Meas. Control* **1999**, *121*, 462–470. [CrossRef]
42. He, K.; Zhang, X.; Ren, S.; Sun, J. Deep Residual Learning for Image Recognition. In Proceedings of the 2016 IEEE Conference on Computer Vision and Pattern Recognition (CVPR), Las Vegas, NV, USA, 26 June–1 July 2016, pp. 770–778. [CrossRef]
43. 2getthere. GRT Vehicle. Available online: https://www.2getthere.eu/technology/vehicle-types/grt-vehicle-automated-minibus/ (accessed on 18 November 2020).
44. Navya. Self-Driving Shuttle for Passenger Transportation. Available online: https://navya.tech/en/solutions/moving-people/self-driving-shuttle-for-passenger-transportation/ (accessed on 18 November 2020).
45. Chen, S.; Leng, Y.; Labi, S. A deep learning algorithm for simulating autonomous driving considering prior knowledge and temporal information. *Comput. Aided Civ. Infrastruct. Eng.* **2019**, *41*, 39. [CrossRef]
46. Keskar, N.S.; Mudigere, D.; Nocedal, J.; Smelyanskiy, M.; Tang, P.T.P. On Large-Batch Training for Deep Learning: Generalization Gap and Sharp Minima. In Proceedings of the 5th International Conference on Learning Representations, ICLR 2017, Toulon, France, 24–26 April 2017.
47. Codevilla, F.; Santana, E.; López, A.M.; Gaidon, A. Exploring the Limitations of Behavior Cloning for Autonomous Driving. In Proceedings of the 2019 IEEE International Conference on Computer Vision, Seoul, Korea, 27 October–2 November 2019.
48. Rajput, N.S. Measurement of IEEE 802.11p Performance for Basic Safety Messages in Vehicular Communications. In Proceedings of the 2018 IEEE International Conference on Advanced Networks and Telecommunications Systems (ANTS), Indore, India, 16–19 December 2018; pp. 1–4. [CrossRef]

Article

Hybrid Path Planning Combining Potential Field with Sigmoid Curve for Autonomous Driving

Bing Lu [1], Hongwen He [1,*], Huilong Yu [2], Hong Wang [3], Guofa Li [4], Man Shi [1] and Dongpu Cao [2]

[1] National Engineering Laboratory for Electric Vehicles, Beijing Institute of Technology, Beijing 100081, China; 3120160195@bit.edu.cn (B.L.); 3120185272@bit.edu.cn (M.S.)

[2] Department of Mechanical and Mechatronics Engineering, Waterloo University, Waterloo, ON N2L 3G1, Canada; huilong.yu@uwaterloo.ca (H.Y.); dongpu.cao@uwaterloo.ca (D.C.)

[3] Tsinghua Intelligent Vehicle Design and Safety Research Institute, Tsinghua University, Beijing 100084, China; hong_wang@tsinghua.edu.cn

[4] College of Mechatronics and Control Engineering, Shenzhen University, Shenzhen 518060, China; guofali@szu.edu.cn

* Correspondence: hwhebit@bit.edu.cn

Received: 10 November 2020; Accepted: 8 December 2020; Published: 16 December 2020

Abstract: The traditional potential field-based path planning is likely to generate unexpected path by strictly following the minimum potential field, especially in the driving scenarios with multiple obstacles closely distributed. A hybrid path planning is proposed to avoid the unsatisfying path generation and to improve the performance of autonomous driving by combining the potential field with the sigmoid curve. The repulsive and attractive potential fields are redesigned by considering the safety and the feasibility. Based on the objective of the shortest path generation, the optimized trajectory is obtained to improve the vehicle stability and driving safety by considering the constraints of collision avoidance and vehicle dynamics. The effectiveness is examined by simulations in multiobstacle dynamic and static scenarios. The simulation results indicate that the proposed method shows better performance on vehicle stability and ride comfortability than that of the traditional potential field-based method in all the examined scenarios during the autonomous driving.

Keywords: potential field; sigmoid curve; path planning; autonomous vehicles

---

1. Introduction

Path planning as an essential part of the autonomous driving has been widely researched in recent years. The path planning layer of autonomous vehicles (AVs) can be classified into the global and local path planners according to the planning horizon [1,2]. The global path planners are mainly focused on the navigation with the optimal economic, the least congestion and the highest average speed by considering the entire configurable space from the start point to the target point [3]. Different from the global path planners, local path planners usually pay more attention to the improvements on the driving safety and the vehicle stability in the process of dynamic obstacle avoidance by considering the constraints of kinematics and dynamics, during autonomous driving [4].

Many planning algorithms of AVs are inherited from wheeled-robotics principles because of their similarities in structure and control. In the wheeled robotics community, path planning methodologies can be classified into four groups, including graph search-based, sampling-based, interpolation-based and numerical optimization-based [5]. The idea behind graph search-based methods is to construct a configurable state-space based on graph theory and then use different search strategies (e.g., Voronoi diagrams [6], the Dijkstras algorithm [7], the A* algorithm [8] or the State Lattice algorithm [9]) to generate a discrete route with grid or lattice occupancy. Being different from graph search-based methods, sampling-based methods can be further categorized into stochastic sampling and deterministic sampling depending on the sampling space used. Deterministic sampling-based methods [10] require less computation cost than the stochastic sampling-based methods [11], as they sample in a semistructured space instead of an entire configuration action-space or state-space. There are some common features between graph-based and sampling-based methods. For example, the paths generated from both methods are connected by a series of discrete waypoints [12], and these paths need further smoothing for practical application in AVs or wheeled-robotics. Since continuous curvature is a necessary requirement of drivable paths, interpolation-based path planning methods have been developed to generate smooth and continuous-curvature routes based on different curve models, such as spline curves model [13], clothoid curve model [14], etc.

The feasible solutions of satisfying the smooth and drivable constraints are usually not unique. Thus, numerical optimization-based path planning methods are developed to obtain the optimal route based on designing an objective function [15], e.g., the shortest-distance, the highest-efficiency, the shortest-time, etc. A potential field-based path planning method (PFBM), as a typical numerical optimization approach, was proposed by establishing the attractive potential field (PF) around a target point and the repulsive potential fields around obstacles to realize the obstacle avoidance of a robot in [16]. A composite PF is established with the constructed repulsive and attractive PFs, to automatically guide a robot to the destination by searching the gradient descent direction. Up to now, the PFBM has been applied both in structured [17,18] and unstructured [19] environments for AV path planning. Traditional PFBMs are usually based on a known target point and span the entire discrete space, which makes them reasonable and efficient for robotics control in indoor or simple environments. However, these requirements are difficult to accurately determine for autonomous driving in practical road environments. Furthermore, there is no qualitative assessment of the reasonability of the paths generated using PFBM, especially in driving scenarios with multiple obstacles. Besides, the planned paths are very sensitive to the configuration of the parameters [16]. For example, if the parameters of the PF functions are inappropriately configured or the obstacles are located with short distances, the route generated using PFBM is likely to fall into local minima, resulting in rough and unexpected routes.

To address the above mentioned problems, in this paper we propose a hybrid potential field sigmoid curve method (HPFSM) as shown in Figure 1, which aims to optimize the planned path of PFBM and to achieve an expected collision-free trajectory to improve the performance of autonomous driving. Firstly, the PF functions are designed based on the geometric shape, relative position and distance, etc. Then, the potential fields are established for the obstacles, the target lane and the road boundaries according to the defined PF functions, respectively. A collision-free route with the minimum PF can be achieved by fusing the established PFs. A constrained nonlinear optimization problem is constructed based on the collision-free path and the sigmoid curves. Finally, an optimal path of satisfying the constraints of collision avoidance and vehicle dynamics can be achieved by solving with the interior point algorithm. The main contributions of this study include:

(1) A novel hybrid path planning method is proposed to get better collision-free path for improvements on vehicle stability and ride comfort during autonomous driving by combining potential field with sigmoid curve.

(2) Based on the distribution function of two-dimensional joint probability density, an improved potential field of the obstacle is designed to mimic more realistic distribution of collision risk by decoupling the PF in longitudinal and lateral directions.

(3) With the designed objective of the shortest path generation, the trajectory is optimized to improve the vehicle stability and the ride comfort during autonomous driving by considering the constraints of collision avoidance and vehicle dynamics.

This paper is organized as follows: Section 2 designs a potential field-based path planning and shows how PFBM can generate unexpected paths. Section 3 introduces the proposed HPFSM. Section 4 presents the validation and evaluation results by comparing the proposed method with PFBM both in the static and dynamic driving scenarios. Finally, Section 5 presents our conclusions.

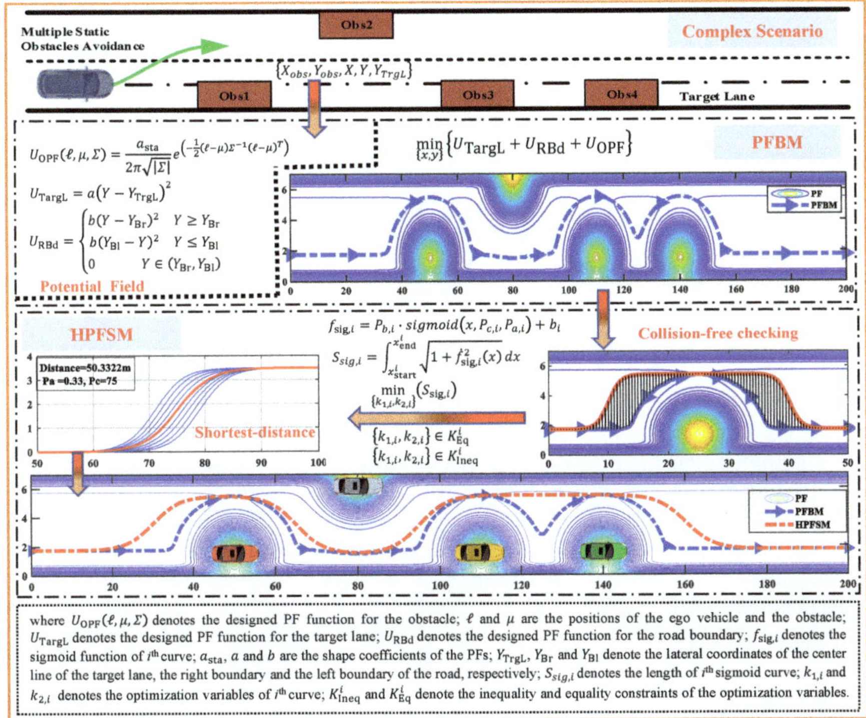

Figure 1. Framework of the hybrid path planning method by combining potential fields with sigmoid curves.

2. An Improved Potential Field-Based Path Planning

Because of the good collision-free performance, PFBM has been widely used as a path planning approach for AVs [20]. The path planning process of PFBM can be mainly divided into two parts, namely, the part for designing PF functions and the collision-free route generation part for obstacle avoidance.

2.1. The Design of PF Functions

The PF is affected by obstacle properties including the obstacle's physical characteristics (e.g., geometric shape and structure), and its dangerous degree is affected by the mass and motion state of the obstacle [21,22].

2.1.1. Road PFs

Road PFs include the PF of road boundary and the PF of target lane. Since the potential field-based path planning is likely to fall into the local minima, especially in an unknown environment [23]. To avoid the local minima problem in our proposed path planning method, both the driving environment and the obstacles (vehicles) are assumed to be known, thus the potential fields can be designed and established more appropriately. In real intelligent transportation systems, these information can be obtained via the vehicle-to-vehicle and vehicle-to-infrastructure technologies [24]. Besides, we design the attractive potential field with the center line of the target lane to attract the ego vehicle driving to the target lane instead of with only one target point [25,26], which will reduce the chance to trap into the local minima by solving a series of optimization subproblems. In this study, the attractive PF is defined as:

$$U_{\text{TrgL}} = a \left(Y - Y_{\text{TrgL}} \right)^2.$$ (1)

The road boundary is designed as a repulsive PF in Equation (2), which is used for preventing the ego vehicle driving out of the road.

$$U_{\text{RBd}} = \begin{cases} b \left(Y - Y_{\text{Br}} \right)^2 & Y \leq Y_{\text{Br}} \\ b \left(Y_{\text{Bl}} - Y \right)^2 & Y \geq Y_{\text{Bl}} \\ 0 & Y \in (Y_{\text{Br}}, Y_{\text{Bl}}) \end{cases},$$ (2)

where $a, b \in \mathbb{R}$, $Y_{\text{Br}} < Y_{\text{Bl}}$ and $Y_{\text{TrgL}} \in (Y_{\text{Br}}, Y_{\text{Bl}})$; a and b respectively denote the shape coefficients of the target-lane PF and the road-boundary PF, which are used to adjust the amplitude of PF; Y denotes the lateral position of the ego vehicle; Y_{TrgL} denotes the lateral position of the center line of the target lane; Y_{Bl} and Y_{Br} denote the lateral positions of the left and right road boundaries, respectively.

Two examples of the successful applications without the local minima problem are shown in Figure 2. In the situation without obstacle vehicles, Figure 2a shows that the ego vehicle will always drive along the path (the central line of the target lane) with the minimum potential field when using our proposed method. In the situation with an obstacle vehicle in the target lane, Figure 2b shows that the planned path with minimum PF will lead the ego vehicle to overtake the obstacle vehicle and then drive back to the target lane.

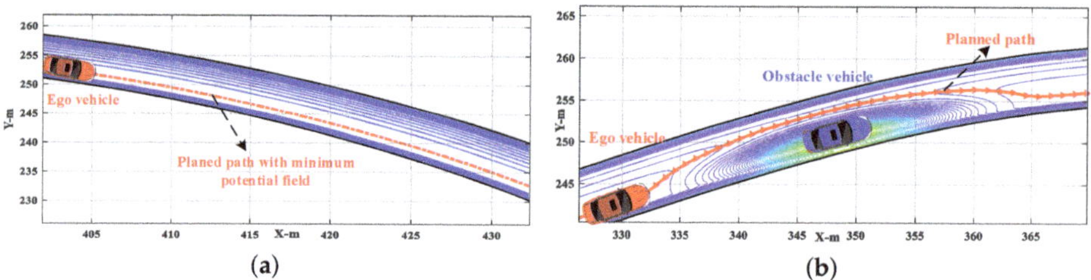

(a)   **(b)**

Figure 2. Examples without trapping into local minima: (**a**) scenario without obstacle vehicle; (**b**) scenario with an obstacle vehicle.

2.1.2. Obstacle Potential Field

The obstacle potential field (OPF) function is defined to construct the repulsive PF according to the longitudinal and lateral safe distances. The calculations of the safe distances are based on the relative speed (between the ego vehicle and the obstacle) and the maximum longitudinal/lateral deceleration of the ego vehicle [12], which means the velocities of the ego vehicle and the obstacle are required. The longitudinal and lateral safe distances $(X_{\text{s}}(t), Y_{\text{s}}(t))$ are calculated as :

$$\begin{cases} X_{\text{s}}(t) = \dfrac{X_{\text{o}}}{2} + \dfrac{\left(V_{\text{x}}(t) - V_{\text{obs,x}}(t) \right)^2}{2a_{\text{x,max}}(t)} \\ \\ Y_{\text{s}}(t) = \dfrac{Y_{\text{o}}}{2} + \dfrac{\left(V_{\text{y}}(t) - V_{\text{obs,y}}(t) \right)^2}{2a_{\text{y,max}}(t)} \end{cases},$$ (3)

where $a_{\text{x,max}}(t) \neq 0$ and $a_{\text{y,max}}(t) \neq 0$ are the maximum longitudinal and lateral decelerations of the ego vehicle; X_{o} and Y_{o} are the length and width of the obstacle, respectively; $V_{\text{obs,x}}(t)$ and $V_{\text{obs,y}}(t)$ represent the longitudinal and lateral velocities of the obstacle, respectively; $V_{\text{x}}(t)$ and $V_{\text{y}}(t)$ are the longitudinal and lateral velocities of the ego vehicle, respectively.

The OPF can be decomposed along the longitudinal and lateral directions of the road coordinate system, and the definition domains of the two directions are usually independent and different [18]. Considering the above characteristics, a two-dimensional (2D) joint probability density distribution function is used as the basic function to define the OPF as:

$$U_{\mathrm{OPF}}\left(\ell, \mu, \Sigma\right) = \frac{a_{\mathrm{sta}}}{2\pi\sqrt{|\Sigma|}} e^{\left(-\frac{1}{2}(\ell-\mu)^T \Sigma^{-1}(\ell-\mu)\right)}, \tag{4}$$

where

$$\mu = (X_{\mathrm{obs}}(t), Y_{\mathrm{obs}}(t))^T, \Sigma = \begin{bmatrix} X_s^2(t) & 0 \\ 0 & Y_s^2(t) \end{bmatrix}, \ell = (X(t), Y(t))^T,$$

where μ and Σ denote the mean and covariance matrix; $(X_{\mathrm{obs}}(t), Y_{\mathrm{obs}}(t))$ and $(X(t), Y(t))$ denote the positions of the obstacle and the ego vehicle at time t, respectively; $a_{\mathrm{sta}} \in \mathbb{R}$ is the shape coefficient used to adjust the amplitude of OPF; $X_s(t)$ and $Y_s(t)$ are the calculated safe distances along the longitudinal and lateral directions of the road coordinate system at time t, respectively. Figure 3 is shown to illustrate that the OPF is adaptive to the safe distance, i.e., the OPF will vary with the velocities of the ego vehicle and the obstacles.

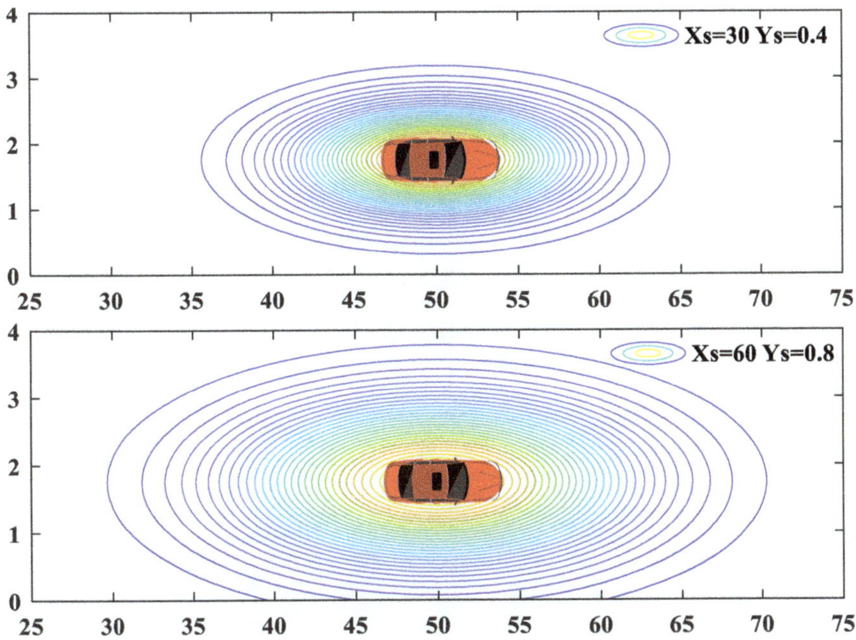

Figure 3. Potential fields with different longitudinal and lateral safe distances.

2.2. Collision-Free Path Generation

The idea behind PFBM is to generate a collision-free path occupied the minimum PF along the driving direction. The related attractive and repulsive PFs can be constructed and integrated according to the parameters described in Table 1.

Table 1. Parameters for PF construction.

Parameter	Value	Parameter	Value	Parameter	Value
X (m)	0~200	Y_{TrgL} (m)	1.75	Y_{obs} (m)	1.5
Y (m)	0~7	Y_{Bl} (m)	6	X_{s} (m)	20
a	0.5	Y_{Br} (m)	1	Y_{s} (m)	1.5
b	100	X_{obs} (m)	50	a_{sta}	1×10^4

The fused PF is shown in Figure 4. Based on this, the minimum PF path is obtained along the longitudinal direction of the road coordinate (X direction). Obviously, the generated collision-free path (the blue trajectory) is optimal subject to the defined PFs.

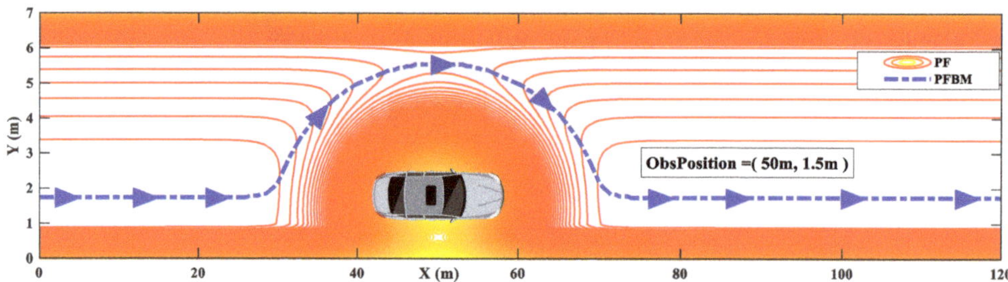

Figure 4. PFBM path planning for collision avoidance.

However, the path obtained using PFBM is not always smooth and expected, especially in these driving scenarios with multiple closely distributed obstacles. The blue path in Figure 5 shows the trajectory planned using PFBM [27] in a driving scenario with two closely distributed obstacles. Although the planned path is collision-free, it involves undesired driving maneuvering, which will affect the efficiency of obstacle avoidance and the tracking performance. An expected trajectory to mimic the real driver (e.g., the red route) is required for AV path planning to ensure the efficiency of the obstacle avoidance and to improve tracking performance.

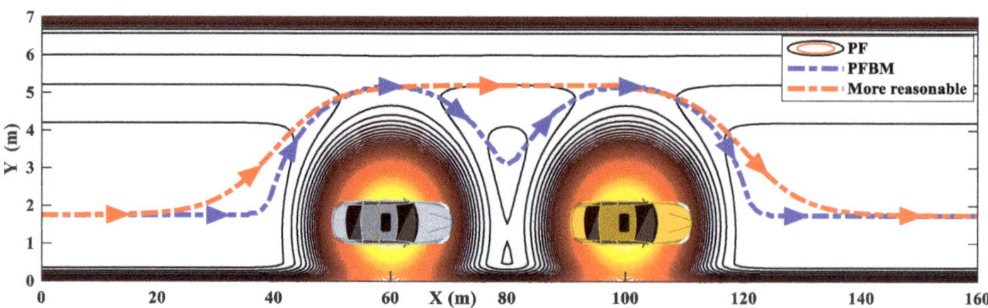

Figure 5. A driving scenario with two obstacle vehicles.

The unexpected maneuvers of the planned path are more evident when the number of obstacles increases, as shown in Figure 6. In practical applications, the AVs would result in frequent unnecessary steering maneuvers when tracking this unexpected path. The planned paths in Figures 5 and 6 indicate that the PFBMs are disadvantaged to be applied in the driving scenarios with multiple closely distributed obstacles. Therefore, a novel hybrid path planning method is proposed, which combines PFs with sigmoid curves to solve the problem of generating unexpected trajectory.

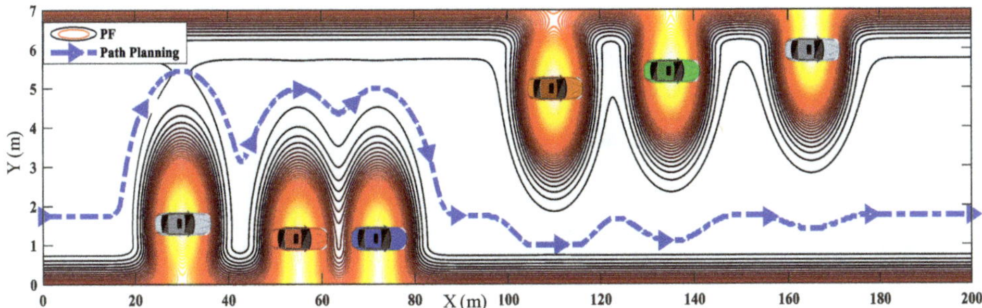

Figure 6. A driving scenario with multivehicle.

3. A Hybrid Path Planning Method

Based on the deterministic curve models, e.g., splines [28], clothoid curves [14], and polynomials [29], etc, the smooth candidate routes can be generated quickly and efficiently. However, it is difficult to shape desiring driving trajectory using the B-spline, C-spline, clothoid and even quintic polynomial models because of the strong coupling relationship among the tunable parameters. Considering the tunable feature of the parameters in the sigmoid curve model [30], which the amplitude, slope, and central symmetry point of the sigmoid curve can be adjusted independently. Therefore, a hybrid path planning method is proposed by combining the PFBM with sigmoid curves to obtain a smooth collision-free and efficient expected route.

3.1. Definition of the Sigmoid Curve

The process of obstacle avoidance is similar to that of vehicle lane change. The trajectory is tangent to the center lines of the related lanes at the start and end points according to the standard lane change path [31]. In this paper, the sigmoid function is introduced as an essential function for generating obstacle avoidance paths. The definition is presented in Equation (5):

$$
\begin{cases}
f_{\text{sig}}(x) = P_{\text{b}} \cdot sigmoid\,(x, P_{\text{c}}, P_{\text{a}}) \\[2mm]
sigmoid\,(x, P_{\text{c}}, P_{\text{a}}) = \frac{1}{1+e^{-P_{\text{a}}(x-P_{\text{c}})}}
\end{cases}
\tag{5}
$$

where P_{a}, P_{b} and P_{c} are the related parameters to shape the sigmoid curve. The parameter P_{a} represents the maximum slope, P_{b} is the amplitude coefficient, P_{c} denotes the centrosymmetric point and the point of maximum slope. These parameters can be used to determine a sigmoid curve uniquely.

3.2. Tunable Features of the Sigmoid Curve

Figure 7 shows the tunable features of the sigmoid curve including the maximum slop, the amplitude and the central point. Figure 7a shows that the centrosymmetric point can be adjusted independently using the parameter of P_c, which can be used to move the sigmoid curve in longitudinal direction. Figure 7b shows that the amplitude can be adjusted using the parameter of P_b, which can be applied to compress or stretch the sigmoid curve in the lateral direction. Furthermore, Figure 7c indicates that the maximum slope is also independently tunable using the parameter of P_a, which can be used to adjust the maximum slope of the sigmoid curve at the centrosymmetric point.

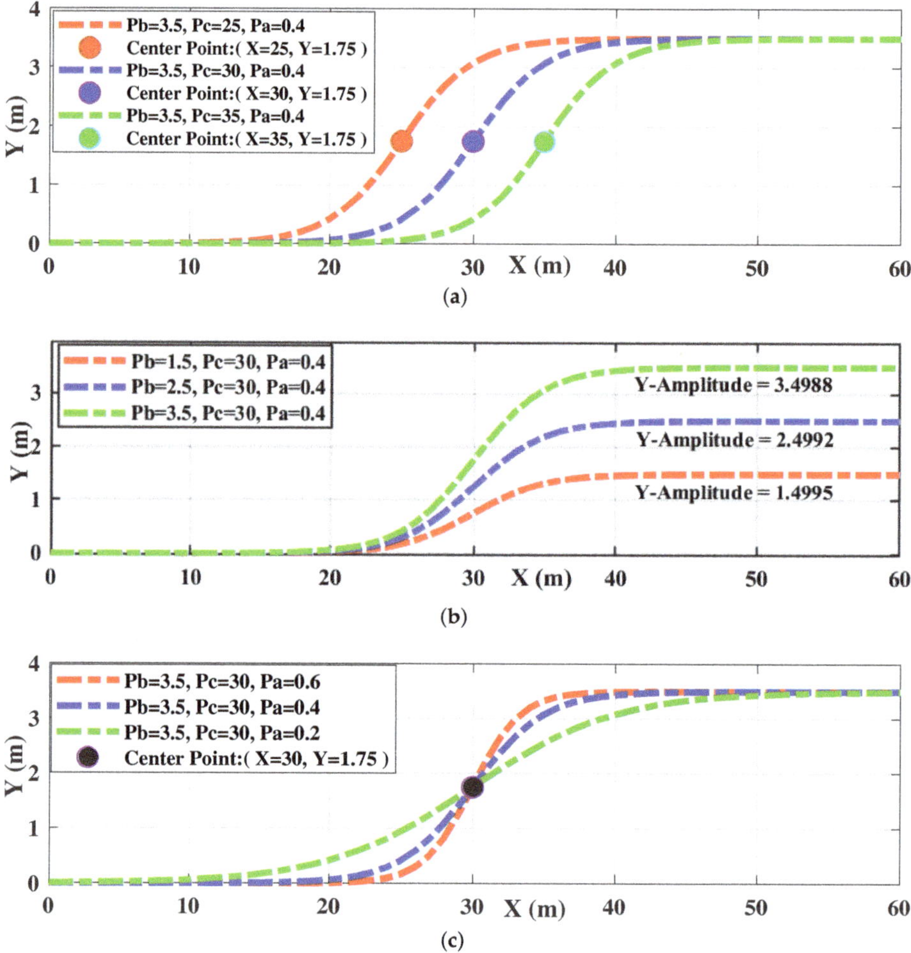

Figure 7. Tunable features of the sigmoid curve: (**a**) Sigmoid curves with different central point; (**b**) Sigmoid curves with different central point; (**c**) Sigmoid curves with different maximum slope.

3.3. Configuration of the Sigmoid Curve

3.3.1. Collision-Free Path Generation of PFBM

The potential fields are integrated according to Equation (6):

$$U_{\mathrm{PF}}(t) = U_{\mathrm{TargL}}(t) + U_{\mathrm{RBd}}(t) + U_{\mathrm{OPF}}(t), \tag{6}$$

where $U_{\mathrm{PF}}(t)$ is the integrated PF, $U_{\mathrm{TargL}}(t)$, $U_{\mathrm{RBd}}(t)$ and $U_{\mathrm{OPF}}(t)$ are the corresponding target lane PF, the road boundary PF and the obstacle PF at time t, respectively.

The PFBM collision-free path is obtained through Equation (7):

$$\{X_{\min}, Y_{\min}\} = \min_{\{x_{\min}(t), y_{\min}(t)\}} U_{\mathrm{PF}}(t), \tag{7}$$

where $\{X_{\min}, Y_{\min}\}$ represent the collision-free path with the minimum PF along the longitudinal direction.

The corresponding lateral positions of the obstacles mapping to the collision-free path are obtained by interpolation through Equation (8):

$$\begin{cases} \forall : x^j_{obs} \in [x_i, x_{i+1}], j \in M \\ \exists : y^j_{obs} = interp1\left(X_{min}, Y_{min}, x^j_{obs}\right) \end{cases} \Rightarrow y^j_{obs} = \frac{y_{i+1} - y_i}{x_{i+1} - x_i}\left(x^j_{obs} - x_i\right) + y_i, \tag{8}$$

where $(x_i, y_i) \in \{X_{min}, Y_{min}\}$ and $(x_{i+1}, y_{i+1}) \in \{X_{min}, Y_{min}\}$ are two known waypoints in the collision-free path; $interp1$ denotes the one-dimensional linear interpolation function for calculating the corresponding lateral coordinates to the collision-free path; i and j denote the index of the waypoints and the index of the obstacles, respectively; M is the amount of the obstacle; x^j_{obs} and y^j_{obs} are the longitudinal and lateral coordinates corresponding to the PFBM path, respectively.

The planned collision-free path is composed of several sigmoid curves, and the definition domains are varying with the positions of obstacles. The definition domains of the sigmoid curves are determined using Equation (9):

$$\Omega_{x,i} = \begin{cases} [x_{start}, x^i_{obs}] & i = 1 \\ \left[x^{i-1}_{obs}, x^i_{obs}\right] & i \in (1, n) \\ \left[x^{i-1}_{obs}, x_{end}\right] & i = n \end{cases}, \tag{9}$$

where $\Omega_{x,i}$ denotes the definition domain of the ith sigmoid curve, x_{start} and x_{end} indicate the start and end points of the planning horizon, $n = M + 1$ represents the amount of sigmoid curves. Considering the detection ranges of on-board sensors [32], the planning horizon is limited to 200 m.

3.3.2. Parameter Configuration

Some key way-points of the collision-free path can be obtained using Equations (7) and (8). Since the amplitude of sigmoid curve is related to the lateral coordinates of the target lane (Y_{TrgL}) and the key waypoints (y_{obs}), the amplitude of the curve is determined in Equation (10):

$$P_{b,i} = \begin{cases} y^i_{obs} - Y_{TrgL} & i = 1 \\ y^i_{obs} - y^{i-1}_{obs} & i \in (1, n) \\ Y_{TrgL} - y^{i-1}_{obs} & i = n \end{cases}. \tag{10}$$

The slope parameter can be determined using Equation (11):

$$P_{a,i} = k_{1,i} \cdot sign\left(P_{b,i}\right), \tag{11}$$

where $k_{1,i}$ denotes the maximum slope at the centrosymmetric point and $sign$ is the sign function.

The centrosymmetric point of the sigmoid curve is defined in Equation (12):

$$P_{c,i} = \begin{cases} x^i_{obs} - k_{2,i}X_s & i = 1 \\ x^{i-1}_{obs} + \frac{k_{2,i}\left(x^i_{obs} - x^{i-1}_{obs}\right)}{2} & i \in (1, n) \\ x^{i-1}_{obs} + k_{2,i}X_s & i = n \end{cases}, \tag{12}$$

where $k_{2,i} \geq 1$ is a tunable coefficient.

When the above three parameters and the bias have been obtained in the definition domain $\Omega_{x,i}$, the sigmoid curve can be determined uniquely using Equation (13):

$$f_{sig,i} = P_{b,i} \cdot sigmoid(x, P_{c,i}, P_{a,i}) + b_i, \tag{13}$$

$$x \in \Omega_{x,i}, \quad b_i = \begin{cases} Y_{TrgL} & i = 1 \\ y^{i-1}_{obs} & i \neq 1 \end{cases},$$

where $f_{sig,i}$ denotes the ith sigmoid curve function, b_i is the corresponding bias. Figure 8 shows the parameters that shape the sigmoid curve in detail.

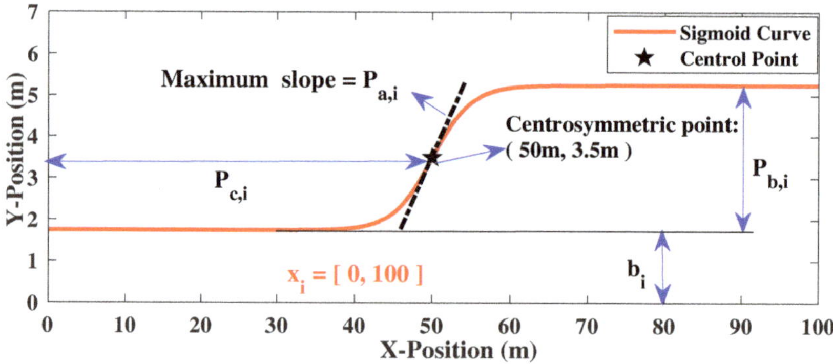

Figure 8. The parameters of sigmoid curve.

3.4. Trajectory Optimization with Sigmoid Curves

Since the coefficients of $k_{1,i}$ and $k_{2,i}$ are not determined yet, a series of sigmoid curves can be generated by the above configurations. An optimization objective function is designed to obtain the shortest path subject to the constraints of the lateral acceleration and the yaw rate to ensure collision avoidance and to improve the vehicle stability of the autonomous driving. The distance of sigmoid curve is calculated according to Equation (14):

$$S_{sig,i} = \int_{x_{start}^i}^{x_{end}^i} \sqrt{1 + \dot{f}_{sig,i}^2(x)}\,dx. \tag{14}$$

3.4.1. Collision Avoidance Constraint

The planned path generated by PFBM is collision-free, which can be used as the constraints of collision avoidance to assist configuring the collision-free sigmoid curves. The collision-free feature can be determined if the sigmoid curves are always farther to the obstacle than that of the collision-free path of PFBM. As Figure 9a indicates, the collision feature cannot be deduced directly and an additional check is required. Therefore, the constraints of collision avoidance should be considered to ensure the collision-free feature of the candidate sigmoid curves.

Instead of combining the geometric information of obstacles [33], the collision avoidance can be ensured by comparing the lateral positions of the candidate paths to that of the PFBM. As illustrated in Figure 9b, collision avoidance is ensured when the red line is completely above the blue line under this situation.

The constraints to ensure collision avoidance based on the path of PFBM are shown in Equation (15):

$$K_{Ineq,i}^{Cons}(1) := \begin{cases} f_{sig,i}(x_i) \geq Y_{min}, & y_{obs}^i \geq Y_{obs} \\ f_{sig,i}(x_i) \leq Y_{min}, & y_{obs}^i < Y_{obs} \end{cases}, \tag{15}$$

where $x_i \in \Omega_{x,i}$ denotes the longitudinal range of the road, and $K_{Ineq,i}^{Cons}$ denotes the inequality constraints of $k_{1,i}$ and $k_{2,i}$.

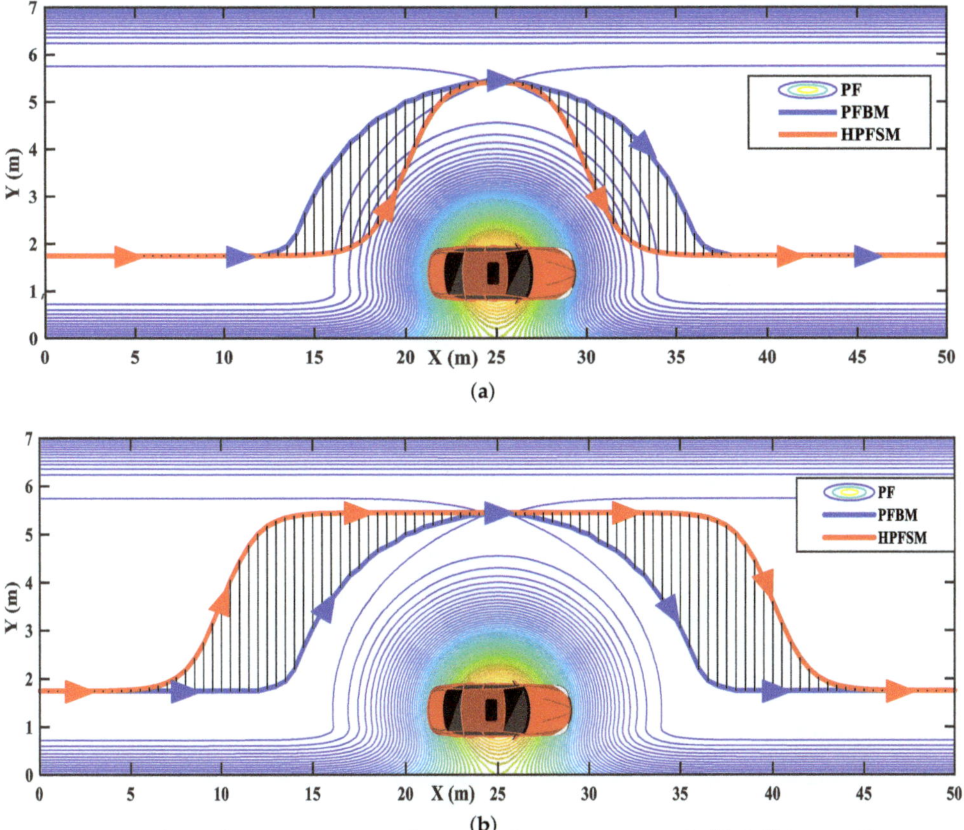

Figure 9. Collision-free checking: (**a**) Collision avoidance is not ensured; (**b**) Collision avoidance is ensured.

3.4.2. The Constraints of Vehicle Dynamics

The constraints of vehicle dynamics should also be considered in path planning module to improve the vehicle stability [34,35] during path tracking. The vehicle stability and ride comfort can be well evaluated based on the lateral acceleration and yaw rate during the path tracking. Assuming that the target velocity is invariant during path tracking, the yaw rate is considered as shown in Equation (16):

$$\begin{cases} \omega_{v,i} = \rho_i V \\ |\omega_{v,i}| \le \omega_s \end{cases},$$ (16)

where $\omega_{v,i}$ (rad/s) is the yaw rate of the ith curve at a speed of V (m/s), ρ_i is the curvature of the ith curve and ω_s (rad/s) denotes the yaw rate constraint to ensure path tracking stability.

The lateral acceleration is considered as follows in Equation (17):

$$\begin{cases} a_{y,i} = V cos\theta_i \omega_{v,i} \\ |a_{y,i})| \le a_s \end{cases},$$ (17)

where V (m/s) denotes the target speed for path tracking, a_s (m/s^2) denotes the constraints of lateral acceleration in path planning module, $a_{y,i}$ is the lateral acceleration of the ith curve.

The constraints of vehicle dynamics can be transformed into a constraint of path curvature in Equation (18):

$$
K^{Cons}_{Ineq,i}(2) := \begin{cases} \rho_i = \dfrac{|\ddot{f}_{sig,i}(x)|}{\left(1+\dot{f}_{sig,i}(x)^2\right)^{3/2}} \\ |\rho_i| \le \rho_{cos} \\ \rho_{cos} = min\left(\frac{a_s}{V^2}, \frac{w_s}{V}\right) \end{cases} ,
$$
(18)

where ρ_{cos} denotes the curvature constraint of planned path considering the ride comfort and vehicle stability in path planning module.

3.4.3. Geometric Constraints

The geometric constraints include amplitude, start point, endpoint and central symmetric point constraints. The end point constraint is defined as:

$$
K^{Cons}_{Eq,i}(1) := \begin{cases} x^{end}_i = \Omega_{x,i}(end) \\ y^{end}_i = interp1\left(X_{min}, Y_{min}, x^{end}_i\right) \\ f_{sig,i}\left(x^{end}_i\right) = y^{end}_i \\ 0 < \dot{f}_{sig,i}\,|_{x^{end}_i} \le \epsilon \end{cases} ,
$$
(19)

where ϵ denotes the infinitesimal value, and $K^{Cons}_{Eq,i}$ denotes the equality constraints of $k_{1,i}$ and $k_{2,i}$.

The start point constraint is defined as:

$$
K^{Cons}_{Eq,i}(2) := \begin{cases} x^{start}_i = \Omega_{x,i}(start) \\ y^{end}_i - f_{sig,i}\left(x^{start}_i\right) = P_{b,i} \\ 0 < \dot{f}_{sig,i}\,|_{x^{start}_i} \le \epsilon \end{cases} .
$$
(20)

The constraint of the centrosymmetric point $P_{c,i}$ is defined as:

$$
K^{Cons}_{Ineq,i}(3) := \begin{cases} x^{end}_i - P_{c,i} \ge X_s \\ x^{start}_i < P_{c,i} < x^{end}_i \end{cases} .
$$
(21)

The inequality constraints are thus summarized as:

$$
K^i_{Ineq} = \left\{ K^{Cons}_{Ineq,i}(1), K^{Cons}_{Ineq,i}(2), K^{Cons}_{Ineq,i}(3) \right\},
$$
(22)

where K^i_{Ineq} refers to the inequality constraints of the ith curve, including the constraints of the collision avoidance , the constraints of the lateral acceleration, the constraints of the yaw rate and the constraints of the geometric.

The constrained nonlinear optimization problem is formulated as:

$$
\min_{\{k_{1,i},k_{2,i}\}} \int_{x^i_{start}}^{x^i_{end}} \sqrt{1 + \dot{f}^2_{sig,i}(x)}\,dx
$$
(23)

$$s.t.$$
$$x_i^{\text{end}} = \Omega_{x,i}(end)$$
$$y_i^{\text{end}} = interp1\left(X_{\min}, Y_{\min}, x_i^{\text{end}}\right)$$
$$f_{sig,i}\left(x_i^{\text{end}}\right) = y_i^{\text{end}}$$
$$x_i^{\text{start}} = \Omega_{x,i}(start)$$
$$y_i^{\text{end}} - f_{sig,i}\left(x_i^{\text{start}}\right) = P_{b,i}$$
$$0 < \dot{f}_{sig,i}\,|_{x_i^{\text{end}}} \leq \epsilon$$
$$0 < \dot{f}_{sig,i}\,|_{x_i^{\text{start}}} \leq \epsilon, \epsilon > 0$$
$$\{k_{1,i}, k_{2,i}\} \in K_{\text{Ineq}}^i$$

A driving scenario with one static obstacle is proposed to analyze the planned path of HPFSM. The relevant parameters for realizing HPFSM are described in Table 2. Figure 10 shows that there are several trajectories satisfying the collision-free constraints, e.g., the blue solid and dotted curves. With a target speed of 20 m/s, the optimal trajectory among the candidate curves is the shortest trajectory (composed with the red and black curves) satisfying the constraints of vehicle dynamics, which require the yaw rate and lateral acceleration are within 25 deg/s and 2 m/s^2, respectively.

Table 2. Initialization parameters.

Parameter	Value	Parameter	Value	Parameter	Value
μ	$(75, 1.5)$	b_1	0	x_2	$[75, 150]$
Σ	$diag([10, 1.5])$	$P_{c,1}$	$[0, 75]$	b_2	3.5
$P_{b,1}$	3.5	$P_{a,1}$	$[0, 1]$	$P_{c,2}$	$[75, 150]$
x_1	$[0, 75]$	$P_{b,2}$	-3.5	$P_{a,2}$	$[-1, 0]$
V	$20\ (\text{m/s})$	a_s	$2\ (\text{m/s}^2)$	ω_s	$25\ (\text{deg/s})$

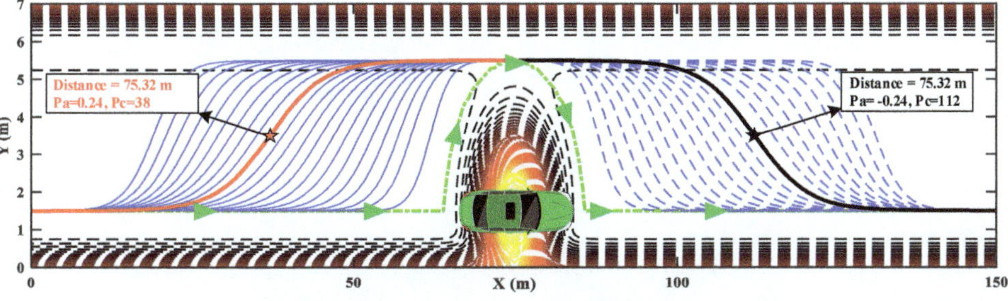

Figure 10. The optimal trajectory with sigmoid curves in a static scenario.

Figure 11a shows the curvature of the optimal trajectory generated using HPFSM. It shows the curvature is continuous, which means the optimal trajectory is drivable. Figure 11b shows the yaw rate and lateral acceleration calculated under the target tracking speed of 20 m/s. It indicates that both the constraints of the lateral acceleration and the yaw rate are effectiveness during the path planning. The results of Figure 11 illustrate that the trajectory can be optimized to satisfy the constraints of vehicle dynamics with the parameters optimization of sigmoid curves.

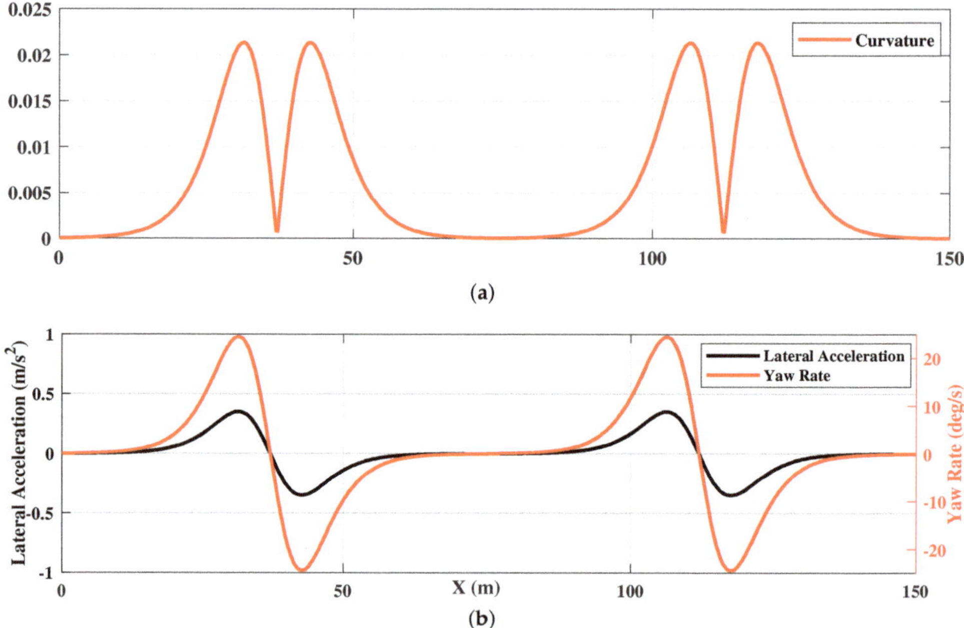

Figure 11. HPFSM path planning with constraints: (**a**) curvature of the planned trajectory; (**b**) yaw rate and lateral acceleration.

4. Verification and Discussion

To further examine and evaluate the proposed approach, a static and a dynamic driving scenario are designed for simulation, respectively. The parked vehicles are considered as the obstacles in the static scenario, and vehicles with short intervals are introduced as the overtaking objects in the dynamic scenario.

4.1. Driving Scenarios for Simulation and Evaluation

A static scenario is with three cars parked on the roadsides as shown in Figure 12. There are two parked cars located in the target lane, while the ego vehicle is approaching to the parked car with a speed of 20 m/s. The positions of the three parked vehicles are ($X_{\mathrm{obs},1} = 80$ m, $Y_{\mathrm{obs},1} = 1.5$ m), ($X_{\mathrm{obs},2} = 180$ m, $Y_{\mathrm{obs},2} = 6.2$ m) and ($X_{\mathrm{obs},3} = 280$ m, $Y_{\mathrm{obs},3} = 1.5$ m), respectively. The constraints of the lateral acceleration and yaw rate are designed within 2 m/s^2 and 25 deg/s, respectively.

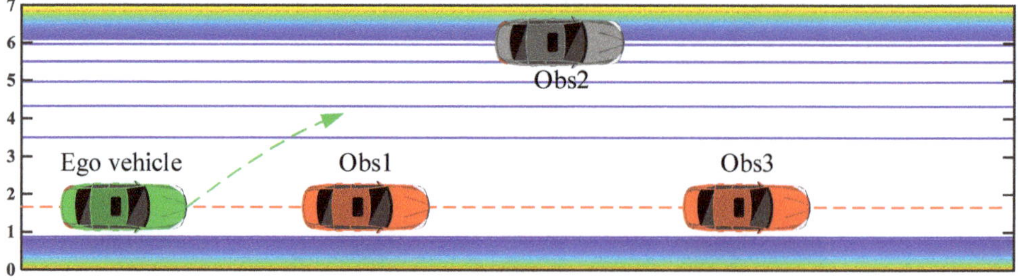

Figure 12. A static driving scenario with three vehicles parked on roadsides.

A dynamic overtaking scenario is designed with three leading vehicles located with short distances, as shown in Figure 13. The red dotted line denotes the target lateral position, i.e., the center line of the target lane. Three leading vehicles are driving with a constant speed ($V_{obs,1} = V_{obs,2} = V_{obs,3} = 15$ m/s) from different initial positions ($X_{obs,1} = 50$ m, $X_{obs,2} = 70$ m, $X_{obs,3} = 85$ m). The initial position and speed of the ego vehicle are set as $X_{ego} = 0$ m and $V_{ego} = 15$ m/s, respectively. The target speed and lateral position of the ego vehicle are set as 20 m/s and 1.75 m, respectively. Meanwhile, the constraints of yaw rate and lateral acceleration are designed within 2 m/s^2 and 25 deg/s, respectively.

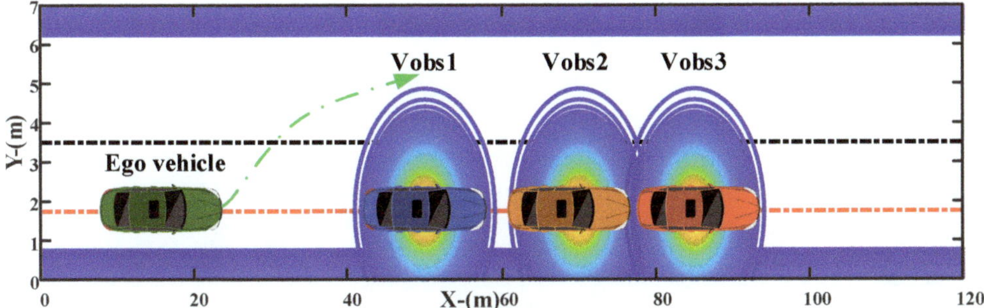

Figure 13. Leading vehicles driving with short interval distance.

4.2. Path Tracking Controller for Validation

Since the main purpose of path planning is to provide an expected reference trajectory for path tracking, it is more meaningful to evaluate the proposed path planning method with the combination of a path tracking controller. Based on these, a linear time-varying model predictive tracking controller (LTV-MPC) [20] is used to evaluate the HPFSM by comparing with PFBM. The 3-DOF bicycle model, including the longitudinal, lateral and yaw directions, is used as the prediction model of the LTV-MPC. The dynamics equations of the 3-DOF dynamics model are presented in Equation (24):

$$\begin{cases} m\left(\dot{v}_x - \omega v_y\right) = F_x \cos \delta \\ m\left(\dot{v}_y + \omega v_x\right) = F_{y,r} + F_{y,f} \cos \delta \\ I_z \dot{\omega} = F_{y,f} L_f \cos \delta - F_{y,r} L_r \end{cases} \tag{24}$$

The motion equations of the vehicle are shown in Equation (25):

$$\begin{cases} \dot{X} = v_x \cos \varphi - v_y \sin \varphi \\ \dot{Y} = v_x \sin \varphi + v_y \cos \varphi \end{cases} \tag{25}$$

where v_x, v_y and ω are the longitudinal velocity, lateral velocity and yaw rate of the vehicle, respectively; X, Y and φ denote the vehicle longitudinal, lateral position and the heading angle; m and F_x represent the vehicle mass and longitudinal force of the front-driving tire; $F_{y,f}$ and $F_{y,r}$ denote the lateral force of front and rear tires; L_f, L_r and I_z represent the front, rear wheelbase and the vehicle inertia around vertical axis, respectively; δ is the steering angle of the front wheel. The relevant vehicle parameters are the same as Table I in [17]. The configuration parameters of the LTV-MPC are presented in Table 3.

Table 3. Parameters of the MPC Controller.

Symbol	Description	Value [Units]
N_p	Prediction horizon	20 [unitless]
N_u	Control variable's number	2 [unitless]
N_s	State variable's number	6 [unitless]
T_s	Sampling period	0.05 [s]
δ_w	Limitation of steering wheel angle	$[-540, 540]\ [°]$
$\Delta\delta_w$	Steering wheel angle rate	$[-5, 5]\ [°]$
F_x	Longitudinal tire force limitation	$[-2000, 2000]\ [N]$
ΔF_x	Tire force rate limitation	$[-50, 50]\ [N]$
Q	Weights matrix of states tracking	$diag([1 \times 10^{-7}, 1 \times 10^2, 1 \times 10^{-7}, 0, 1 \times 10^{-7}, 0])$
R	Weights matrix of control variables	$diag([1 \times 10^{-7}, 1 \times 10^{-5}])$

4.3. Results and Discussion

4.3.1. Static Scenario

The comparisons of trajectories between HPFSM and PFBM are shown in Figure 14a, respectively. The green and red trajectories are the generated paths of PFBM and HPFSM, respectively. It shows that the optimized path by HPFSM is more feasible to be an expected driving trajectory, because the path is smoother than that of PFBM without increasing in length (400.87 m vs. 403.88 m). The shaded part illustrates that the red trajectory is collision-free by comparing the lateral positions of the two paths. The comparison of the curvatures between the two trajectories is shown in Figure 14b, which further illustrates the red path is smoother than the green one.

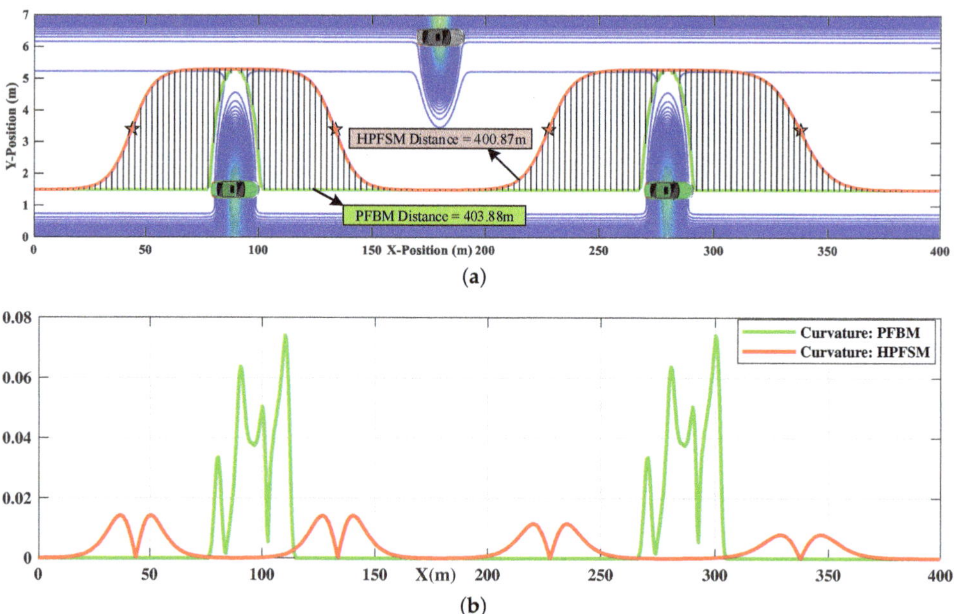

Figure 14. The path comparisons between HPFSM and PFBM in a static scenario: (**a**) Trajectory comparison: HPFSM vs PFBM; (**b**) Curvature comparison: HPFSM vs PFBM.

The planned paths of HPFSM and PFBM are tracked by the LTV-MPC with a target speed of 20 m/s, respectively. The comparisons of the yaw rate and lateral acceleration between the two methods are shown in Figure 15. The comparison of lateral acceleration shows that the instantaneous and average values of HPFSM are smaller than that of PFBM in Figure 15a, which illustrates the ride comfort is improved with HPFSM. Meanwhile, Figure 15b shows the comparison of yaw rate, which illustrates that the yaw rate based on PFBM does not satisfy the designed constraint of 25 deg/s; however, the yaw rate based on HPFSM can be well constrained. This implies that the stability of the ego vehicle is also better while tracking the path of HPFSM.

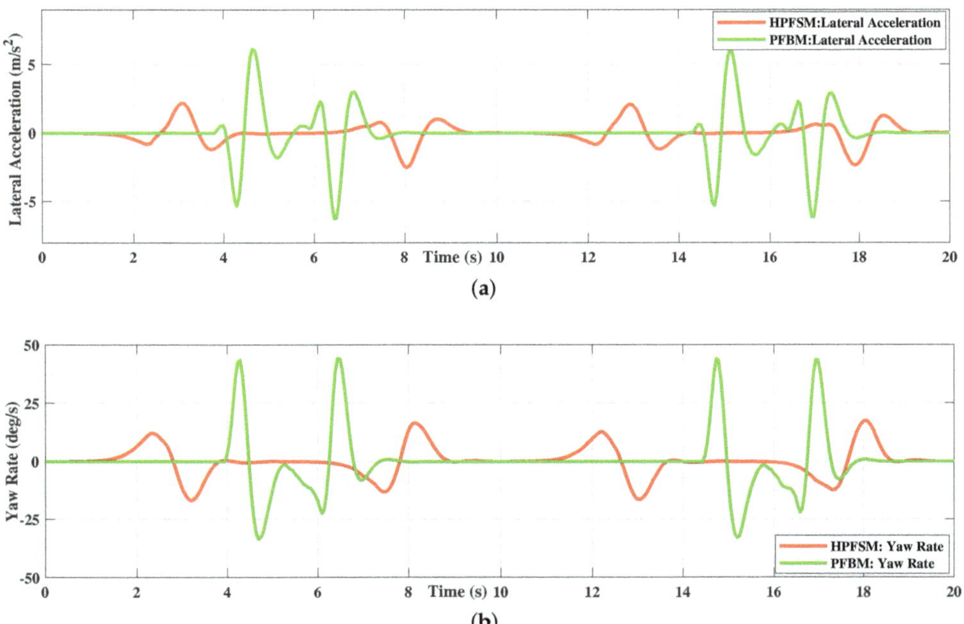

Figure 15. Path tracking comparisons between the two methods in a static scenario: (**a**) LTV-MPC tracking: Lateral acceleration comparison; (**b**) LTV-MPC tracking: Yaw rate comparison.

The improvements with HPFSM in the static scenario is analyzed in Table 4. It shows that the maximum and average lateral accelerations are decreased almost 60% (6.254 m/s² vs. 2.504 m/s²) and 40.6% (0.475 m/s² vs. 0.282 m/s²) comparing to PFBM, respectively. Meanwhile, the yaw rate is also optimized in the maximum and average values, respectively. The maximum and average yaw rates with HPFSM are improved 60.47% (44.17 deg/s vs. 17.459 deg/s) and 28.2% (3.517 deg/s vs. 2.524 deg/s), respectively.

Table 4. Results comparisons in the static scenario.

Symbol	Description	HPFSM	PFBM	$-(\%)$
$a_{y,max}$ (m/s²)	Maximum lateral acceleration	2.504	6.254	59.9
$a_{y,mean}$ (m/s²)	Average lateral acceleration	0.282	0.475	40.6
ω_{max} (deg/s)	Maximum yaw rate	17.459	44.170	60.47
ω_{mean} (deg/s)	Average yaw rate	2.524	3.517	28.2

4.3.2. Dynamic Scenario

The tracking velocities of HPFSM and PFBM are shown in Figure 16a to ensure a consistent speed environment during the overtaking task. The trajectories of the two methods are shown in Figure 16b; they show that the ego vehicle can finish the overtaking task with both of these two methods. However, the trajectory with PFBM shows an sudden fluctuation at the position around $X = 300$ m, which will result in sharp steering maneuvers as shown in Figure 16c. These unexpected steering maneuvers will further affect both the driving safety and stability.

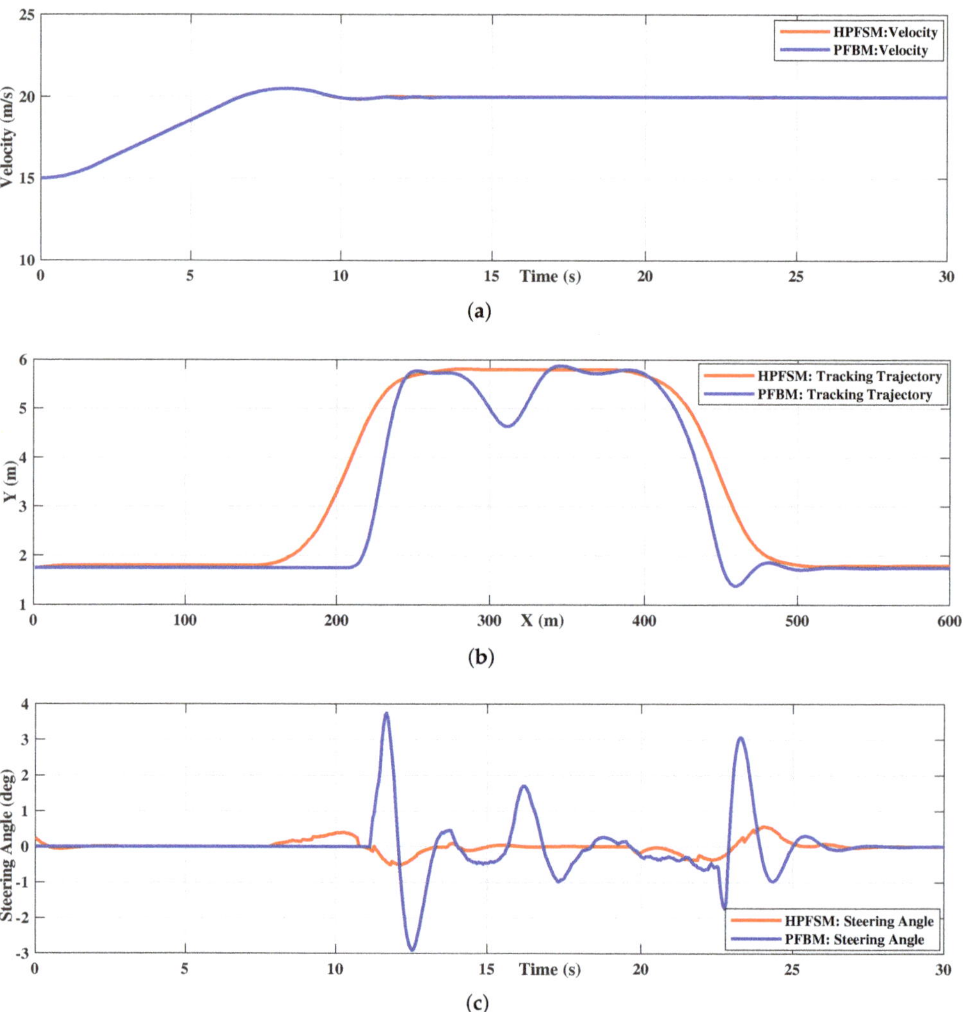

Figure 16. An overtaking scenario with multivehicle distributed in short interval distance: (**a**) HPFSM vs. PFBM: Tracking velocity; (**b**) HPFSM vs. PFBM: Tracking trajectory; (**c**) HPFSM vs. PFBM: Steering angle of front wheel.

The yaw rate and the lateral acceleration of the two methods are compared in Figure 17. The yaw rate is constrained within 10 deg/s while tracking the path of HPFSM; however, the yaw rate is beyond the designed constraint of 25 deg/s while tracking the path of PFBM, as shown in Figure 17a. Meanwhile, the comparison of the lateral acceleration in Figure 17b shows the lateral acceleration based on HPFSM is much smaller than that of PFBM during tracking. These illustrate that both the vehicle stability and the ride comfort of the ego vehicle are improved with the proposed HPFSM comparing to the PFBM. The improvements with HPFSM in the dynamic scenario are shown in Table 5. This indicates that the maximum and average lateral accelerations with HPFSM are decreased 87.8% (2.4 m/s^2 vs. 0.29 m/s^2) and 83.9% (0.18 m/s^2 vs. 0.029 m/s^2), respectively. Meanwhile, the yaw rate is also optimized in the maximum and average values compared to the PFBM. The maximum and average yaw rates are improved 82.8% (20.4 deg/s vs. 3.5 deg/s) and 72.2% (1.7 deg/s vs. 0.47 deg/s), respectively.

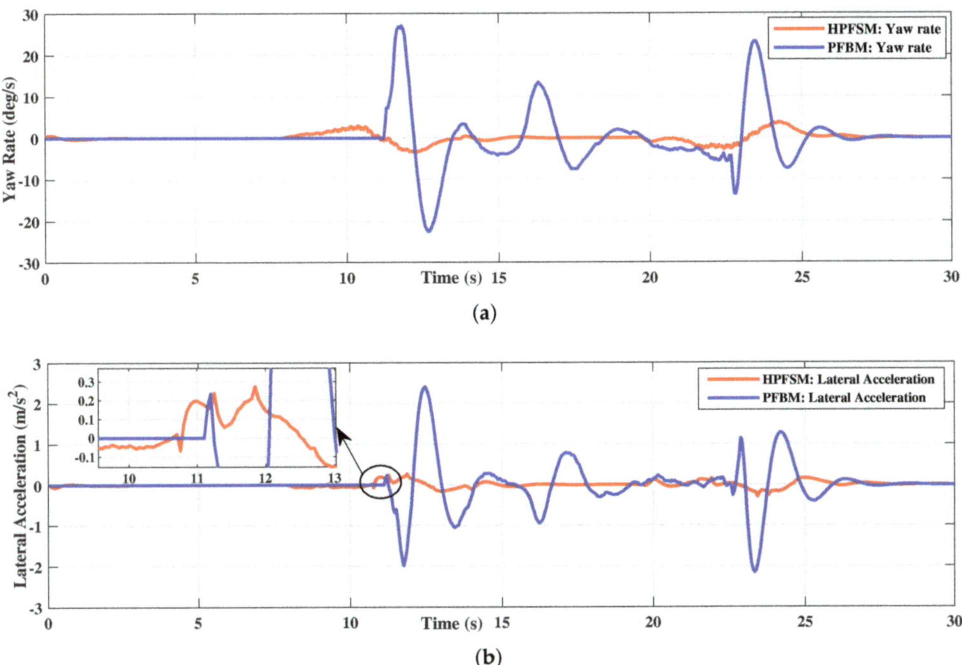

Figure 17. The comparisons of yaw rate and lateral acceleration: (**a**) HPFSM vs. PFBM: Yaw rate; (**b**) HPFSM vs. PFBM: Lateral acceleration.

Table 5. Results comparisons in the dynamic scenario.

Symbol	Description	HPFSM	PFBM	$-(\%)$
$a_{y,max}$ (m/s²)	Maximum lateral acceleration	0.293	2.410	87.8
$a_{y,mean}$ (m/s²)	Average lateral acceleration	0.029	0.180	83.9
ω_{max} (deg/s)	Maximum yaw rate	3.508	20.430	82.8
ω_{mean} (deg/s)	Average yaw rate	0.477	1.713	72.2

5. Conclusions

A hybrid path planning is proposed to achieve an expected path generation and to improve the vehicle stability and the ride comfort during autonomous driving by combining the potential field with the sigmoid curve. The collision avoidance and the vehicle dynamics are considered to obtain the shortest collision-free trajectory composed by sigmoid curves. The multiobstacle static and dynamic scenarios are designed to examine the effectiveness of HPFSM, respectively. To evaluate the performance of autonomous driving with HPFSM, an LTV-MPC is used to track the planned paths of HPFSM and PFBM, respectively. The simulation results of the static scenario show that the maximum and average lateral accelerations are decreased 60% and 40%, and the maximum and average yaw rates are decreased almost 60.47% and 28.2%, respectively. The results of the simulated dynamic scenario show the same trend as the static scenario with a decrease of almost 80% in the indexes of both the lateral acceleration and the yaw rate. However, these improvements are achieved on the basis of the analysis of the simulation results; the figures are likely to be more modest in the practical application. These simulation results indicate that the vehicle stability and the ride comfort are well improved with the proposed method during autonomous driving. How the local minima problem can be completely or sufficiently avoided in more complex and unknown driving scenarios with more traffic participants is still a challenging task and should be further addressed in the future work. Meanwhile, our future work will present experimental applications of the proposed method under real driving scenarios.

Author Contributions: The presented work was carried out in collaboration with all authors. B.L. conceived the idea, realized the algorithms and wrote the manuscript. H.H. and D.C. reviewed the manuscript and provided important suggestions. H.Y., H.W. and G.L. gave some useful advice to organize the work. M.S. organized the figures. All authors have read and agreed to the published version of the manuscript.

Funding: This work was supported by the National key R&D program of China (2018YFB0106200) in part, and by the National Science Foundation of China Joint Fund Project (U1864205).

Conflicts of Interest: The authors declare no conflict of interest.

Abbreviations

The following abbreviations are used in this manuscript:

AVs	Autonomous Vehicles
PF	Potential Field
OPF	Obstacle Potential Field
PFBM	Potential Field-based Path Planning Method
HPFSM	Hybrid Potential Field Sigmoid Curve Method
LTV-MPC	Linear Time-varying Model Predictive Tracking Controller

References

1. Rimon, E.; Koditschek, D.E. Exact robot navigation using artificial potential functions. *IEEE Trans. Robot. Autom.* **1992**, *8*, 501–518. [CrossRef]
2. Yu, X.; Liu, L. Distributed formation control of nonholonomic vehicles subject to velocity constraints. *IEEE Trans. Ind. Electron.* **2016**, *63*, 1289–1298. [CrossRef]
3. Huang, Y.; Wang, H.; Khajepour, A.; Ding, H.; Yuan, K.; Qin, Y. A Novel Local Motion Planning Framework for Autonomous Vehicles Based on Resistance Network and Model Predictive Control. *IEEE Trans. Veh. Technol.* **2020**, *69*, 55–66. [CrossRef]
4. Wang, H.; Huang, Y.; Khajepour, A.; Zhang, Y.; Rasekhipour, Y.; Cao, D. Crash mitigation in motion planning for autonomous vehicles. *IEEE Trans. Intell. Transp. Syst.* **2019**, *20*, 3313–3323. [CrossRef]
5. González, D.; Pérez, J.; Milanés, V.; Nashashibi, F. A review of motion planning techniques for automated vehicles. *IEEE Trans. Intell. Transp. Syst.* **2016**, *17*, 1135–1145. [CrossRef]
6. Anderson, S.J.; Karumanchi, S.B.; Iagnemma, K. Constraint-based planning and control for safe, semi-autonomous operation of vehicles. In Proceedings of the 2012 IEEE Intelligent Vehicles Symposium, Alcala de Henares, Spain, 3–7 June 2012; pp. 383–388.
7. Hwang, J.Y.; Kim, J.S.; Lim, S.S.; Park, K.H. A fast path planning by path graph optimization. *IEEE Trans. Syst. Man Cybern. Part A Syst. Hum.* **2003**, *33*, 121–128. [CrossRef]
8. Ziegler, J.; Werling, M.; Schröder, J. Navigating car-like robots in unstructured environments using an obstacle sensitive cost function. In Proceedings of the 2008 IEEE Intelligent Vehicles Symposium, Eindhoven, The Netherlands, 4–6 June 2008; pp. 787–791.
9. Ziegler, J.; Stiller, C. Spatiotemporal state lattices for fast trajectory planning in dynamic on-road driving scenarios. In Proceedings of the IEEE/RSJ International Conference on Intelligent Robots and Systems, St. Louis, MO, USA, 11–15 October 2009; pp. 1879–1884.
10. Chen, L.; Shan, Y.; Tian, W.; li, B.; Cao, D. A fast and efficient double-tree RRT*-like sampling-based planner applying on mobile robotic systems. *IEEE/Asme Trans. Mechatronics* **2018**, *23*, 2568–2578. [CrossRef]
11. Kavraki, L.E.; Švestka, P.; Latombe, J.C.; Overmars, M.H. Probabilistic roadmaps for path planning in high-dimensional configuration spaces. *IEEE Trans. Robot. Autom.* **1996**, *12*, 566–580. [CrossRef]

12. Zhang, Y.; Chen, H.; Waslander, S.L.; Gong, J.; Xiong, G.; Yang, T.; Liu, K. Hybrid trajectory planning for autonomous driving in highly constrained environments. *IEEE Access* **2018**, *6*, 32800–32819. [CrossRef]

13. Mercy, T.; Van Parys, R.; Pipeleers, G. Spline-based motion planning for autonomous guided vehicles in a dynamic environment. *IEEE Trans. Control Syst. Technol.* **2018**, *26*, 2182–2189. [CrossRef]

14. Brezak, M.; Petrovic, I. Real-time approximation of clothoids with bounded error for path planning applications. *IEEE Trans. Robot.* **2014**, *30*, 507–515. [CrossRef]

15. Li, X.; Sun, Z.; Cao, D.; He, Z.; Zhu, Q. Real-Time Trajectory Planning for Autonomous Urban Driving: Framework, Algorithms, and Verifications. *IEEE/ASME Trans. Mechatronics* **2016**, *21*, 740–753. [CrossRef]

16. Khatib, O., The Potential Field Approach And Operational Space Formulation In Robot Control. In *Adaptive and Learning Systems: Theory and Applications*; Narendra, K.S., Ed.; Springer: Boston, MA, USA, 1986; pp. 367–377.

17. Ji, J.; Khajepour, A.; Melek, W.W.; Huang, Y. Path planning and tracking for vehicle collision avoidance based on model predictive control with multiconstraints. *IEEE Trans. Veh. Technol.* **2017**, *66*, 952–964. [CrossRef]

18. Rasekhipour, Y.; Khajepour, A.; Chen, S.K.; Litkouhi, B. A potential field-based model predictive path-planning controller for autonomous road vehicles. *IEEE Trans. Intell. Transp. Syst.* **2017**, *18*, 1255–1267. [CrossRef]

19. Krogh, B.H.; Thorpe, C.E. Integrated path planning and dynamic steering control for autonomous vehicles. In Proceedings of the IEEE International Conference on Robotics and Automation, San Francisco, CA, USA, 7–10 April 1986; pp. 1664–1669.

20. Li, M.; Song, X.; Cao, H.; Wang, J.; Huang, Y.; Hu, C.; Wang, H. Shared control with a novel dynamic authority allocation strategy based on game theory and driving safety field. *Mech. Syst. Signal Proc.* **2019**, *124*, 199–216. [CrossRef]

21. Park, D.H.; Hoffmann, H.; Pastor, P.; Schaal, S. Movement reproduction and obstacle avoidance with dynamic movement primitives and potential fields. In Proceedings of the IEEE-RAS International Conference on Humanoid Robots, Daejeon, Korea, 1–3 December 2008; pp. 91–98.

22. Li, G.; Yang, Y.; Zhang, T.; Qu, X.; Cao, D.; Cheng, B.; Li, K. Risk assessment based collision avoidance decision-making for autonomous vehicles in multi-scenarios. *Transp. Res. Pt. C Emerg. Technol.* **2021**, *122*, 102820.

23. Lee, J.; Nam, Y.; Hong, S. Random force based algorithm for local minima escape of potential field method. In Proceedings of the 2010 11th International Conference on Control Automation Robotics and Vision, Singapore, 7–10 December 2010; pp. 827–832.

24. G. Li, Y. Yang, X.Q. Deep learning approaches on pedestrian detection in hazy weather. *IEEE Trans. Ind. Electron.* **2020**, *67*, 8889–8899. [CrossRef]

25. Montiel, O.; Orozco-Rosas, U.; Sepúlveda, R. Path planning for mobile robots using bacterial potential field for avoiding static and dynamic obstacles. *Expert Syst. Appl.* **2015**, *42*, 5177–5191. [CrossRef]

26. Kovács, B.; Szayer, G.; Tajti, F.; Burdelis, M.; Korondi, P. A novel potential field method for path planning of mobile robots by adapting animal motion attributes. *Robot. Auton. Syst.* **2016**, *82*, 24–34. [CrossRef]

27. Hwang, Y.K.; Ahuja, N. A potential field approach to path planning. *IEEE Trans. Robot. Autom.* **1992**, *8*, 23–32. [CrossRef]

28. Elbanhawi, M.; Simic, M.; Jazar, R. Randomized bidirectional B-spline parameterization motion planning. *IEEE Trans. Intell. Transp. Syst.* **2016**, *17*, 406–419. [CrossRef]

29. Resende, P.; Nashashibi, F. Real-time dynamic trajectory planning for highly automated driving in highways. In Proceedings of the International IEEE Conference on Intelligent Transportation Systems, Funchal, Portugal, 19–22 September 2010; pp. 653–658.

30. Finney, D.J. *Probit Analysis: A Statistical Treatment of the Sigmoid Response Curve*; Cambridge University Press: Cambridge, UK, 1952.

31. You, F.; Zhang, R.; Lie, G.; Wang, H.; Wen, H.; Xu, J. Trajectory planning and tracking control for autonomous lane change maneuver based on the cooperative vehicle infrastructure system. *Expert Syst. Appl.* **2015**, *42*, 5932–5946. [CrossRef]

32. Li, S.E.; Li, G.; Yu, J.; Liu, C.; Cheng, B.; Wang, J.; Li, K. Kalman filter-based tracking of moving objects using linear ultrasonic sensor array for road vehicles. *Mech. Syst. Signal Proc.* **2017**, *98*, 173–189. [CrossRef]

33. Bialkowski, J.; Otte, M.; Karaman, S.; Frazzoli, E. Efficient collision checking in sampling-based motion planning via safety certificates. *Int. J. Robot. Res.* **2016**, *35*, 767–796. [CrossRef]

34. Liu, W.; He, H.; Sun, F.; Lv, J. Integrated chassis control for a three-axle electric bus with distributed driving motors and active rear steering system. *Veh. Syst. Dyn.* **2017**, *55*, 601–625. [CrossRef]
35. Ataei, M.; Khajepour, A.; Jeon, S. Reconfigurable Integrated Stability Control for Four- and Three-wheeled Urban Vehicles With Flexible Combinations of Actuation Systems. *IEEE/ASME Trans. Mechatronics* **2018**, *23*, 2031–2041. [CrossRef]

Publisher's Note: MDPI stays neutral with regard to jurisdictional claims in published maps and institutional affiliations.

 © 2020 by the authors. Licensee MDPI, Basel, Switzerland. This article is an open access article distributed under the terms and conditions of the Creative Commons Attribution (CC BY) license (http://creativecommons.org/licenses/by/4.0/).

Article

A Novel Path Planning Algorithm for Truck Platooning Using V2V Communication

Yongki Lee [1], Taewon Ahn [1], Chanhwa Lee [2], Sangjun Kim [2] and Kihong Park [1,*]

[1] Graduate School of Automotive Engineering, Kookmin University, Seoul 02707, Korea;
    yklee0731@kookmin.ac.kr (Y.L.); atw6754@gmail.com (T.A.)
[2] Research and Development Division, Hyundai Motor Company, Gyeonggi-do 18280, Korea;
    chanhwa.lee@gmail.com (C.L.); rixoest@hyundai.com (S.K.)
*   Correspondence: kpark@kookmin.ac.kr

Received: 18 November 2020; Accepted: 7 December 2020; Published: 8 December 2020

Abstract: In truck platooning, the leading vehicle is driven manually, and the following vehicles run by autonomous driving, with the short inter-vehicle distance between trucks. To successfully perform platooning in various situations, each truck must maintain dynamic stability, and furthermore, the whole system must maintain string stability. Due to the short front-view range, however, the following vehicles' path planning capabilities become significantly impaired. In addition, in platooning with articulated cargo trucks, the off-tracking phenomenon occurring on a curved road makes it hard for the following vehicle to track the trajectory of the preceding truck. In addition, without knowledge of the global coordinate system, it is difficult to correlate the local coordinate systems that each truck relies on for sensing environment and dynamic signals. In this paper, in order to solve these problems, a path planning algorithm for platooning of articulated cargo trucks has been developed. Using the Kalman filter, V2V (Vehicle-to-Vehicle) communication, and a novel update-and-conversion method, each following vehicle can accurately compute the trajectory of the leading vehicle's front part for using it as a target path. The path planning algorithm of this paper was validated by simulations on severe driving scenarios and by tests on an actual road. The results demonstrated that the algorithm could provide lateral string stability and robustness for truck platooning.

Keywords: TROOP; truck platooning; path planning; kalman filter; V2V communication; string stability; off-tracking; articulated cargo trucks; kabsch algorithm

---

1. Introduction

Truck platooning refers to a form in which a number of trucks run as a fleet with short inter-vehicle distance using V2V (Vehicle-to-Vehicle) communication. Figure 1 shows the architecture of a truck platooning system. The leading vehicle (LV) is driven manually by an experienced driver, and the following vehicles (FVs) run by autonomous driving. The following vehicle (FV) uses environment sensors such as radar and camera to perceive vehicles and lanes ahead and perform autonomous driving by longitudinal and lateral vehicle control. The autonomous driving algorithm of the FV does not rely on GPS since the vehicle cannot receive correct GPS signals in some conditions, like when driving through tunnels. In the case of platooning of large cargo trucks, the length of the fleet can easily reach 100 m. Thus, the number of trucks in one platoon is usually limited to 3 or 4, considering the safety of the nearby vehicles.

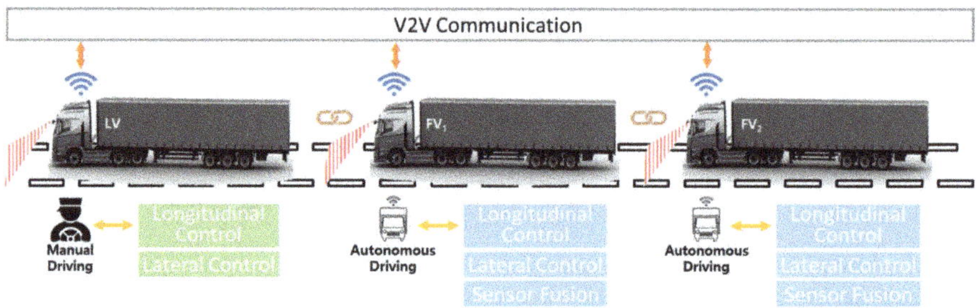

Figure 1. Overall architecture of truck platooning system.

Figure 1. Overall architecture of truck platooning system.

Many studies are being conducted worldwide on truck platooning since it can bring improvement in driving safety, driver convenience, traffic throughput, fuel economy, and emission reduction. In Europe, truck manufacturers have been establishing consortiums for collaborative research on truck platooning. In 2016, they hosted the European Truck Platooning Challenge [1], and in 2018, they launched a large-scale inter-country project ENSEMBLE for multi-brand truck platooning [2]. In the US, the legislation necessary for platooning is actively being prepared to spur truck platooning to practical use in the near future [3].

In Korea, a first government project on truck platooning-TROOP (TRuck platOOning Project)-was launched in 2018 [4]. It will last until the end of 2021, with the final goal of developing the most advanced truck platooning system covering not just the control technologies but also the operational and management technologies based on the C-ITS services that the government has already established.

Figure 2 shows the photo of the trucks developed in the TROOP project. They are an articulated cargo truck with two bodies-tractor and trailer-linked by a kingpin where all the longitudinal and lateral controls are performed only at the tractor. The total length of each truck is 16.66 m. In the TROOP project, the ODD (Operational Design Domain) of truck platooning includes highway driving with a radius of 460R or more at a design speed of 90 kph. This paper introduces the research on the path planning algorithm of the FV, which has been developed as part of the TROOP project.

Figure 2. Photo of trucks in the TRuck platOOning Project (TROOP) Project. (Pictured by Hyundai Motor Company).

The longitudinal control of the FV aims at maintaining a short distance to the front truck, and this is basically done by adopting the well-established ACC (Adaptive Cruise Control) algorithm, which relies on radar. Short inter-vehicle distance is important in truck platooning since it gives fuel economy to the FV by reducing aerodynamic drag. In TROOP, the final target is 0.5 s time-gap at 90 kph, or 12.5 m, which is shorter than the truck's length. The FV also uses V2V communication, thus, it can immediately respond to critical situations such as when the LV makes sudden braking.

The lateral control of the FV aims at following the driving path of the LV while staying on its own lane. The FV uses a camera to perform lane keeping, but the well-established LKS (Lane Keeping System) algorithm [5,6] cannot be used since the front-view range of the camera is severely limited by the preceding truck [7]. The path following control of the FV sets as its target the trajectory that the LV has gone through. Therefore, in truck platooning, a higher level of control technology is required than the lateral control method used in general autonomous driving. There are two main methods for lateral control of platooning, 'Direct vehicle-following' and 'Vehicle path-following' [8].

In the direct vehicle-following method, the following vehicle directly follows the preceding vehicle by calculating the steering angle based on a geometrical principle using the relative longitudinal and lateral distance with the preceding vehicle [9,10]. Alternatively, using the relative position and relative angle between the subject vehicle CG (Center of Gravity) and the rear center of the preceding vehicle, a virtual curved path to the rear of the preceding vehicle can be computed [11,12]. However, since these methods use the relative position information of the rear of the preceding vehicle, not the trajectory of the steering wheel of the preceding vehicle, there may be a problem of driving inside the actual trajectory of the preceding vehicle during turning. In addition, when driving on a highway with a small curvature, since the relative yaw angle with the preceding vehicle is quite small, the reliability of the virtual curved path for following the preceding vehicle cannot be guaranteed if the accuracy of perception is low or the resolution of the measured value is small.

On the other hand, vehicle path-following is a method of following the trajectory of the preceding vehicle. The trajectory of the preceding vehicle can be obtained using motion parameters of the subject vehicle and storing the position coordinates of the rear of the preceding vehicle [8,13,14]. As the look-ahead distance within the trajectory of the preceding vehicle can be controlled, the performance of path tracking can be improved. However, there is a problem that enough look-ahead distance cannot be obtained at high speed due to a short inter-vehicle distance during platooning. In case of a semi-trailer truck, off-tracking, which is the difference in the path between the tractor's steering axle and the trailer's rear bumper, occurs during turning, causing a tracking error when following the preceding vehicle.

Figure 3 shows the off-tracking phenomenon, which occurs when a tractor-and-trailer type truck runs on a curved road. It displays different patterns at low speed and high speed for the same truck. At low speed, the trajectory of the rear bumper of the trailer is formed inside the trajectory of the tractor, while at high speed, the trajectory of the trailer travels outward than that of the tractor due to the increase in lateral acceleration [15]. Off-tracking is a major factor that harms the stability of the lateral dynamics of the platoon, and the stability gets worse as it propagates towards the tail of the platoon.

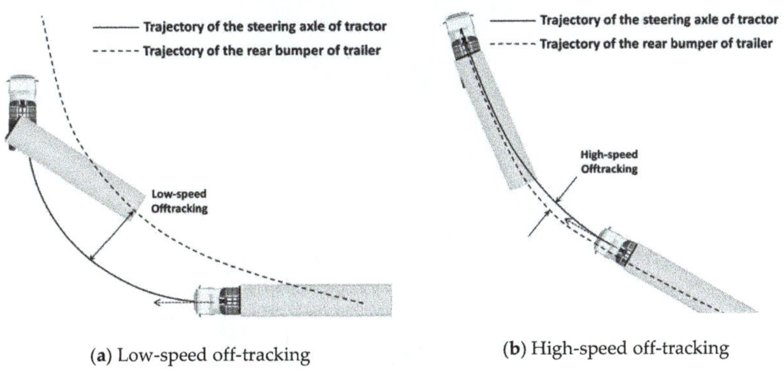

(**a**) Low-speed off-tracking

(**b**) High-speed off-tracking

Figure 3. Schematic of off-tracking [15].

To overcome the off-tracking problem, the FV needs to use the trajectory of the tractor—not the trailer—of the preceding vehicle for its own target path [16]. However, it is not possible for the FV to perceive the position of the preceding vehicle's tractor only by camera. A study has been proposed to take advantage of the curvature of the trajectory of the preceding vehicle's trailer position [17], but it is effective only when the yaw angle between the two vehicles is small. Another study has been proposed to use the DRTK (Dynamic Based Real-Time Kinematic) and V2V to access the global position of the tractor of the preceding vehicle [18,19], but platooning trucks generally do not employ GPS since GPS signal cannot be received in conditions such as when driving through tunnels.

In this study, a path planning algorithm has been developed for lateral control of the FV in truck platooning formed by articulated cargo trucks. The algorithm uses camera/radar fusion data, IVN (In-Vehicle Network) chassis signals, and V2V communication. Using the Kalman filter and a novel coordinate conversion method, the FV is now able to figure out the trajectory of the LV's tractor position to use as its own target path. The algorithm of this paper was validated by simulations on severe driving scenarios and by tests on an actual road. The results demonstrated that the algorithm can provide lateral string stability and robustness in truck platooning.

In addition, the proposed path planning algorithm can be expanded by generating a target path in the interchange and junction of the highway with a small turning radius and can be applied to various specially equipped vehicles as well as trucks. Furthermore, since the target path is generated based on the trajectory of the preceding vehicle using V2V communication, it can be applied even on an unpaved road without a lane. As a result, this study is expected to improve stability and fuel economy through platooning by applying it to various specially equipped vehicles in various road environments as well as large cargo trucks.

2. System Architecture

Figure 4 shows Hyundai Xcient 6 × 2 tractor used in the TROOP project. The actuating mechanisms for steering, braking, and acceleration have been modified to enable autonomous driving. A mono camera and a radar are mounted on the dashboard and front bumper, respectively, to perceive the center point of the preceding vehicle's bumper. The V2V module uses dual antennas, and they are installed inside the left and right side-mirrors to minimize the area of communication blind spot. Computation of the proposed path planning algorithm is carried out using MicroAutoBox II, which also serves as a CPU (Central Processing Unit) for implementing platooning control logics.

Figure 4. Hyundai Xcient 6 × 2 tractor.

Figure 5 shows the specification of the truck used in the TROOP project. The total length of the vehicle is 16.66 m. The tractor and trailer are connected by a kingpin, but all the longitudinal and lateral controls are performed only at the tractor.

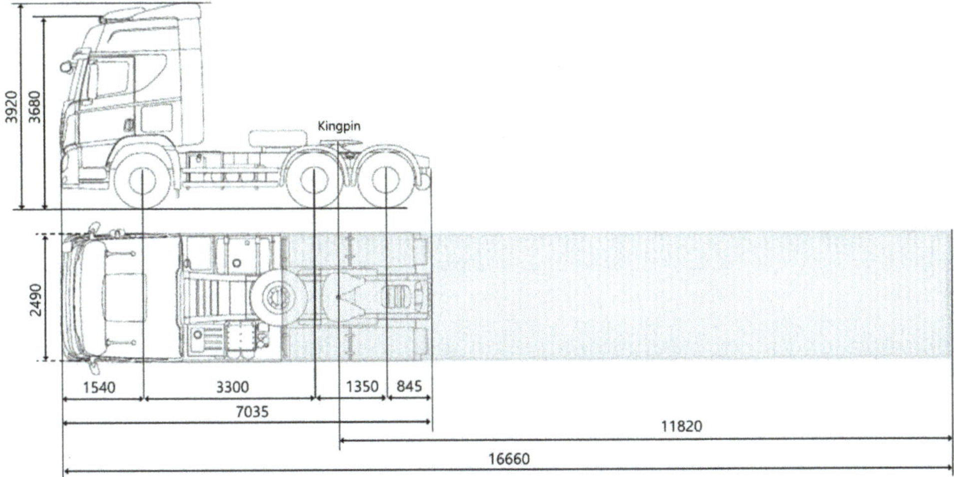

Figure 5. Specification of whole truck (unit: mm).

In the TROOP project, the truck platooning system was developed in three parts: Platooning operation control system, longitudinal control system, and lateral control system. The platooning operation control system performs the function of join, maintain, leave, and gap change of the platoon vehicle. In this paper, we only deal with the path planning method for lateral control, not the platooning operation system and the longitudinal control system.

Figure 6 shows the overall architecture of the truck platooning lateral controller. All vehicles participating in the fleet perform V2V communication among each other using DSRC/802.11p WAVE (Wireless Access in Vehicular Environment) protocol [20].

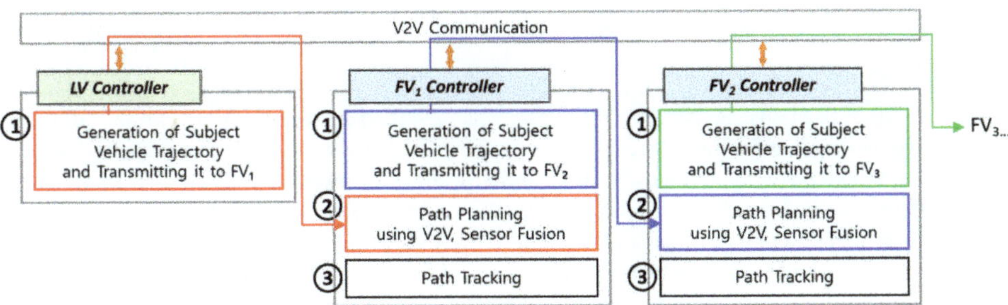

Figure 6. Overall architecture of truck platooning lateral controller.

In Figure 6, the LV controller creates its driven trajectory (①) and transmits it to the FV via V2V communication. Using the LV's trajectory, the FV performs path planning (②), i.e., calculates its own target path to follow. Finally, the FV implements path tracking control to follow the target path (③). In the same way, target paths are created between any adjacent FVs, so in essence, all FVs can follow the trajectory of the LV. Although the path tracking algorithm was developed as well in the TROOP project, this paper will cover only the path planning algorithm.

In order for the FV to create its own target path using the proposed path planning algorithm, longitudinal speed, lateral speed, yaw rate, and kingpin angle are required. Among them, vehicle speed, yaw rate, and kingpin angle can be measured, but their values are vulnerable to sensor noise. In addition, lateral speed cannot be directly measured. In order to solve this problem, the Kalman filter was designed in this study, and its details will be covered in Chapter 3.

Since all signals measured by each truck are measured in its own coordinate system, the driving trajectory of the preceding vehicle received via V2V communication must be converted to fit the local coordinate system of the recipient truck. This is also necessary since truck platooning in the TROOP project does not rely on any global positioning equipment. To solve this problem, a novel point matching method was developed in this study, and its details will be covered in Chapter 4.

If V2V communication is disconnected during platooning, platooning operation, and longitudinal/lateral control cannot be performed. The supervisor controller cancels platooning, and the control mode of each vehicle is changed to the independent autonomous driving mode, and the situation is notified to the driver. For example, the longitudinal control mode is changed to the ACC mode and lateral control mode to LKS. In this paper, we are dealing with the path planning method in the situation where V2V communication is operating normally.

3. Generation of Subject Vehicle Trajectory

Described in this chapter is how each truck creates its own driving trajectory, and the overall architecture is shown in Figure 7. Explanation in this chapter may be based on the LV, but the same method equally applies to all FVs.

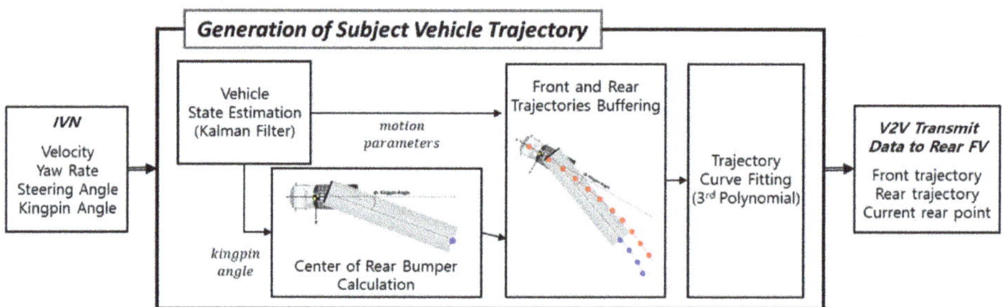

Figure 7. Architecture of subject vehicle trajectory generation.

3.1. Vehicle State Estimation by Kalman Filter

This section explains the Kalman filter that was designed in this study to estimate the state variables needed to generate the subject vehicle trajectory. In previous studies, since it is impossible to measure the lateral speed of a vehicle, only general highway driving scenarios with a small lateral speed were considered, and the lateral speed was assumed to be zero [8,13,14]. However, the large cargo truck, which is the target vehicle of this study, has sensitive dynamic characteristics depending on the load weight and the road environment, and reliability of the buffered trajectory is important to consider not only driving within a lane but also a lane change scenario. Therefore, it is necessary to generate a trajectory more accurately in consideration of the lateral speed of the vehicle. There is a method of using a kinematic model of a vehicle [21], but as described in Chapter 1: Introduction, the kinematic model cannot represent the off-tracking characteristics of a truck. Thus, a 3 DOF (Degrees of Freedom) articulated vehicle model was selected to represent the dynamic characteristics of a truck properly. Figure 8 shows the 3 DOF articulated vehicle model from which the Kalman filter has been built. All variables and parameters of the model are defined with respect to the local coordinate system of each truck, which has its origin at the tractor CG, x-axis facing front, and y-axis facing left.

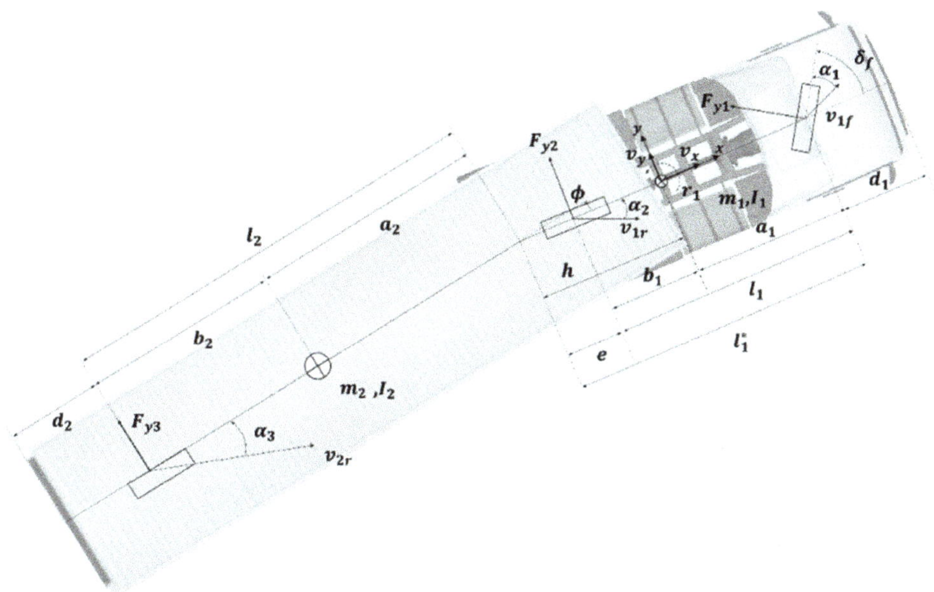

Figure 8. 3 degrees of freedom articulated vehicle model.

The articulated vehicle model in Figure 8 is represented by the state vector $\left[v_{y1},\gamma_1,\dot\phi,\phi\right]^T$. Here, v_{y1} is the lateral speed of the tractor, γ_1 is the yaw rate of the tractor, and $\dot\phi$ and ϕ are the angular velocity and angle of the kingpin, respectively. Equation (1) shows the equations of motion of this articulated vehicle model.

$$
\begin{aligned}
&(m_1+m_2)\left(\dot v_{y1}+v_x\gamma_1\right)-m_2(h_1+a_2)\dot\gamma_1-m_2a_2\ddot\phi\\
&=-\frac{1}{v_x}\left[Cv_{y1}+\left\{C_{s_1}-C_3(h_1+a_2+b_2)\right\}\gamma_1-C_3(a_2+b_2)\dot\phi\right]+C_1\delta_f\\[4pt]
&-h_1m_2\left(\dot v_{y1}+v_x\gamma_1\right)+\{I_1+m_2h_1(h_1+a_2)\}\dot\gamma_1+m_2h_1a_2\ddot\phi\\
&=-\frac{1}{v_x}\left[C_{s_1}v_{y1}+\left\{C_{q_1^2}+C_3h_1(h_1+a_2+b_2)\right\}\gamma_1+C_3h_1(a_2+b_2)\dot\phi\right]+C_1a_1\delta_f\\[4pt]
&-m_2a_2\left(\dot v_{y1}+v_x\gamma_1\right)+\{I_2+m_2a_2(h_1+a_2)\}\dot\gamma_1+\left(I_2+m_2a_2^2\right)\ddot\phi\\
&=-\frac{1}{v_x}\left[-C_3(a_2+b_2)v_{y1}+\{C_3(a_2+b_2)(h_1+a_2+b_2)\}\gamma_1+C_3(a_2+b_2)^2\left(\dot\phi+v_x\phi\right)\right]
\end{aligned}
\tag{1}
$$

In Equation (1), the subscripts $()_1$ and $()_2$ are used to denote the tractor and the trailer, respectively; m is the vehicle mass; I is the yaw moment of inertia; l is the wheelbase; a is the distance from CG to the front axle and b the distance from CG to the rear axle; h is the distance between the tractor's CG and the kingpin; e is the distance between the tractor's rear axle and the kingpin; and l^* is the distance between the tractor's front axle and the kingpin; v_x is the longitudinal vehicle speed; v_y is the lateral vehicle speed; and γ is the yaw rate. C_i, α_i and F_{y_i} are the tire cornering stiffness, the wheel slip angle, and the lateral force of the ith wheel axle, respectively; and δ_f is the front steer angle.

Equation (1) can be made into a matrix-type state equation as below.

$$
\dot X = M^{-1}A\,X + M^{-1}B\,u
$$
$$
\text{where } X=\left[v_{y1},\gamma_1,\dot\phi,\phi\right]^T,\, u=\delta_f
\tag{2}
$$

In the above, X is the state and u is the input, which is the front steer angle of the tractor. The parameter matrices in Equation (2) are defined as below.

$$M = \begin{bmatrix} m_1 + m_2 & -m_2(h_1 + a_2) & -m_2 a_2 & 0 \\ -m_2 h_1 & I_1 + m_2 h_1(h_1 + a_2) & m_2 h_1 a_2 & 0 \\ -m_2 a_2 & I_2 + m_2 a_2(h_1 + a_2) & I_2 + m_2 a_2^2 & 0 \\ 0 & 0 & 0 & 1 \end{bmatrix}$$

$$A = -\frac{1}{v_x} \begin{bmatrix} C + C_3 & C_{S_1} - C_3(h_1 + l_2) + (m_1 + m_2)v_x^2 & -C_3 l_2 & -C_3 v_x \\ C_{S_1} - C_3 h_1 & C_{q_1^2} + C_3 h_1(h_1 + l_2) - m_2 h_1 v_x^2 & C_3 h_1 l_2 & C_3 h_1 v_x \\ -C_3 l_2 & C_3 l_2(h_1 + l_2) - m_2 a_2 v_x^2 & C_3 l_2^2 & C_3 l_2 v_x \\ 0 & 0 & -v_x & 0 \end{bmatrix} \tag{3}$$

$$B = \begin{bmatrix} C_1 \\ a_1 C_1 \\ 0 \\ 0 \end{bmatrix}$$

where $C = C_1 + C_2, C_{s1} = a_1 C_1 - b_1 C_2, C_{q_1^2} = a_1^2 C_1 + b_1^2 C_2$

The Kalman filter of this study has been designed to estimate the state every 10 ms, which is equal to the CAN communication period in each vehicle. To do this, Equation (2) was converted into its discrete-time form as in Equation (4).

$$X_{k+1} = A_d \cdot X_k + B_d \cdot \delta_f \tag{4}$$

where

$$A_d = \left(I + \frac{\Delta t}{2} M^{-1} A\right)\left(I - \frac{\Delta t}{2} M^{-1} A\right)^{-1}, B_d = \Delta t \cdot M^{-1} B \text{ with } \Delta t = 0.01$$

In this study, the yaw rate and the kingpin angle were measured and used as input to the Kalman filter. Equation (5) shows the measurement model for the Kalman filter where z_k is the measurement variable.

$$z_k = H \cdot X_k$$

$$\text{where } H = \begin{bmatrix} 0 & 1 & 0 & 0 \\ 0 & 0 & 0 & 1 \end{bmatrix} \tag{5}$$

The Kalman filter operates by repeating a series of two stages: Prediction and update [22]. Using the system model, it predicts the state variables, compensates for the difference between the measured variables and their predicted values, and outputs a new estimation of the state variables

In the prediction stage, the predicted state estimate $\hat{x}_k^-$ is computed together with the predicted error covariance P_k^- by the following equation.

$$\hat{x}_k^- = A_d \hat{x}_{k-1} + B_d u_{k-1}$$
$$P_k^- = A_d P_{k-1}^+ A_d^T + Q \tag{6}$$

In Equation (6), the overstrike $\hat{}$ means an estimate of the corresponding variable, and the superscripts $()^-$ and $()^+$ denote the predicted estimate and updated estimate, respectively. Q is a diagonal matrix that represents the covariance of the process noise. Q is used as a tuning parameter with the influence that: A larger value for a certain diagonal element of Q makes estimation of the corresponding state variable more affected by the measurement variables. In this study, Q was chosen as Equation (7) for the reason that the lateral velocity v_{y1}, which is not directly measurable, should depend more heavily on the measured variables for its estimation than the remaining state variables do.

$$Q = \begin{bmatrix} 45 & 0 & 0 & 0 \\ 0 & 1 & 0 & 0 \\ 0 & 0 & 1 & 0 \\ 0 & 0 & 0 & 1 \end{bmatrix} \tag{7}$$

In the update stage, the algorithm computes the measurement residual $\widetilde{y}_k$ and the Kalman gain K_k using Equation (8). The measurement residual is the difference between the actual measurement z_k and its estimate $H\hat{x}_k^-$, and the Kalman gain is a weight matrix for updating the prediction in Equation (6).

$$\widetilde{y}_k = z_k - H\hat{x}_k^-$$
$$K_k = P_k^- H^T \left(H P_k^- H^T + R \right)^{-1} \tag{8}$$

In Equation (8), R is a 2×2 diagonal matrix representing the covariance of the measurement noise. A larger diagonal element implies a higher reliability of the corresponding measurement signal. In this study, R was chosen as Equation (9), reflecting the characteristics of the sensors used in the trucks of the TROOP project. The yaw rate sensor gave fairly precise measurement, while the kingpin angle sensor had notable hysteresis property and not very high resolution.

$$R = \begin{bmatrix} 1 & 0 \\ 0 & 1.5 \end{bmatrix} \tag{9}$$

Finally, the state estimate is updated to $\hat{x}_k^+$ using Equation (10), and at the next time step, it is used as $\hat{x}_{k-1}$ in Equation (6). Likewise, the error covariance matrix is updated to P_k^+, and it is used as P_{k-1}^+ at the next time step.

$$\hat{x}_k^+ = \hat{x}_k^- + K_k\widetilde{y}$$
$$P_k^+ = (I - K_k H) P_k^- \tag{10}$$

Figure 9 shows the results of the simulation, which was conducted to verify the performance of the Kalman filter. TruckSim was used for the vehicle model [23], and Matlab/Simulink was used to implement the filter algorithm. As input, a sinusoidal front steer angle with 90 deg amplitude and 0.125 Hz frequency was applied to the truck model running at 90 kph.

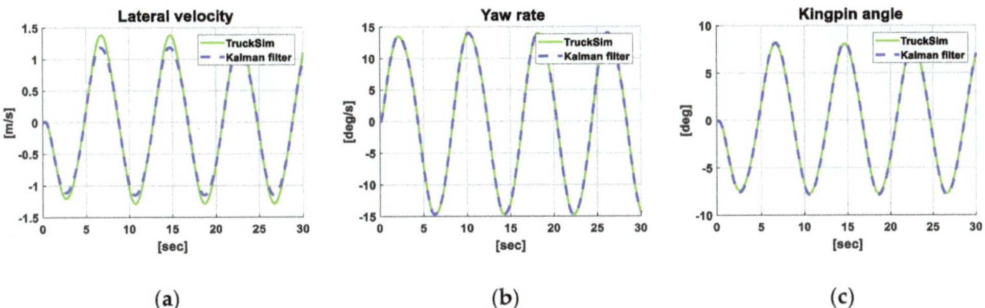

(a) **(b)** **(c)**

Figure 9. Comparison of simulation results of TruckSim and Kalman filter. (a) Lateral velocity; (b) Yaw rate; (c) Kingpin angle.

Figure 9 shows that in the graph of lateral velocity, the estimation deviates from the true value (TruckSim) by as much as 12.6% in magnitude, but more importantly, there was little phase delay between the estimation and the true value. In the case of the yaw rate and kingpin angle, Figure 9 shows that their estimations are very close to their true values both in magnitude and phase.

Although there are more simulation results of the designed Kalman filter, they are not shown here since they all show similar credibility as above. The robustness of the Kalman filter was not actively examined in the simulation environment since the path planning test results with actual vehicles will demonstrate those properties anyway. The state estimation by the Kalman filter in this section is used for the path planning in Section 4.2 as well as for the subject vehicle trajectory generation in the following section.

3.2. Generation of Front and Rear Trajectories of Subject Vehicle

In the proposed path planning algorithm, the LV must generate its own driving trajectory since this trajectory makes it possible for the FV to create its target paths. This trajectory is composed of two parts - front trajectory $T_{F.LV}$ and rear trajectory $T_{R.LV}$-each of which is formed by accumulating into a buffer a total of 300 samples as below.

$$T_{F.LV} = \begin{bmatrix} X_{F.LV} \\ Y_{F.LV} \end{bmatrix} = \begin{bmatrix} x_{F.LV_1} \ x_{F.LV_2} \ \cdots \ x_{F.LV_n} \\ y_{F.LV_1} \ y_{F.LV_2} \ \cdots \ y_{F.LV_n} \end{bmatrix} 2 \times 300 \text{ matrix}$$
$$T_{R.LV} = \begin{bmatrix} X_{R.LV} \\ Y_{R.LV} \end{bmatrix} = \begin{bmatrix} x_{R.LV_1} \ x_{R.LV_2} \ \cdots \ x_{R.LV_n} \\ y_{R.LV_1} \ y_{R.LV_2} \ \cdots \ y_{R.LV_n} \end{bmatrix} 2 \times 300 \text{ matrix}$$

$$(11)$$

Throughout this paper, the subscript "F" (short for "Front") refers to the center of the steering axle of the LV tractor, and the subscript "R" (short for "Rear") refers to the center of the rear bumper of the LV trailer. In addition, the second subscript "LV" in $(\ )_{.LV}$ refers to the coordinate system in which the value is defined. For example, $x_{R.LV}$ means x values of "R" (= LV's rear point) in terms of the LV local coordinate, $x_{R.FV}$ (which will be introduced later) means x values of "R" (= LV's rear point) in terms of the FV local coordinate. Care must be taken not to be confused with the absolute concept of the first subscripts F and R and the relative concept of the second subscripts LV and FV, both used in the same variable.

The numbering among 300 samples was made thus that the number increases from the most recent one to the past. The buffer is a pipeline with FIFO (First In First Out) property. Thus, $(\ )_1$ is the value at the current sample and $(\ )_2$ is the value at one sample before. When time passes to the next sample, all the elements of the buffer are shifted by one to the past, with the new sample added at the front. Figure 10 shows how the trajectory looks like from the view point of the truck.

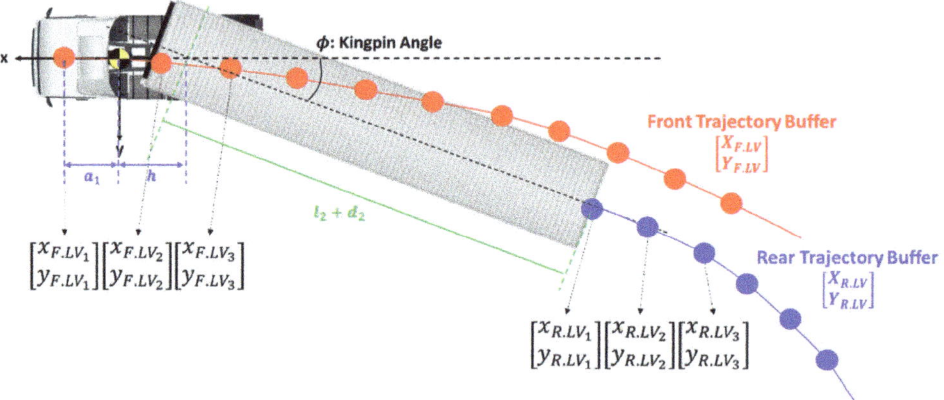

Figure 10. Subject vehicle trajectory buffer.

Equation (12) shows the geometric equations of the "Front" and "Rear" points with respect to the LV's local coordinate system. Since the local coordinate system on a truck changes as the truck moves, the trajectories $T_{F.LV}$ and $T_{R.LV}$ are updated at every sample with the information of translation and rotation made during the last sampling period.

$$\begin{bmatrix} x_{F.LV_1} \\ y_{F.LV_1} \end{bmatrix} = \begin{bmatrix} a_1 \\ 0 \end{bmatrix}$$
$$\begin{bmatrix} x_{R.LV_1} \\ y_{R.LV_1} \end{bmatrix} = \begin{bmatrix} -h \\ 0 \end{bmatrix} + \begin{bmatrix} \cos\phi & -\sin\phi \\ \sin\phi & \cos\phi \end{bmatrix} \cdot \begin{bmatrix} -l_2 - d_2 \\ 0 \end{bmatrix} \tag{12}$$

As seen above, the subject vehicle trajectory is represented by a total of 1200 points (counting front and rear trajectories and x-y values for each sample), which will be sent to the rear vehicles. The TROOP project mandates V2V communication to occur every 20 ms, but 1200 data points are too big for this purpose. To solve this problem, curve fitting to a third-order polynomial was performed, which reduces 600 data points to 4 coefficient values. Curve fitting also gives the side benefit of making the trajectory smooth even in the presence of outliers or random noises in the raw data of $T_{F.LV}$ and $T_{R.LV}$.

Table 1 shows the message that the LV transmits to the FV via V2V communication. Along with $T_{F.LV}$ and $T_{R.LV}$ in the form of the coefficients for their 3rd order polynomial, the LV sends to the FV $[x_{R.LV_1}, y_{R.LV_1}]$, which is the center point of the LV's rear bumper at the current sample. The is because the FV requires a reference point when implementing conversion between the local coordinates, which is explained in Chapter 4.

Table 1. The message that the leading vehicle (LV) sends to the following vehicle (FV) via V2V communication.

Message	Notation
coefficients of 3rd order polynomial for LV's "Front" trajectory	$\begin{bmatrix} C_{F_3}, & C_{F_2}, & C_{F_1}, & C_{F_0} \end{bmatrix}$
coefficients of 3rd order polynomial for LV's "Rear" Trajectory	$\begin{bmatrix} C_{R_3}, & C_{R_2}, & C_{R_1}, & C_{R_0} \end{bmatrix}$
coordinate of LV's "Rear" point at the current sample	$\begin{bmatrix} x_{R.LV_1}, & y_{R.LV_1} \end{bmatrix}$

4. Proposed Path Planning Algorithm

Figure 11 shows the architecture of the FV lateral controller. Using the received message in Table 1, the FV performs path planning, i.e., calculates its target path, and this enters the path tracking control module as input.

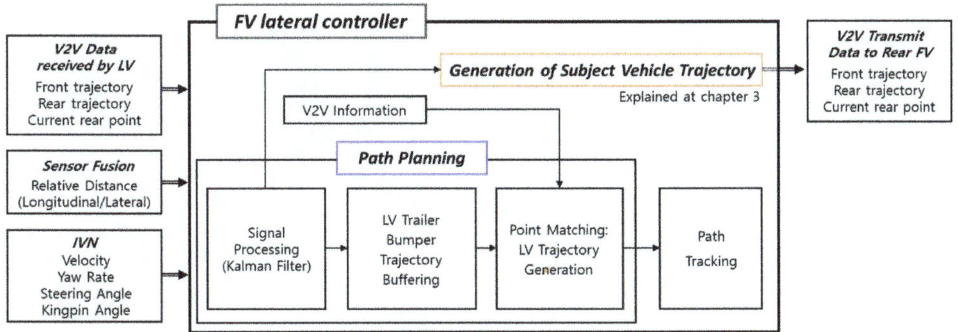

Figure 11. The architecture of the FV lateral controller.

The target path of the FV is basically the past trajectory of the LV, $T_{F.LV}$, but since $T_{F.LV}$ is defined from the viewpoint of the LV, it must be converted into the local coordinate system of the FV. This conversion is difficult without having any knowledge of the global coordinate system, or perhaps the most challenging part in the whole truck platooning. This chapter describes how this problem was solved in this research.

4.1. Concept of Coordinate Matching

Figure 12 illustrates how the coordinate matching algorithm in this paper works, with an example of representing a dataset P in terms of the coordinate system of a dataset Q, when the two datasets P and Q have the same shape but are dislocated in the 2D plane.

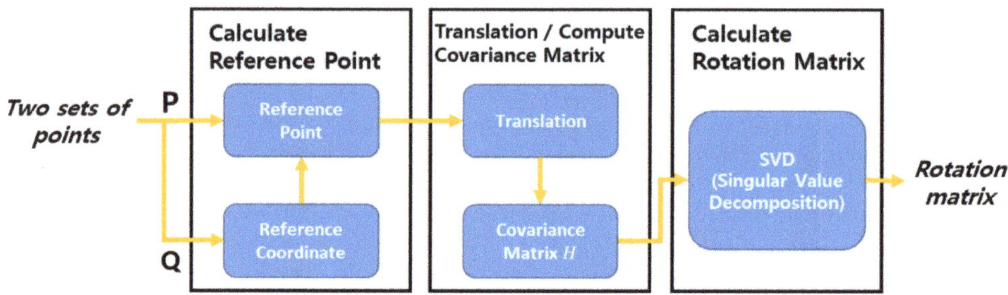

Figure 12. Workflow of point matching algorithm.

Coordinate matching is equivalent to point matching among two datasets P and Q. First, a reference point is selected from P and Q, respectively. They must represent an identical point in an identical 2D shape. From the difference of their locations, a translation vector can be found, and all points in Q are translated accordingly. Finally, a rotation matrix is found that makes P and Q coincide.

To apply the above concept to path planning in the truck platooning, $T_{R.LV}$ and $\overline{T}_{R.FV}$ were chosen as P and Q in Figure 12, respectively. As previously explained, $T_{R.LV}$ is the trajectory of the LV's rear point in the LV's coordinate system. $\overline{T}_{R.FV}$, which appears for the first time here, represents the trajectory of the LV's rear point in the FV's coordinate system. As Figure 13 shows, the FV can generate $\overline{T}_{R.FV}$ since the FV can perceive the rear end of the LV with a camera and radar.

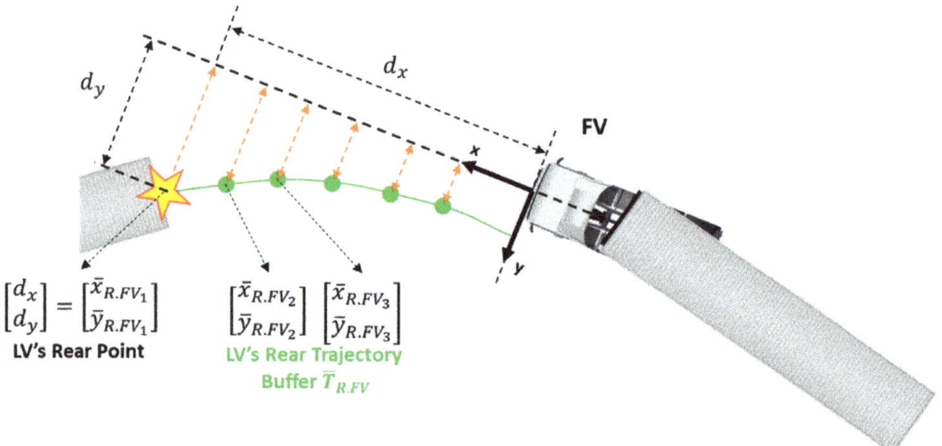

Figure 13. LV's rear bumper trajectory from the viewpoint of FV.

For the reference point in the two datasets $T_{R.LV}$ and $\overline{T}_{R.FV}$, their first elements $\left[x_{R.LV_1}, \; y_{R.LV_1} \right]^T$ and $\left[\overline{x}_{R.FV_1}, \; \overline{y}_{R.FV_1} \right]^T$ were used since they represent an identical point at an identical time. Figure 14 shows the schematics of the coordinate matching between LV and FV.

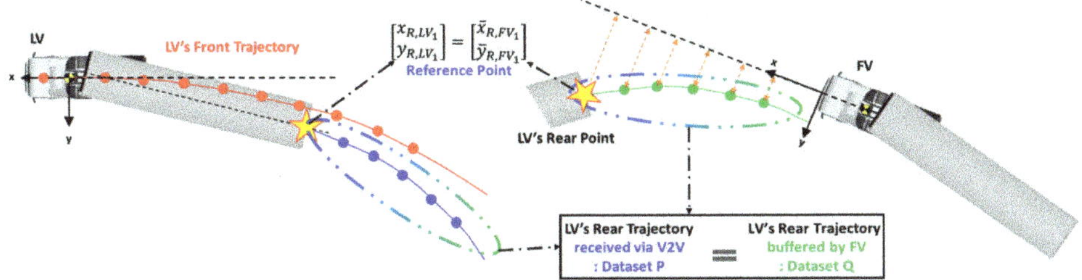

Figure 14. Schematics of coordinate matching in path planning of truck platooning.

4.2. Steps for Path Planning

This section shows how the FV computes its target path. Although the explanation is based on the LV–FV relationship, the same argument equally applies to any two adjacent trucks in the platoon, like in the FV_1-FV_2 relationship. First, Table 2 shows four trajectories involved in path planning. It is reminded that the "Front" point refers to the center of the steering axle of the LV tractor, and the "Rear" point refers to the center of the rear bumper of the LV trailer.

Table 2. Trajectories involved in path planning.

Mathematical Notation	Description
$T_{F.LV} = \begin{bmatrix} X_{F.LV} \\ Y_{F.LV} \end{bmatrix} = \begin{bmatrix} x_{F.LV_1} \; x_{F.LV_2} \; \cdots \; x_{F.LV_n} \\ y_{F.LV_1} \; y_{F.LV_2} \; \cdots \; y_{F.LV_n} \end{bmatrix}$	trajectory of "Front" point in LV's coordinate system
$T_{R.LV} = \begin{bmatrix} X_{R.LV} \\ Y_{R.LV} \end{bmatrix} = \begin{bmatrix} x_{R.LV_1} \; x_{R.LV_2} \; \cdots \; x_{R.LV_n} \\ y_{R.LV_1} \; y_{R.LV_2} \; \cdots \; y_{R.LV_n} \end{bmatrix}$	trajectory of "Rear" point in LV's coordinate system
$T_{F.FV} = \begin{bmatrix} X_{F.FV} \\ Y_{F.FV} \end{bmatrix} = \begin{bmatrix} x_{F.FV_1} \; x_{F.FV_2} \; \cdots \; x_{F.FV_n} \\ y_{F.FV_1} \; y_{F.FV_2} \; \cdots \; y_{F.FV_n} \end{bmatrix}$	trajectory of "Front" point in FV's coordinate system
$T_{R.FV} = \begin{bmatrix} X_{R.FV} \\ Y_{R.FV} \end{bmatrix} = \begin{bmatrix} x_{R.FV_1} \; x_{R.FV_2} \; \cdots \; x_{R.FV_n} \\ y_{R.FV_1} \; y_{R.FV_2} \; \cdots \; y_{R.FV_n} \end{bmatrix}$	trajectory of "Rear" point in FV's coordinate system
$\overline{T}_{R.FV} = \begin{bmatrix} \overline{X}_{R.FV} \\ \overline{Y}_{R.FV} \end{bmatrix} = \begin{bmatrix} \overline{x}_{R.FV_1} \; \overline{x}_{R.FV_2} \; \cdots \; \overline{x}_{R.FV_n} \\ \overline{y}_{R.FV_1} \; \overline{y}_{R.FV_2} \; \cdots \; \overline{y}_{R.FV_n} \end{bmatrix}$	trajectory of "Rear" point generated by FV

Figure 15 illustrates the process of how the FV performs path planning. First, the LV generates $T_{F.LV}$ and $T_{R.LV}$ and concurrently, the FV generates $\overline{T}_{R.FV}$. Through V2V communication, the FV receives $T_{F.LV}$ and $T_{R.LV}$ (in the form of the 3rd order polynomial coefficients) from the LV. Using $T_{R.LV}$ and $\overline{T}_{R.FV}$, the FV performs a coordinate conversion, which is possible since they represent an identical trajectory. Coordinate conversion is to find the translational and rotational relationship between LV and FV as in Equation (13).

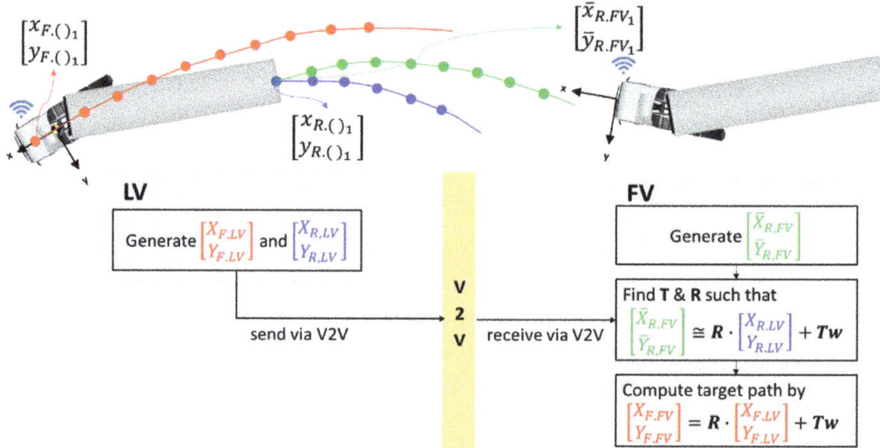

Figure 15. Schematics of path planning.

Find 2×1 translation vector $T = \begin{bmatrix} x_T \\ y_T \end{bmatrix}$ and 2×2 rotation matrix R such that

$$\overline{T}_{R.FV} \cong R \cdot T_{R.LV} + T \cdot w \ \textit{where } w = 1 \times n \text{ vector with ones} \tag{13}$$

Using the same relationship between the coordinate systems of LV and FV in the above, the FV can compute $T_{F.FV}$ by Equation (14).

$$T_{F.FV} = R \cdot T_{F.LV} + T \cdot w \tag{14}$$

$T_{F.FV}$ is the trajectory of the center of the steering axle of the LV's tractor from the viewpoint of the FV, as seen in Figure 16. $T_{F.FV}$ is important because it can serve as the target path to the FV. However, the FV cannot generate $T_{F.FV}$ using only camera and radar since the tractor of the LV is blocked by the trailer of the LV most of the time. $T_{F.FV}$ is the final output of the path planning algorithm of this research.

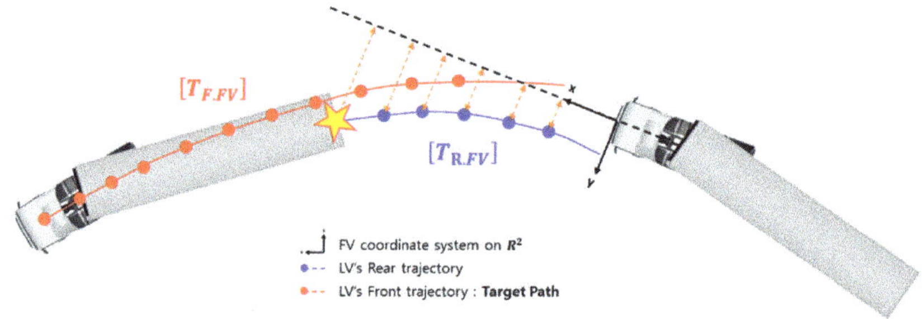

Figure 16. LV trajectory defined in the FV coordinate system.

4.3. Kabsch Algorithm

The rotation matrix in Equation (13) is computed after $T_{R.LV}$ is translated thus that its first point coincides with the first point of $\overline{T}_{R.FV}$, and this is illustrated in Figure 17.

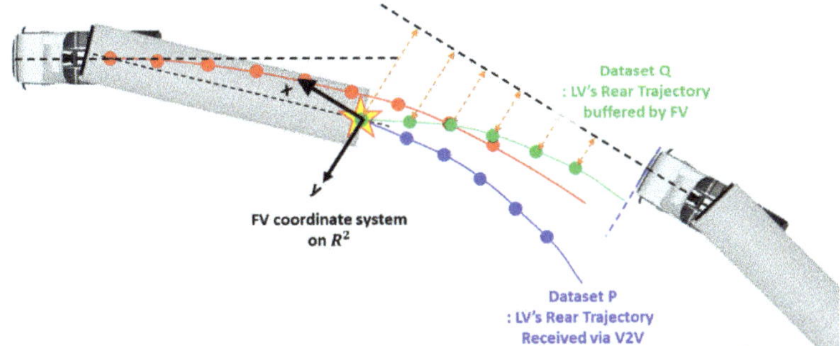

Figure 17. Reference point matching before finding the rotational relationship.

In the study, the Kabsch algorithm [24,25] was adopted to find the optimal rotation matrix after the reference point matching. The Kabsch algorithm is a method to compute the optimal rotation matrix by minimizing the RMSD (Root Mean Squared Deviation) between the two datasets to be matched. This algorithm is used to convert the LV's trajectory received by V2V communication into the FV's coordinate system, and it is explained below in the context of path planning.

First, Equation (15) shows the two datasets and their reference points.

$$
\begin{aligned}
P = T_{R.LV} = \begin{bmatrix} X_{R,LV} \\ Y_{R,LV} \end{bmatrix} = \begin{bmatrix} x_{R.LV_1} & x_{R.LV_2} & \cdots & x_{R.LV_n} \\ y_{R.LV_1} & y_{R.LV_2} & \cdots & y_{R.LV_n} \end{bmatrix}, \quad p_0 = \begin{bmatrix} x_{R.LV_1} \\ y_{R.LV_1} \end{bmatrix} : \text{ reference point of } P \\
Q = \overline{T}_{R.FV} = \begin{bmatrix} \overline{X}_{R,FV} \\ \overline{Y}_{R,FV} \end{bmatrix} = \begin{bmatrix} \overline{x}_{R.FV_1} & \overline{x}_{R.FV_2} & \cdots & \overline{x}_{R.FV_n} \\ \overline{y}_{R.FV_1} & \overline{y}_{R.FV_2} & \cdots & \overline{y}_{R.FV_n} \end{bmatrix}, \quad q_0 = \begin{bmatrix} \overline{x}_{R.FV_1} \\ \overline{y}_{R.FV_1} \end{bmatrix} : \text{ reference point of } Q
\end{aligned}
\tag{15}
$$

Next, the reference point matching in Figure 17 is done by translating both trajectories, thus that their reference points are located at the origin of the FV's local coordinate system.

$$
\begin{aligned}
\overline{P} &= P - p_0 \cdot w \\
\overline{Q} &= Q - q_0 \cdot w
\end{aligned}
\tag{16}
$$

Next, the rotation matrix is computed thus that the root mean squared error between the two datasets $\overline{P}$ and $\overline{Q}$ are minimized. To do this, their covariance matrix H is formed and singular value decomposition is done to this matrix (Equation (17)).

$$
\begin{aligned}
H &= \overline{P} \cdot \overline{Q}^T \\
H &= U \cdot S \cdot V^T
\end{aligned}
\tag{17}
$$

Finally, the rotation matrix and the translation matrix T are computed by Equation (18), and using them the target path for the FV can be computed by Equation (14).

$$
\begin{aligned}
d &= sign\left(\det\left(V \cdot U^T\right)\right) \\
R &= V \cdot \begin{pmatrix} 1 & 0 \\ 0 & d \end{pmatrix} \cdot U^T \\
T &= q_0 - R \cdot p_0
\end{aligned}
\tag{18}
$$

Algorithm 1 shows the proposed path planning process.

Algorithm 1. Path planning algorithm

Input:

$T_{F.LV}$: Trajectory of "Front" point in LV's coordinate system

$T_{R.LV}$: Trajectory of "Rear" point in LV's coordinate system, Dataset P

$T_{R.FV}$: Trajectory of "Rear" point in FV's coordinate system, Dataset Q

$[x_{R.LV_1}, y_{R.LV_1}]$: Coordinate of LV's "Rear" point at current sample

Output:

$T_{F.FV}$: Trajectory of "Front" point in FV's coordinate system

Find the reference points p_0, q_0 using $[x_{R.LV_1}, y_{R.LV_1}]$

Translate the trajectories to coincide with the origin of FV's local coordinate system

$\overline{P} = P - p_0 \cdot w$, $\overline{Q} = Q - q_0 \cdot w$, *w: weight matrix*

Find the rotation matrix and translation matrix

$H = \overline{P} \cdot \overline{Q}^T$ ← *covariance matrix*

$[U, S, V] = svd(H)$ ← *singular value decomposition*

$D \leftarrow$ *2-by-2 diagonal matrix*

if $\left| V \cdot U^T \right| < 0$, **then**

$D(2,2) = -1$, *return D;*

end

$R = V \cdot D \cdot U^T$ ← *rotation matrix*

$T = q_0 - R \cdot p_0$ ← *translation matrix*

Compute the target path

$T_{F.FV} \leftarrow T_{F.LV}$ *using R and T*

return $T_{F.FV}$;

---

5. Results of Simulation and Road Test Experiments

The proposed path planning algorithm of this paper was validated by both simulation and road test experiments, and their results are shown and analyzed in this chapter.

5.1. Simulation Result

A simulation environment was constructed to validate the path planning algorithm of this paper. The plant model for truck platooning was made with TruckSim, and the path planning logic was implemented with Matlab/Simulink. Table 3 shows three test scenarios used in the simulation. Scenarios S1 and S2 are the cases of driving on a curved road at low and high speeds, and Scenario S3 is a double lane change.

Table 3. Test scenarios for simulation.

No.	Speed	Time-Gap	Method	Radius [m]
Scenario S1	40 kph [1]	0.7 s [1]	Driving on a curved road	100R
Scenario S2	90 kph [1]	0.7 s	Driving on a curve road	250R
Scenario S3	90 kph	0.7 s	Double lane change	Straight road

[1] In the TROOP project, the minimum speed of platooning is 40 kph, the maximum speed is 90 kph, and the target time-gap for demonstration of platooning in 2020 is 0.7 s.

5.1.1. Scenario S1–Curved Road with 100R, 40 kph, 0.7 s Time Gap

Figure 18 shows the simulation results for Scenario S1.

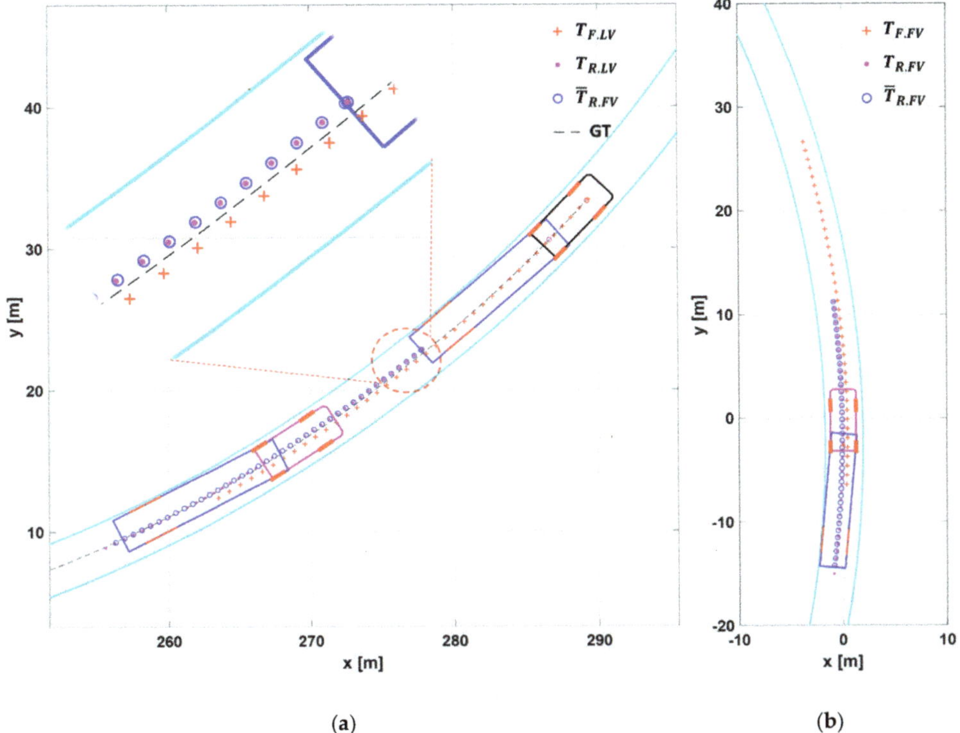

Figure 18. The simulation results for Scenario S1 (a) $T_{F.LV}$, $T_{R.LV}$, and $\overline{T}_{R.FV}$ in LV and FV; (b) $T_{F.FV}$, $T_{R.FV}$ and $\overline{T}_{R.FV}$ in FV. The black dash is the actual travel path (GT) of the front trajectory of LV. The cyan line next to the vehicles means the left/right lane.

Figure 18a shows that in the LV, the tractor trajectory $T_{F.LV}$ is formed inside the trailer trajectory $T_{R.LV}$, which verifies the pattern of the low-speed off-tracking illustrated in Figure 3a. An almost exact agreement of $T_{R.LV}$ and $\overline{T}_{R.FV}$ in Figure 18a indicates that the point matching algorithm using the two trajectories for the LV's rear point was working successfully.

Comparing $T_{F.LV}$ and the trajectory of its true value (GT), the graph shows that they started at the same point, but $T_{F.LV}$ gradually deviates from GT as the point goes afterward. This is due to the estimation error in the Kalman filter, which was used in the generation of the subject vehicle trajectory. This kind of error cannot be completely avoided in any case. However, since the FV's actual target path begins at some distance ahead—typically 16 m ahead at 40 kph—to secure enough look-ahead distance, this estimation error does not cause any significant trouble in path planning. In addition, this trajectory error can be reduced by giving more weight to the latest trajectory point than the past, when performing point matching.

Figure 18b shows the target path of the FV, $T_{F.FV}$, which is the final output of the proposed path planning algorithm. It can be observed that this target path is located outside of $\overline{T}_{R.FV}$ (or $T_{R.FV}$). This is an effort to overcome the off-tracking phenomenon which produces fairly large positive kingpin angle in both trucks in this case.

In Figure 18b, the LV truck was removed from the plot to demonstrate that the FV can generate a long-range target path even when its front view is severely impaired by the trailer of the LV. This is important since it can provide lateral string stability to the fleet. In fact, the conditions of Scenario S1 are too harsh to be found in actual driving situations.

5.1.2. Scenario S2-Curved Road with 250R, 90 kph, 0.7 s Time Gap

Figure 19 shows the simulation results for Scenario S2.

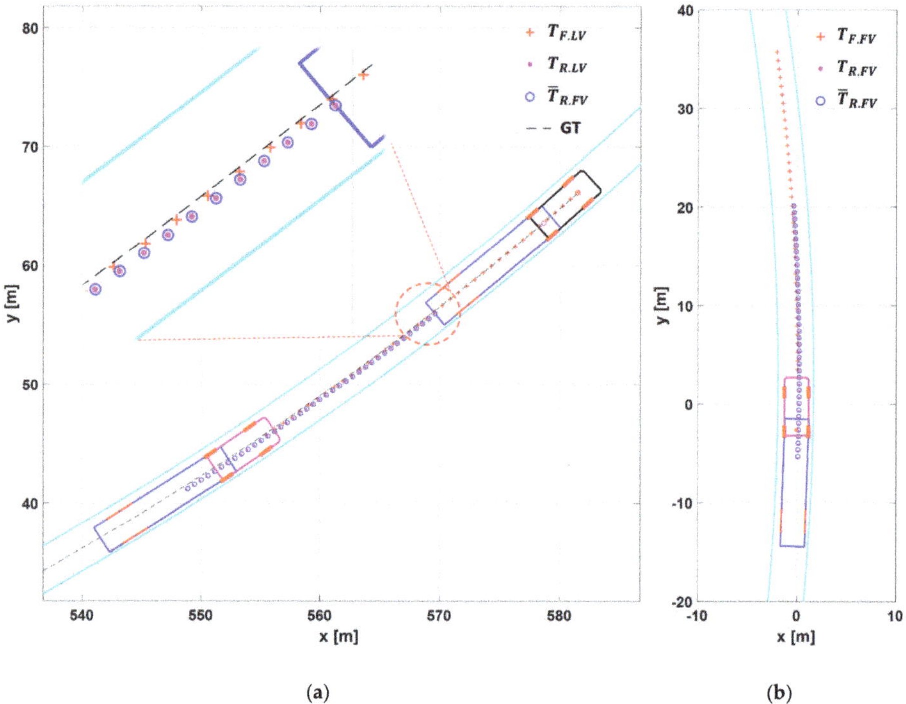

Figure 19. The simulation results for Scenario S2 (**a**) $T_{F.LV}$, $T_{R.LV}$, and $\overline{T}_{R.FV}$ in LV and FV; (**b**) $T_{F.FV}$, $T_{R.FV}$ and $\overline{T}_{R.FV}$ in FV.

Figure 19a shows that in the LV, the tractor trajectory $T_{F.LV}$ was formed outside the trailer trajectory $T_{R.LV}$, which was the opposite of what happened in Scenario S1. In Scenario S2, the lateral acceleration was 0.25 g—much higher than 0.13g in Scenario S1—and due to this large centrifugal force, the trailer was pushed outside to yield the high-speed off-tracking pattern in Figure 3b. Like in Scenario S1, the two trajectories $\overline{T}_{R.FV}$ and $T_{R.LV}$ matched almost exactly, which indicates the high reliability of the path planning algorithm of this paper. The deviation of $T_{F.LV}$ from its true trajectory (GT) can be explained similarly as in Scenario S1, except that at 90 kph, the look-ahead distance where the actual target path begins was about 35m ahead of the FV.

Figure 19b shows that the target path of the FV, $T_{F.FV}$, was located slightly inside of $\overline{T}_{R.FV}$ (or $T_{R.FV}$). With such configuration, the FV can prevent the vehicle from leaving its lane outward. This path must be very close to the actual trajectory of the LV truck, which was driven manually by an experienced driver.

For large cargo trucks, if the target path was incorrectly generated at high speeds, the lateral stability of the vehicle may be compromised, raising the risk of rollovers. Since the driver of LV is a professional driver who understands platoon driving well, the trajectory generated by LV driving at high speeds is the target path that can guarantee the stability of the vehicle. FV performs the proposed path planning using the trajectory of LV and directly follows the path, and then rollover can be prevented. In the same principle as feed-forward control using the target longitudinal acceleration of the LV for longitudinal control, using the LV trajectory received via V2V communication as a target path is a key to ensuring the string stability of the truck platoon.

5.1.3. Scenario S3–DLC (Double Lane Change) on a Straight Road, 90 kph, 0.7 s Time Gap

Figures 20 and 21 show the simulation results for Scenario S3 in which the platoon makes double lane change to avoid stopped vehicles ahead in their driving lane. While former scenarios were for verifying the steady-state performance of the proposed algorithm, this scenario is for verifying transient performance.

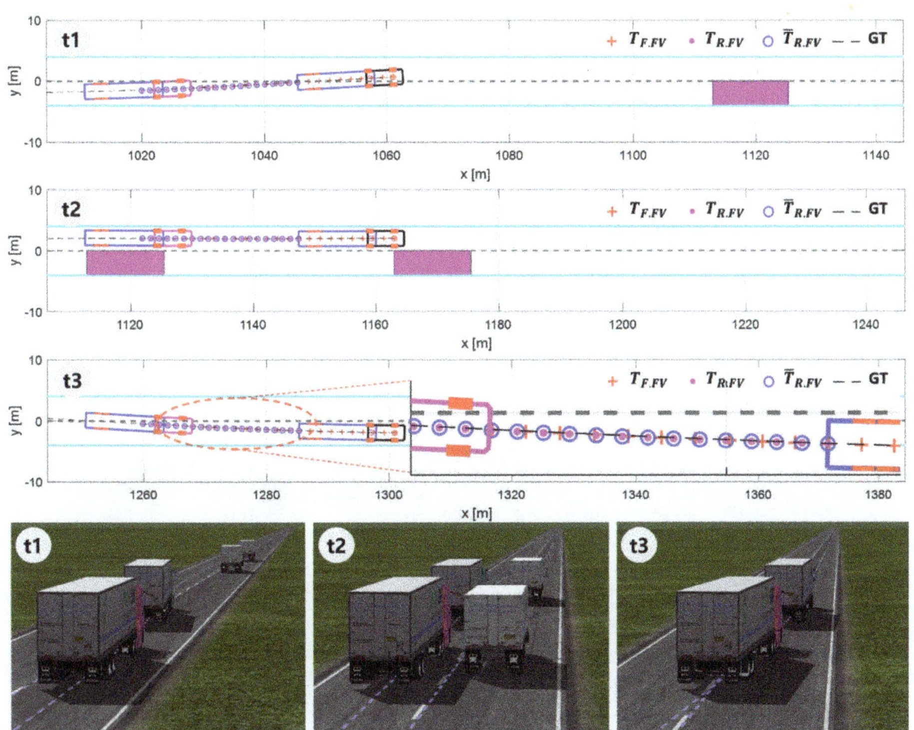

Figure 20. The simulation results for Scenario S3. At t1, LV finds stopping vehicles and begins to change lanes. At t2, the platoon is passing by them. At t3, the platoon has overtaken stopped vehicles and is returning to its original lane.

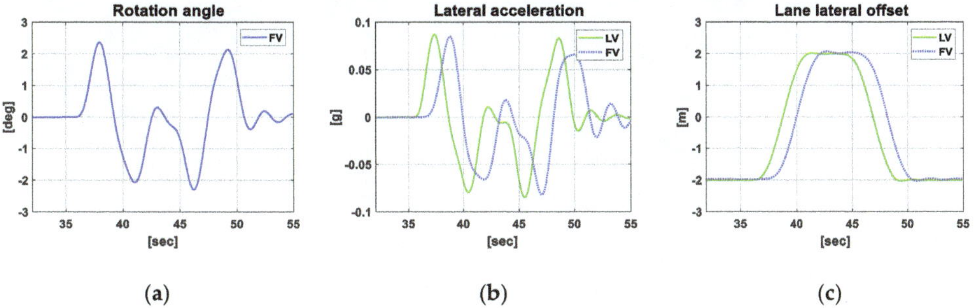

Figure 21. The simulation results for Scenario S3. (**a**) Rotation angle for point matching; (**b**) Lateral acceleration; (**c**) lateral offset of the vehicle with respect to the center of the lane.

Figure 20 shows snapshots of the platoon at three different moments during DLC: t1 is when DLC initiates, t2 is when the platoon is passing the stopped vehicles, and t3 is when the platoon is returning to its original lane after securing enough space past the stopped vehicles.

Figure 20 shows that the FV recognizes the change in the driving path of the LV truck and successfully creates its target path throughout the DLC maneuver: During the transient periods of lane change and during the straight driving when passing the stopped vehicles.

Figure 21 shows that during the DLC the rotation angle computed in the point matching algorithm varies in the range of −2.3 deg to 2.4 deg, and the lateral acceleration varies between −0.085 g and 0.085 g. This is rather a mild variation, whose amount depends on the platooning strategy for the lane change. As mentioned earlier, since the fleet length can reach 100 m in the platooning of articulated cargo trucks, the platoon can cause safety issues to the nearby vehicles. For this reason, lane change of the platoon must be minimized as much as possible, and even when it should happen as in the current scenario, both individual stability and string stability should not be violated. As shown in Figure 21b, the peak to peak of the lateral acceleration of the FV is slightly smaller than that of the LV. This means that the transient response characteristics have improved from the LV to the rear FV of the platoon, indicating that the lateral string stability has been secured. Figure 21c implies that the whole process of DLC is completed in 14 s, which corresponds to 350 m at 90 kph. Figure 21c also shows that the transient response of the platoon is quite stable, with almost negligible overshoot on both lane changes.

From the simulation results thus far, it could be verified that the proposed path planning algorithm provides lateral string stability for various driving conditions of truck platooning.

5.2. Road Test Experiments Result

Along with simulation, actual vehicle tests were conducted to validate the proposed path planning algorithm. It was performed in the Yeoju Smart Highway, which was built by the Korean government in 2014 as a testbed for cooperative autonomous vehicles. It is located right next to the Jungbu Naeryuk Highway near Yeoju Junction and is a two-lane road with a total length of 7.7 km. Figure 22 shows the satellite photograph. It includes a straight road part and two 2000R curved road parts with opposite curvatures.

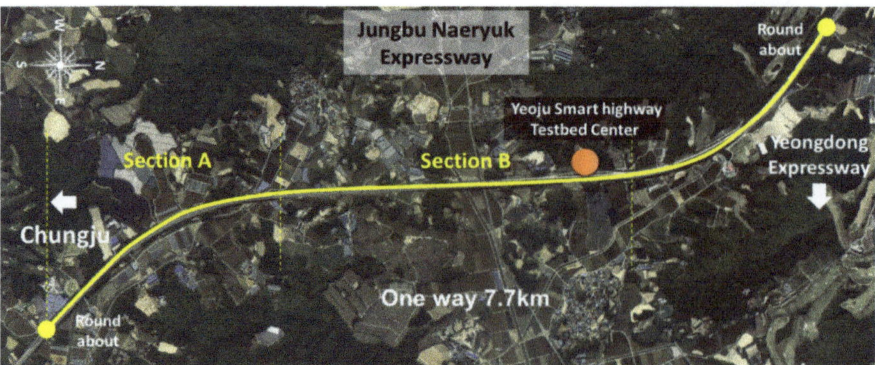

Copyright 2019 NGII. Retrieved from http://map.ngii.go.kr/pd/ctlsSvc/ctlsSvc.do

Figure 22. Yeoju Smart Highway testbed.

5.2.1. Scenario T1–Curved Road with 2000R, 80 kph, 0.7 s Time Gap

Figure 23 shows the test results for Scenario T1. Using a drone, the image of the platooning trucks was recorded (Figure 23a), and the image of the LV's rear view was captured by a camera installed in the FV (Figure 23b).

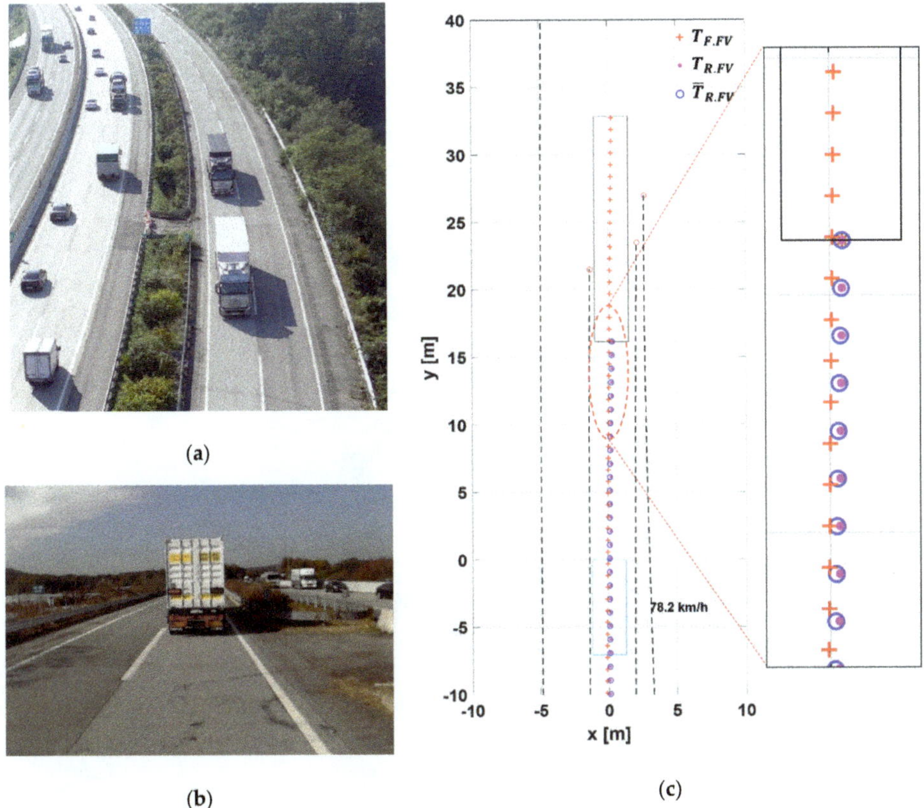

Figure 23. The test results for Scenario T1. (**a**) Top-view of the platooning trucks; (**b**) front-camera view of FV; (**c**) LV trajectories generated by FV in FV coordinate system.

As can be imagined from the snapshots in Figure 23a,b, the platooning trucks could maintain the time gap without causing any lateral stability issues in this test. In Figure 23c, the front box represents the LV, and the rear box represents the FV's tractor, and the black dotted lines on both sides of the vehicles indicate the road lanes that the camera perceived.

The plot in Figure 23c indicates that, with LV blocking the front view area of the FV's camera, the lane detection range of the FV was not more 24 m ahead. Even worse, Figure 23b shows that the lane markings can be missing at some intervals in real situations. However, Figure 23c plot shows that despite these difficulties, the proposed path planning algorithm provides a reliable target path that spans 33 m ahead of the FV.

In Figure 23, it is very interesting to note that, although the road is gently curved to the right and the trucks are running at high speeds, the off-tracking is occurring in the opposite direction of the high-speed off-tracking pattern seen in Figure 3b. This is because the test road has a slight bank with a downside inside the curve, which is typical on a curved road of every highway. To a cargo truck, even a slight bank can notably affect its lateral motion by making the trailer slide down the bank by its own weight. Thus, the pattern of the target path in Figure 23c indicates that the proposed path planning algorithm is robust to environmental disturbances.

5.2.2. Scenario T2–SLC (Single Lane Change) on a Straight Road, 80 kph, 0.7 s Time Gap

Figure 24 shows the test results for Scenario T2. The first picture shows an overlapped image of three drone images shot at three different moments during SLC. The pictures on the second row show the snapshots of the LV at those three different moments.

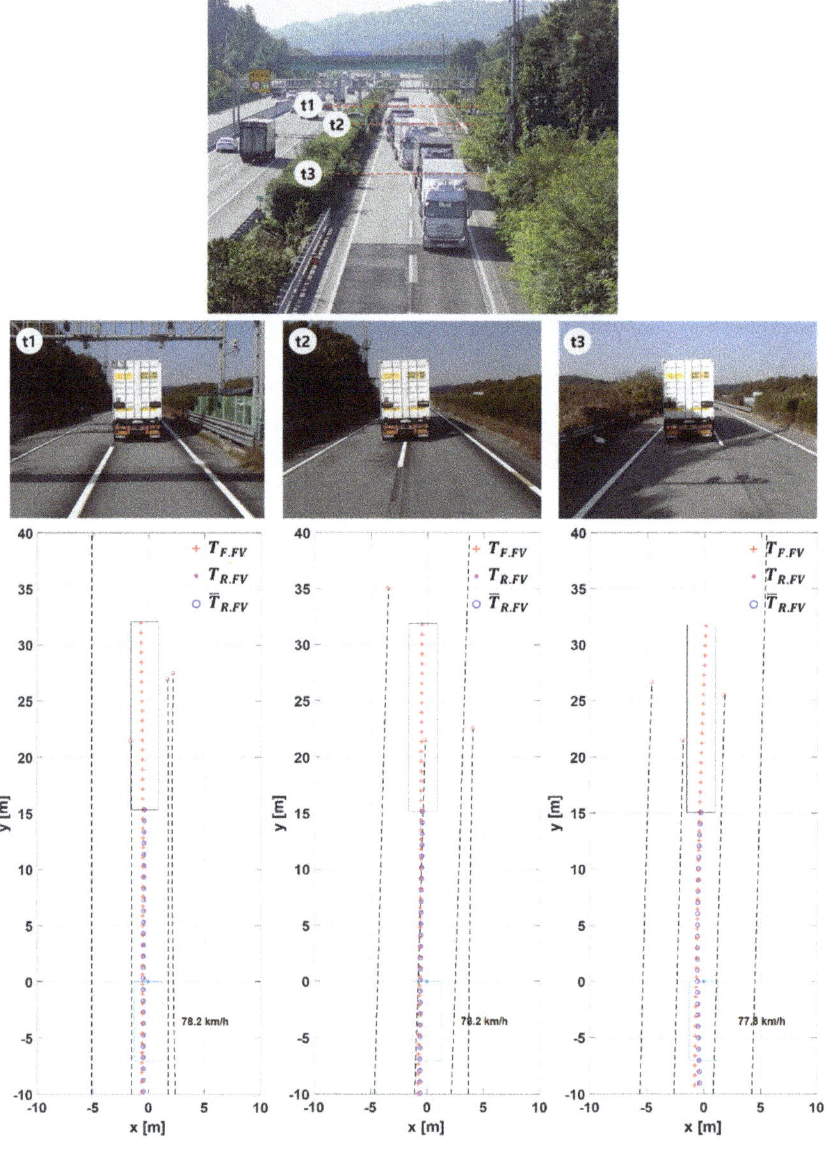

Figure 24. The test results for Scenario T2. At t1, LV initiates single lane change (SLC), at t2, FV changes lanes, and at t3, LV settles in the new lane.

In Figure 24, t1 is the time when the LV initiates SLC to the left lane. The bottom graphs show that at t1 the target path of the FV is properly generated toward the left lane. If the FV was relying only on camera/radar to generate its target path, it would have been hard to tell whether the LV intended to change lanes or else (like when the trailer of the LV unintentionally drifts out of its lane due to a bank). Using V2V communication and the proposed algorithms, however, the path planning method could tell the LV's intention and generate the target path of the FV to agree with the intention.

From t2 to t3, the FV changes lanes and then settles on the new lane. Similarly to the above case, if the FV follows the rear point of the LV that its front camera/radar perceives, the FV would experience significant overshoot to settle on the new lane by knowing the LV's intention only after its result occurs. In a typical single lane change maneuver, an experienced driver is known to perform reverse steering as early as when the vehicle is in the middle of crossing lanes to reduce the overshoot later when the vehicle settles on the new lane. In this test, it was observed that the proposed path planning algorithm worked just like an experienced driver with enough look-ahead distance virtually made.

As shown in the front camera view in the middle of Figure 24, a downward slope is formed in the direction of changing lanes. Heavy-duty trucks are at risk of being pushed down the slope by the load, and the risk will be increased when changing lanes. The FV needs to quickly control the vehicle's attitude through reverse steering at the appropriate point during the lane change. The proposed algorithm enables the FV to immediately respond to the steering intention that the professional LV driver responds to changes in the road, thus that the FV can stably follow the LV in any road environment.

From the results by simulation and experimental tests, the path planning algorithm of this paper has demonstrated capabilities to respond quickly to the LV's steering intention and to act against unexpected road disturbances, which can enable more sophisticated steering control and thus secure lateral string stability of the platoon.

5.2.3. Scenario T3–Unintended Steering Input, 80 kph, 0.7 s Time Gap

Figure 25 shows the test results for Scenario T3. This scenario was considered to validate the reliability of the proposed path planning algorithm in a situation where an unintended steering disturbance is applied to the vehicle. The first graph shows the measured steering wheel angle and steering wheel angle command. The pure pursuit algorithm [26] was used for path tracking. Three photos of the inside of the tractor's cab are shown in chronological order from t1 to t3.

At t1, the FV driver turned the steering wheel counterclockwise to make an unintended steering disturbance into the vehicle. As shown in the first graph in Figure 25, a steering wheel angle of 10.2 deg was applied to FV. The driver released his hands from the steering wheel immediately after making the steering input. At t2, while moving out of the lane, the FV generates the target path that the LV, which is running normally, traveled based on the coordinate system of the FV. At this time, steering wheel angle command was generated up to −18 deg to follow the trajectory of the LV. At t3, FV is stably converging through a slight opposite steering after reaching the target path. Through scenario T3, it was verified that the FV can follow the LV by stably generating the target path even when steering disturbance occurs.

In road test scenarios T1, T2, and T3, the proposed path planning algorithm is robust and functioning accurately in various driving environments. By receiving the trajectories of the LV via V2V communication, it is possible to effectively solve the problem of the limited perceived distance of the existing path tracking algorithm for platooning. First, it is possible to overcome an error between the LV's tractor trajectory and the FV's trajectory by following the trajectory of the tractor rather than the trailer in consideration of the off-tracking characteristics of the truck. Next, by using the path reflecting the steering intention of the professional driver of the LV, the FV can generate a target path that is robust against changes in road conditions such as banks. Lastly, it was possible to create a target path considering the influence of the driver's unintended steering input as well as the external disturbance factor.

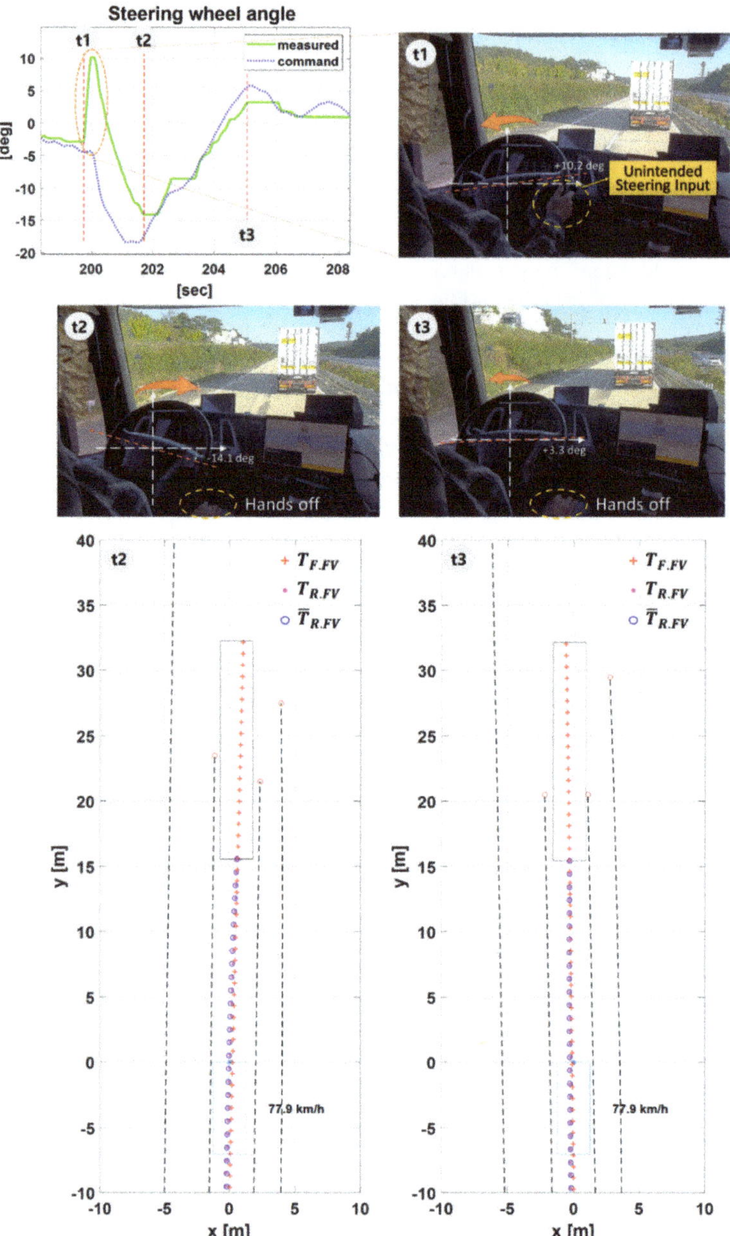

Figure 25. The test results for Scenario T3. At t1, FV's driver make a steering input, at t2, FV is returning to the center of the lane, and at t3, FV settles in the lane.

6. Conclusions

Truck platooning refers to a form in which a number of trucks run as a fleet with short inter-vehicle distance using V2V communication. The leading vehicle is driven manually by an experienced driver, and the following vehicles run by autonomous driving. To successfully perform platooning in various situations, each truck must maintain dynamic stability and at the same time, the whole system must maintain string stability.

Due to the short front-view range, the following vehicles' path planning capabilities become significantly impaired. In addition, in platooning with articulated cargo trucks, which is the case of this study, an off-tracking phenomenon occurring on a curved road makes it hard for the following vehicle to track the trajectory of the preceding truck. Furthermore, without knowledge of the global coordinate system, it is difficult to correlate the local coordinate systems that each truck relies on for sensing environment and dynamic signals.

In this paper, to solve these problems, a path planning algorithm for platooning of articulated cargo trucks has been developed. Using the Kalman filter, V2V communication, and a novel update-and-conversion method, each following vehicle can accurately compute the trajectory of the leading vehicle's front part for using it as a target path. This paper's path planning algorithm was validated by simulations on severe driving scenarios and by tests on an actual road. From the simulation and experimental results, it could be verified that the proposed path planning algorithm provides lateral string stability, even for very harsh driving conditions of truck platooning. The algorithm also demonstrated the capabilities to respond quickly to the leading vehicle's steering intention and to act against unexpected road disturbances, which can enable sophisticated path tracking control.

Author Contributions: Conceptualization, Y.L.; investigation, C.L.; methodology, Y.L.; resources, S.K.; software, Y.L., T.A., C.L.; validation, Y.L., T.A., S.K.; writing—original draft preparation, Y.L.; writing—review and editing, K.P.; visualization, Y.L. All authors have read and agreed to the published version of the manuscript.

Funding: This research was supported by the Academy Industry Research Collaboration funded by Hyundai Motor Group. In addition, this research was supported by the Transportation Logistics Development Program (20TLRP-B147674-03) funded by the Ministry of Land, Infrastructure and Transport (MOLIT Korea); and by Innovative Incubation Center for Autonomous xEV Technology through the National Research Foundation of Korea (NRF) funded by the Ministry of Education (5199990814084).

Conflicts of Interest: The authors declare no conflict of interest.

References

1. Aarts, L.; Feddes, G. European truck platooning challenge. In Proceedings of the HVTT14: International Symposium on Heavy Vehicle Transport Technology, Rotorua, New Zealand, 15–18 November 2016.
2. Konstantinopoulou, L.; Coda, A.; Schmidt, F. Specifications for Multi-Brand Truck Platooning. In Proceedings of the ICWIM8, 8th International Conference on Weigh-In-Motion, Prague, Czech Republic, 19–23 May 2019; p. 8.
3. Scribner, M. *Authorizing Automated Vehicle Platooning: A Guide for State Legislators*; The National Academies of Sciences, Engineering, and Medicine: Washington, DC, USA, 2019.
4. Development of Operation Technology for V2X Truck Platooning. Available online: https://www.kaia.re.kr/portal/landmark/readTskView.do?tskId=147674&yearCnt=1&menuNo (accessed on 17 November 2020).
5. Rajamani, R. *Vehicle Dynamics and Control*; Springer Science & Business Media: Berlin/Heidelberg, Germany, 2011.
6. Khodayari, A.; Ghaffari, A.; Ameli, S.; Flahatgar, J. A historical review on lateral and longitudinal control of autonomous vehicle motions. In Proceedings of the 2010 International Conference on Mechanical and Electrical Technology, Singapore, 10–12 September 2010; IEEE: Piscataway, NJ, USA, 2010; pp. 421–429. [CrossRef]
7. Solyom, S.; Idelchi, A.; Salamah, B.B. Lateral control of vehicle platoons. In Proceedings of the 2013 IEEE International Conference on Systems, Man, and Cybernetics, Manchester, UK, 13–16 October 2013; IEEE: Piscataway, NJ, USA, 2013; pp. 4561–4565. [CrossRef]
8. Jansen, W. Lateral Path-Following Control of Automated Vehicle Platoons. Master's Thesis, Delft University of Technology, Delft, The Netherlands, 2016.
9. Lu, G.; Tomizuka, M. A laser scanning radar based autonomous lateral vehicle following control scheme for automated highways. In Proceedings of the 2003 American Control Conference, Denver, CO, USA, 4–6 June 2003; IEEE: Piscataway, NJ, USA, 2003; pp. 30–35. [CrossRef]

10. Hellstrom, T.; Ringdahl, O. Follow the Past: A path-tracking algorithm for autonomous vehicles. *Int. J. Veh. Auton. Syst.* **2006**, *4*, 216–224. [CrossRef]
11. Tsugawa, S.; Murata, S. Steering Control Algorithm for Autonomous Vehicle. *Trans. Inst. Syst. Control Inf. Eng.* **1989**, *2*, 360–362. [CrossRef]
12. Kato, S.; Tsugawa, S. *Lateral and Longitudinal Control Algorithms for Visual Platooning of Autonomous Vehicles*; No. 2000-05-0373; SAE Technical Paper; SAE: Warrendale, PA, USA, 2000.
13. Gehrig, S.K.; Stein, F.J. A trajectory-based approach for the lateral control of vehicle following systems. In Proceedings of the IEEE International Conference on Intelligent Vehicles, Stuttgart, Germany, 28–30 October 1998; IEEE: Piscataway, NJ, USA, 1998; pp. 156–161.
14. Bandi, V. Thorvaldsson, Reference Path Estimation for Lateral Vehicle Control. Master's Thesis, Chalmers University of Technology, Gothenburg, Sweden, 2015.
15. Isiklar, G.; Besselink, I.I.; Nijmeijer, H.; Veenhuizen, P.; Verbeek, H. Simulation of Complex Articulated Commercial Vehicles for Different Driving Manoeuvres. Master's Thesis, Eindhoven University of Technology, Eindhoven, The Netherlands, 2007.
16. Papadimitriou, I.; Tomizuka, M. Lateral control of platoons of vehicles on highways: The autonomous following based approach. *Int. J. Veh. Des.* **2004**, *36*, 24–37. [CrossRef]
17. White, R.; Tomizuka, M. Autonomous following lateral control of heavy vehicles using laser scanning radar. In Proceedings of the 2001 American Control Conference, Arlington, VA, USA, 25–27 June 2001; Cat. No.01CH37148. IEEE: Piscataway, NJ, USA, 2001; pp. 2333–2338. [CrossRef]
18. Travis, W.; Bevly, D.M. Trajectory duplication using relative position information for automated ground vehicle convoys. In Proceedings of the 2008 IEEE/ION Position, Location and Navigation Symposium, Monterey, CA, USA, 5–8 May 2008; IEEE: Piscataway, NJ, USA, 2008; pp. 1022–1032. [CrossRef]
19. Martin, S.; Bevly, D.M. Comparison of GPS-based autonomous vehicle following using global and relative positioning. *Int. J. Veh. Auton. Syst.* **2012**, *10*, 229–255. [CrossRef]
20. Rezgui, J.; Cherkaoui, S.; Chakroun, O. Deterministic access for DSRC/802.11p vehicular safety communication. In Proceedings of the 2011 7th International Wireless Communications and Mobile Computing Conference, Istanbul, Turkey, 4–8 July 2011; IEEE: Piscataway, NJ, USA, 2011; pp. 595–600. [CrossRef]
21. Minh, V.T. Trajectory Generation for autonomous vehicles. In *Mechatronics 2013*; Springer: Cham, Switzerland, 2014; pp. 615–626. [CrossRef]
22. Kim, Y.; Bang, H. Introduction to Kalman filter and its applications. In *Introduction and Implementations of the Kalman Filter*; IntechOpen: London, UK, 2018. [CrossRef]
23. TruckSim Overview—Mechanical Simulation. Available online: https://www.carsim.com/products/trucksim/ (accessed on 17 November 2020).
24. Kabsch, W. A solution for the best rotation to relate two sets of vectors. *Acta Crystallogr. Sect. A Cryst. Phys. Diffr. Theor. Gen. Crystallogr.* **1976**, *32*, 922–923. [CrossRef]
25. Kabsch, W. A discussion of the solution for the best rotation to relate two sets of vectors. *Acta Crystallogr. Sect. A Cryst. Phys. Diffr. Theor. Gen. Crystallogr.* **1978**, *34*, 827–828. [CrossRef]
26. Implementation of the Pure Pursuit Path Tracking Algorithm. Available online: https://www.ri.cmu.edu/pub_files/pub3/coulter_r_craig_1992_1/coulter_r_craig_1992_1.pdf (accessed on 17 November 2020).

Publisher's Note: MDPI stays neutral with regard to jurisdictional claims in published maps and institutional affiliations.

© 2020 by the authors. Licensee MDPI, Basel, Switzerland. This article is an open access article distributed under the terms and conditions of the Creative Commons Attribution (CC BY) license (http://creativecommons.org/licenses/by/4.0/).

MDPI

St. Alban-Anlage 66

4052 Basel

Switzerland

Tel. +41 61 683 77 34

Fax +41 61 302 89 18

www.mdpi.com

Sensors Editorial Office

E-mail: sensors@mdpi.com

www.mdpi.com/journal/sensors

www.ingramcontent.com/pod-product-compliance
Lightning Source LLC
LaVergne TN
LVHW070156170726
843515LV00007B/1935